Advances in Aquaculture Feed Additives

Advances in Aquaculture Feed Additives

Guest Editors

Adolfo Jatobá
Delano Dias Schleder

MDPI
Basel • Beijing • Wuhan • Barcelona • Belgrade • Novi Sad • Cluj • Manchester

Guest Editors
Adolfo Jatobá
Aquaculture Laboratory
Catarinense Federal Institute of Education, Science and Technology
Araquari
Brazil

Delano Dias Schleder
Aquaculture Laboratory
Catarinense Federal Institute of Education, Science and Technology
Araquari
Brazil

Editorial Office
MDPI AG
Grosspeteranlage 5
4052 Basel, Switzerland

This is a reprint of the Special Issue, published open access by the journal *Fishes* (ISSN 2410-3888), freely accessible at: https://www.mdpi.com/journal/fishes/special_issues/Q8794T30X3.

For citation purposes, cite each article independently as indicated on the article page online and as indicated below:

Lastname, A.A.; Lastname, B.B. Article Title. *Journal Name* **Year**, *Volume Number*, Page Range.

ISBN 978-3-7258-6836-0 (Hbk)
ISBN 978-3-7258-6837-7 (PDF)
https://doi.org/10.3390/books978-3-7258-6837-7

Contents

About the Editors

Adolfo Jatobá

Adolfo Jatobá is a researcher and professor at the Federal Institute Catarinense, Campus Araquari, Brazil. His work focuses on aquaculture nutrition, feed additives, probiotics and postbiotics, biofloc technology, and integrated multitrophic aquaculture systems. He has coordinated and participated in numerous research projects addressing sustainable production strategies, fish health, water quality management, and circular economy approaches in aquaculture. His scientific contributions include peer-reviewed articles, editorial activities, and the supervision of undergraduate and graduate students. He has extensive experience in experimental aquaculture systems and actively collaborates with national and international research groups, contributing to the advancement of sustainable and functional aquafeed development.

Delano Dias Schleder

Delano Dias Schleder is a researcher and professor affiliated with the Federal Institute Catarinense, Brazil, with expertise in aquaculture, animal health, applied nutrition and microbiology. His research interests include immune responses, stress physiology, functional ingredients, and sustainable management practices in aquatic production systems. He has been involved in interdisciplinary research projects aimed at improving animal welfare, production efficiency, and environmental performance in aquaculture. His academic achievements include scientific publications, student supervision, and participation in collaborative research initiatives that bridge fundamental knowledge and practical applications for the aquaculture sector.

Preface

This Reprint addresses recent advances in aquaculture feed additives, highlighting their role in improving efficiency, health, and sustainability. Prepared by the Guest Editors, it is intended for researchers and professionals seeking innovative, science based nutritional strategies for modern aquaculture.

Adolfo Jatobá and Delano Dias Schleder
Guest Editors

fishes

MDPI

Editorial

Advances in Aquaculture Feed Additives: Insights from the Special Issue

Adolfo Jatobá * and Delano Dias Schleder

Aquaculture Laboratory, Federal Institute Catarinense–Campus Araquari, BR 280-km 27, Araquari 89245-000, SC, Brazil; delano.schleder@ifc.edu.br
* Correspondence: jatobaadolfo@gmail.com; Tel.: +55-4799970-9988

1. Introduction

Aquaculture continues to expand globally and is now responsible for providing more than half of the aquatic products consumed worldwide. It continues to grow as a vital contributor to global food security and the blue economy. Alongside this expansion, there is an increasing demand for feeding strategies that optimize performance while minimizing environmental impacts. Feed additives have emerged as key nutritional tools capable of improving animal health, digestive efficiency, immune competence, and overall resilience, supporting a transition toward more sustainable and responsible aquaculture practices [1,2]. These compounds—ranging from probiotics, prebiotics, enzymes, and plant extracts to organic acids and functional proteins—have been widely recognized for their potential to enhance nutrient utilization, modulate the gut microbiome, and reduce reliance on chemotherapeutic agents [3,4].

Over the past decade, technological and conceptual advances have deepened our understanding of how bioactive dietary compounds influence fish physiology. New insights from molecular biology, metabolomics, and microbiome research have provided evidence that the interactions between diet, host, and environment are highly dynamic and context-dependent [5,6]. Within this evolving landscape, the Special Issue "Advances in Aquaculture Feed Additives" compiles studies addressing nutritional, physiological, and practical dimensions of feed additive use, offering perspectives that contribute to both scientific and applied innovation in aquaculture nutrition (Articles 1–9).

2. SI Contributions and Challenges in the Development and Application of Feed Additives

The articles gathered in this Special Issue reflect the diversity and complexity of current research on aquaculture feed additives. Rather than summarizing each study individually, the collection highlights several overarching trends shaping the field. One enduring challenge in aquaculture nutrition is the reduction in dependence on finite marine resources such as fishmeal and fish oil, while maintaining or improving growth performance and product quality. Studies in this SI, together with recent reviews [7,8], demonstrate the feasibility of incorporating sustainable ingredients and functional additives that enhance nutrient use, health, and environmental outcomes.

Beyond ingredient substitution, the SI emphasizes the multifaceted role of additives as modulators of gut health, immune function, and stress resilience—attributes increasingly vital in intensive production systems [3,4,9]. However, biological responses to additives remain highly variable and species-specific. Their effects depend on numerous factors,

 https://doi.org/10.3390/fishes11010044

including age, diet composition, rearing conditions, and environmental stressors. Moreover, the mechanisms underlying additive-induced physiological changes are often only partially understood, making it difficult to predict outcomes in large-scale production environments [5]. This complexity reinforces the need for integrative approaches that combine nutritional, physiological, and molecular assessments.

Another key challenge concerns the translation of research into practice. Although many additives show consistent benefits under experimental conditions, the transition from laboratory to industry is often slow and fragmented [2]. Scientific development in this field frequently outpaces industry adoption; by the time an additive has demonstrated robust efficacy and safety, newer compounds or combinations are already under evaluation. Consequently, promising innovations risk being underutilized before reaching commercial maturity. Logistical limitations, regulatory frameworks, cost-effectiveness, and the heterogeneity of production systems further contribute to this gap between scientific advances and field implementation. Bridging this divide is essential to ensure that academic progress translates into tangible benefits for producers and consumers.

By addressing these challenges collectively, this Special Issue contributes to a more comprehensive and operational understanding of feed additive development and application. The studies emphasize that successful additive strategies require a multidisciplinary perspective—encompassing nutrition, physiology, microbiology, environmental management, and economics—to ensure solutions that are not only effective and sustainable but also feasible for large-scale aquaculture.

3. Future Perspectives for Aquaculture Feed Additives

Despite substantial progress, several research and practical gaps remain. Optimizing inclusion levels and understanding potential interactions among multiple additives are recurrent challenges [8]. Future studies should explore synergistic effects among bioactive compounds and the ways in which these combinations influence metabolic and immunological pathways. Multi-omics tools—including transcriptomics, metabolomics, and microbiomics—offer valuable opportunities to clarify mechanisms of action and to design precision feeding strategies tailored to specific species, production stages, and rearing conditions [5,6].

Sustainability will continue to be a central theme. Increasing attention is being given to valuing agro-industrial by-products, microbial biomass, and marine macroalgae as promising sources of functional ingredients for aquaculture feeds [1]. Circular economy principles encourage the identification of additive sources that provide both nutritional and ecological value. In parallel, reducing the use of chemotherapeutic agents in aquaculture is an urgent global priority. Functional feed additives that improve immune competence, gut integrity, and stress tolerance represent viable and increasingly validated alternatives, supporting antimicrobial resistance mitigation efforts and aligning with United Nations Sustainable Development Goals (SDG 2—Zero Hunger, SDG 12—Responsible Consumption and Production, SDG 14—Life Below Water, SDG 15—Life on Land).

Ultimately, progress in feed additive research will depend not only on innovation but also on effective collaboration between academia, industry, and regulatory bodies. Strengthening knowledge transfer mechanisms, promoting transparency in additive evaluation, and fostering co-development with producers are key to accelerating adoption. The collective findings presented in this Special Issue, in concert with ongoing global research efforts, offer valuable guidance toward the responsible and sustainable use of feed additives in aquaculture.

https://doi.org/10.3390/fishes11010044

Informed Consent Statement: Not applicable.

Data Availability Statement: The data related to this research are available upon request.

Acknowledgments: The authors thank the National Council for Scientific and Technological Development (CNPq) for the grants to A. Jatobá (308661/2023-0) and D. D. Schleder (310247/2025-0); FAPESC (2023/TR001478).

Conflicts of Interest: The authors declare no conflicts of interest.

List of Contributions:

1. Xu, R.; Wang, Z.; Liu, H.; Li, R.; Wang, X.; Yang, H.; Ding, J.; Chang, Y.; Zuo, R. Effect of Dietary Proline on the Growth Performance, Collagen Deposition, and Texture Quality of Sea Cucumbers' Body Wall (*Apostichopus japonicus*). *Fishes* **2025**, *10*, 482. https://doi.org/10.3390/fishes10100482.
2. Candela-Maldonado, Y.; Megder, I.; Tefal, E.; Peñaranda, D.S.; Martínez-Llorens, S.; Tomás-Vidal, A.; Jover-Cerdá, M.; Jauralde, I. Four Organic Protein Source Alternatives to Fish Meal for Pacific White Shrimp (*Penaeus vannamei*) Feeding. *Fishes* **2025**, *10*, 384. https://doi.org/10.3390/fishes10080384.
3. Dias, M.; Marmelo, I.; António, C.; Rodrigues, A.M.; Marques, A.; Diniz, M.S.; Maulvault, A.L. *Asparagopsis taxiformis* Feed Supplementation as a Tool to Improve the Resilience of Farmed Diplodus sargus to Marine Heatwave Events—A Metabolomics Approach. *Fishes* **2025**, *10*, 350. https://doi.org/10.3390/fishes10070350.
4. Di, W.; Zhang, Y.; Zuo, H.; Liu, H.; Wang, L.; Ding, J.; Chang, Y.; Zuo, R. Effects of Dietary β-Carotene on the Gonadal Color, Pigmentation, and Regulation Mechanisms in Sea Urchin Strongylocentrotus intermedius. *Fishes* **2025**, *10*, 304. https://doi.org/10.3390/fishes10070304.
5. Zehra, S.; Abul Kasim, A.S.; Saleh, R.; De Mello, P.; Alshaikhi, A.; Laranja, J.; Alhafedh, Y.; Glencross, B.D.; Alghamdi, M.A.; Mohamed, A.W. Evaluation of Apparent Nutrient Digestibility of Novel and Conventional Feed Ingredients in Sobaity Seabream (*Sparidentex hasta*) for Sustainable Aquaculture. *Fishes* **2025**, *10*, 265. https://doi.org/10.3390/fishes10060265.
6. Wu, C.; Yang, H.; Yang, Y.; Yuan, Q.; Lv, W.; Ayana, G.U.; Li, M.; Su, D.; Zhou, W.; Zhang, Q. Dietary Supplementation of Astragalus Polysaccharides Modulates Growth Physiology, Metabolic Homeostasis, and Innate Immune Responses in Rice Field Eels (*Monopterus albus*). *Fishes* **2025**, *10*, 213. https://doi.org/10.3390/fishes10050213.
7. Athira, A.; Nishiguchi, H.; Hayashi, D.; Otsu, Y.; Miura, C.; Suryadi, I.B.B.; Ali, M.F.Z.; Miura, T. Influence of the Silkworm-Derived (*Bombyx mori*) Functional Substance (Silkrose-BM) on the Fish Meat Quality of Yellowtail (*Seriola quinqueradiata*). *Fishes* **2025**, *10*, 130. https://doi.org/10.3390/fishes10030130.
8. Jatobá, A.; Pereira, M.O.; Jesus, G.F.A.; Dutra, S.A.P.; Mouriño, J.L.P.; Owatari, M.S.; Schleder, D.D. Effects of Blending Curcuma longa Hydrolate and *Lactobacillus plantarum* on the Growth and Health of Nile Tilapia. *Fishes* **2024**, *9*, 503. https://doi.org/10.3390/fishes9120503.
9. Okon, E.; Iyobhebhe, M.; Olatunji, P.; Adeleke, M.; Matekwe, N.; Okocha, R. Feed Additives in Aquaculture: Benefits, Risks, and the Need for Robust Regulatory Frameworks. *Fishes* **2025**, *10*, 471. https://doi.org/10.3390/fishes10090471.

References

1. Midhun, S.J.; Arun, D. Alternative feed technology in aquaculture. In *Recent Advances in Aquaculture Microbial Technology*; Academic Press: Cambridge, MA, USA, 2023; pp. 291–306. [CrossRef]
2. Zuberi, A.; Kamran, M.; Younus, N.; Abdel-Tawwab, M. Functional feed additives: Current trends. *Front. Aquac.* **2024**, *3*, 1385508. [CrossRef]
3. Mustafa, S.A.; Sanaa, A.; Al-Faragi, J.K. Supplementation of feed additives on aquaculture feeds: A review. *Int. J. Pharm. Res.* **2021**, *13*, 561–567.
4. Maurya, N.K.; Dubey, S.; Samal, K.; Ramachandra Naik, A.T.; Lakshmipathi, M.T.; Khalasi, B.R.; Shinglai, C.; Singh, K.; Chaturvedi, P.; Yadav, B. Biotechnology in Fisheries and Aquaculture: A Holistic Review of Innovations in Health, Nutrition, and Production. *J. Adv. Biol. Biotechnol.* **2025**, *28*, 175–204. [CrossRef]

https://doi.org/10.3390/fishes11010044

5. Hu, X.; Ma, W.; Zhang, D.; Tian, Z.; Yang, Y.; Huang, Y.; Hong, Y. Application of natural antioxidants as feed additives in aquaculture: A review. *Biology* **2025**, *14*, 87. [CrossRef] [PubMed]
6. Martin Samuel, A.M.; Król, E. Nutrigenomics and immune function in fish: New insights from omics technologies. *Dev. Comp. Immunol.* **2017**, *75*, 86–98. [CrossRef] [PubMed]
7. Elumalai, P.; Kurian, A.; Lakshmi, S.; Faggio, C.; Esteban, M.A.; Ringø, E. Herbal immunomodulators in aquaculture. *Rev. Fish. Sci. Aquacult.* **2020**, *29*, 33–57. [CrossRef]
8. Osei, S.A.; Ayisi, C.L.; Boamah, G.A.; Mensah, G.D. The Circular Economy in Aquaculture and Fisheries: Enhancing Sustainability and Food Security. *Circ. Econ. Sustain.* **2025**, *5*, 4565–4614. [CrossRef]
9. De Moraes, A.V.; Owatari, M.S.; da Silva, E.; de Oliveira Pereira, M.; Piola, M.; Ramos, C.; Farias, D.R.; Scheleder, D.D.; Jesus, G.A.F.; Jatobá, A. Effects of microencapsulated probiotics-supplemented diet on growth, non-specific immunity, intestinal health and resistance of juvenile Nile tilapia challenged with *Aeromonas hydrophila*. *Anim. Feed. Sci. Technol.* **2022**, *287*, 115286. [CrossRef]

https://doi.org/10.3390/fishes11010044

fishes

MDPI

Article

Effect of Dietary Proline on the Growth Performance, Collagen Deposition, and Texture Quality of Sea Cucumbers' Body Wall (*Apostichopus japonicus*)

Rujian Xu, Zitong Wang, Haijing Liu, Ruixue Li, Xianyu Wang, Hongbing Yang, Jun Ding, Yaqing Chang and Rantao Zuo *

Key Laboratory of Mariculture and Stock Enhancement in North China's Sea (Ministry of Agriculture and Rural Affairs), Dalian Ocean University, Dalian 116023, China

* Correspondence: rtzuo@dlou.edu.cn or zuorantao@163.com; Tel.: +86-411-84763513

Abstract: Sea cucumber (*Apostichopus japonicus*) is an important economically cultured species in the northern coastal regions of China. Its body wall is rich in collagen, which directly determines product quality and market value. However, with the expansion of aquaculture scale, issues such as insufficient collagen deposition have led to inconsistent quality among cultured individuals. Therefore, there is an urgent need to improve growth performance and body wall quality through nutritional regulation. As functional nutrients, amino acids play key roles in collagen synthesis, yet relevant research on *A. japonicus* remains limited. This study was conducted to investigate the effects of dietary proline on the growth performance, body wall collagen deposition and organoleptic quality of sea cucumber (initial body weight: 30.20 ± 2.02 g). Three kinds of feed with equal levels of nitrogen and other lipids, and supplemented with different concentrations of proline (0%, 1.5%, 3%) in the basal diet, were named P0, P1.5, and P3, and the experiment was conducted for 60 days. The results showed that supplementation with 3% proline significantly increased final body weight and weight gain rate ($p < 0.05$), reaching 66.39 g and 115.30%, respectively. Collagen content in the body wall increased by approximately 18.5% compared to the control group. Histological analysis of the body wall showed that the collagen fibers in the P1.5 and P3 groups were thicker, with an increased proportion of type I collagen. Texture profile analysis indicated that hardness, cohesiveness, and chewiness were significantly improved in the P3 group ($p < 0.05$). In summary, supplementation with 1.5% and 3% proline effectively enhanced growth, collagen deposition, and body wall quality. Compared to the P0 group, the relative expression levels of collagen type I alpha 2 chain (*COL1A1*), Sma- and Mad-related protein 1(*SMAD1*), and sp-smad2/3 (*SMAD2/3*) in the body wall tissue were significantly upregulated in both the P1.5 and P3 groups ($p < 0.05$).

Keywords: Amino acid; collagen synthesis; weight gain rate; texture profile; TGF-β/SMAD signaling pathway

Key Contribution: This study provides the first form of evidence that dietary proline supplementation enhances growth performance, collagen deposition, and body wall quality in *Apostichopus japonicus*. Proline promoted collagen fiber development and upregulated *COL1A1* and *SMAD* pathway genes, suggesting its potential as a functional nutrient to improve sea cucumber aquaculture quality and consistency.

1. Introduction

Sea cucumber (*Apostichopus japonicus*) is an invertebrate belonging to the class Holothuroidea of the echinoderms, naturally distributed in the northern part of the West Pacific, including the Far East coast of Russia, Japan, Korea, and the Yellow and Bohai Seas of China [1,2]. In 2022, *A. japonicus* production in China reached 248,508 tons, making it one of the most valuable mariculture species in coastal provinces such as Liaoning, Shandong, and Fujian, with significant contributions to local economies [3]. *A. japonicus*, as a premium sea food product, is characterized by high protein content and rich bioactive compounds including collagen, polysaccharides, and saponins, which has been proven to possess immunomodulatory, anti-inflammatory, antioxidant, and health-promoting properties [4,5]. Due to its high nutritional value, it is gradually receiving more attention. With the enhancement of people's health and wellness awareness, the demand for sea cucumbers is increasing, and sea cucumber farming has become an important part of China's aquaculture industry [6].

The body wall is the main edible part of the sea cucumber, with collagen being the primary component of the body wall [7]. Collagen, as the main component of connective tissue, affects the functional and structural characteristics of muscles [8]. The content and structure of collagen in the body wall not only affect its texture but also determine the soaking ratio of the sea cucumbers. Nutritional regulation technology has proven to be an efficient and effective solution to enhance the quality of fresh and processed aquatic products. It was found that appropriate supplementation of certain amino acids effectively increased the collagen content and texture quality of muscle in several fish species. The addition of taurine in feeds can improve the growth performance, muscle composition, digestive enzyme activity, and stress resistance of white shrimp (*Litopenaeus vannamei*) [9]. The addition of hydroxyproline in feeds can improve the growth performance and muscle hardness of Nibe croaker (*Nibea diacanthus*) [10]. Proline is the only α-imino acid that constitutes collagen and plays an important role in maintaining the structural integrity of the protein framework [11]. Studies have reported that proline, as a functional amino acid, can regulate the synthesis of collagen and promote the repair of damaged tissues of animals [12]. Studies have shown that the combination of proline and vitamin C in feeds can improve the hardness and collagen content in the muscles of the large yellow croaker (*Larimichthys crocea*). Rong et al. [13,14] found that adding proline to the feeds improved the crude protein content and antioxidant capacity in the body of the Yellow Drum (*Nibea coibor*), and affected the expression of genes related to collagen synthesis in the swim bladder tissue, involving type I collagen and prolyl 4-hydroxylase (P4H). Deng et al. [15] found that adding proline to the feeds improved the growth performance and collagen synthesis of juvenile Chinese soft-shelled Turtles (*Pelodiscus sinensis*).

In recent years, increasing attention has been paid to the molecular regulatory mechanisms underlying collagen synthesis in the body wall of sea cucumbers. The biosynthesis of collagen involves the coordinated regulation of key enzymes such as prolyl 4-hydroxylase (P4H) and lysyl oxidase (LOX), which catalyze hydroxylation and cross-linking of procollagen, playing a decisive role in collagen fiber formation and deposition [16]. Among the signaling pathways involved, the transforming growth factor-β (*TGF-β*)/Smads pathway is considered a central regulatory axis. Activation of *TGF-β* receptors promotes the phosphorylation of *Smad2/3*, which subsequently translocate into the nucleus to regulate the transcription of collagen synthesis-related genes, thereby facilitating collagen deposition [17,18]. For example, Wang et al. reported that dietary supplementation with vitamin E significantly increased collagen content in the body wall of *A. japonicus* by enhancing antioxidant capacity, upregulating the expression of *TGF-β1*, *Smad2*, and *Smad3*, and downregulating the inhibitory factor *Smad7*, indicating that vitamin E promotes collagen synthesis and deposition through activation of the *TGF-β/Smads* pathway [19]. These findings not only reveal the molecular mechanisms of collagen deposition but also provide a theoretical basis for nutritional regulation of body wall quality in sea cucumbers. However, there is currently a lack of studies elucidating the effects of dietary proline on collagen synthesis and the transcriptional regulation of collagen-related genes in *A. japonicus*.

Although numerous studies have been conducted extensively on the role of proline in collagen synthesis in aquatic animals, the effects of proline on the quality of the body wall of sea cucumbers are not yet fully understood. Therefore, this experiment aimed to investigate the effects of adding different concentrations of proline to the feeds on the growth performance of sea cucumbers, the content and structure of collagen, the organoleptic quality, and the expression of relevant collagen synthesis genes in the body wall of sea cucumbers.

2. Materials and Methods

2.1. Ethics Statement

In this study, sea cucumbers were successfully bred in a controlled hatchery environment. All experimental procedures strictly adhered to China's national ethical guidelines and the research protocols established by Dalian Ocean University.

2.2. Experimental Diets and Feeding Experiment

In this experiment, defatted fish meal, *Sargassum thunbergii* meal, and fermented soybean meal were used as the main sources of protein. Different ratios of L-Proline and L-Serine (provided by Shanghai Sangon Biotechnology, Shanghai, China purity ≥ 98.5%) were added to make three experimental feeds with graded levels of proline (0, 1.5%, and 3%). The feed formula and nutritional composition are presented in Table 1.

Table 1. Ingredients and proximate analysis of experimental diets (%).

Ingredient	Dietary Proline Levels (% Dry Diet)		
	0	1.5	3
Skim Fish Meal [1]	2	2	2
Sargassum Thunbergii Meal [2]	28	28	28
Fermented Soybean Meal [3]	12	12	12
Proline	0	1.5	3
Serine	3	1.5	0
Vitamin Premix [4]	0.5	0.5	0.5
Mineral Premix [5]	0.5	0.5	0.5
$Ca(H_2PO4)_2$ [6]	2	2	2
Sea Mud	52	52	52
Proximate composition			
Crude Protein (%)	9.52	9.56	9.48
Crude lipid (%)	0.12	0.11	0.13

Note: [1] Skim fish meal: crude protein 68.7% dry matter, crude lipid 2.6% dry matter. Dalian Xinyulong Marine Biological Seed Technology Co., Ltd. (Dalian, Liaoning Province, China). [2] Sargassum thunbergii meal: crude protein 16.8% dry matter, Dalian Xinyulong Marine Biological Seed Technology Co., Ltd. (Dalian, Liaoning Province, China). [3] Fermented Soybean meal: crude protein 51.56% dry matter, crude lipid 0.9% dry matter. Dalian Xinyulong Marine Biological Seed Technology Co., Ltd. (Dalian, Liaoning Province, China). [4] Vitamin premix: provided by Dalian Xinyulong Marine Biological Seed Technology Co., Ltd. (Dalian, Liaoning Province, China). [5] Mineral premix: provided by Dalian Xinyulong Marine Biological Seed Technology Co., Ltd. (Dalian, Liaoning Province, China). [6] $Ca(H_2PO4)_2$: Source of available phosphorus and calcium.

All solid ingredients were carefully ground into a fine powder and then passed through an 80-mesh sieve. L-proline and L-serine were first proportionally mixed. Then, other feed ingredients were weighed, added sequentially and mixed evenly with the mixture of proline and serine. Subsequently, the experimental feeds were sealed in plastic bags and stored at −20 °C until use. A measured portion of the feed was mixed with 30% distilled water to achieve the right consistency. The mixture was then kneaded into spherical solid pellets, each approximately 1.5 cm in diameter.

The feeding trial was conducted at the Key Laboratory of Mariculture and Stock Enhancement in the North China Sea, Ministry of Agriculture and Rural Affairs, Dalian Ocean University. Experimental *A. japonicus* were obtained from Dalian Xinyulong Marine Biotechnology Co., Ltd. Prior to the trial, individuals were acclimated for two weeks in a controlled aquaculture system and provided with a standard basal diet to ensure uniform health status and adaptation to experimental conditions. Following a 36 h fasting period, 180 healthy individuals were randomly assigned to nine 160 L square tanks (20 per tank), with an initial total biomass of approximately 500 ± 1 g per group. The rearing experiment lasted 80 days. *A. japonicus* were fed twice (07:00 and 17:00) every day. For each tank, feeds were allocated to animals at the amount of 2% of their body weight. The *A. japonicus* in each tank were weighed and the feeding amount was accordingly adjusted every week. During the experiment, the water temperature was maintained at 19-21 °C, salinity at 30 ± 1 ‰, pH at 8.0 ± 0.1, dissolved oxygen above 8.0 mg/L, ammonia concentration below 0.2 mg/L, and nitrite concentration below 0.1 mg/L. One-third of the water volume was exchanged daily, and residual feed and feces were removed every other day. Tanks were maintained under dim illumination to simulate the natural habitat.

2.3. Sample Collection

Upon completion of the feeding trial, *A. japonicus* species were subjected to a 48 h fasting period. All individuals in each tank were then counted, measured, and weighed to calculate survival rate, weight gain rate, and body wall mass. Coelomic fluid was extracted from six randomly selected specimens per tank. For histological examination, the excised body wall tissues were fixed in 4% paraformaldehyde. A section from the mid-dorsal region of the body wall was trimmed into approximately 1.5 cm × 1.5 cm × 1.5 cm cubes for texture profile analysis. Body wall and intestinal samples were collected, placed in 1.5 mL RNase-free tubes (Axygen, Union City, CA, USA), and stored at −80 °C for subsequent proximate composition analysis.

2.4. Feed Composition Analysis

The proximate composition of the experimental diets was analyzed according to AOAC (1995) [20]. The crude protein (CP) content of the diets was determined according to the Kjeldahl method. Briefly, approximately 1.0 g of finely ground sample was digested with sulfuric acid. The digest was then neutralized with sodium hydroxide, and the released ammonia was distilled into a boric acid solution and subsequently titrated with standardized hydrochloric acid. The nitrogen content was calculated, and the crude protein was expressed as nitrogen × 6.25. The crude fat (ether extract, EE) was measured following the Soxhlet extraction method. About 2.0 g of dried sample was wrapped in filter paper and extracted with petroleum ether in a Soxhlet apparatus for 6 h. After extraction, the solvent was removed by evaporation, and the residue was dried to a constant weight at 105 °C. The crude fat content was calculated as the percentage of residue weight relative to the sample weight.

2.5. Collagen Contents Analysis

Hydroxyproline (Hyp) content in the body wall was quantified using a commercial assay kit (Art. No. A030-2; Nanjing Jiancheng Bioengineering Institute, Nanjing, China) according to the manufacturer's instructions. Absorbance was measured at 550 nm using a microplate reader (Infinite® Pro 200, Tecan, Männedorf, Switzerland). Collagen content was calculated as Hyp × 8, as described by AOAC (2002) [21].

2.6. Amino Acid Contents Analysis

Amino acid composition of the body wall was determined following the AOAC (2002) standard procedure with minor modifications. Briefly, freeze-dried body wall samples (approximately 100 mg) were hydrolyzed with 6 mol/L HCl at 110 °C for 24 h under a nitrogen atmosphere. After hydrolysis, the samples were neutralized, filtered, and diluted to a fixed volume with ultrapure water. The amino acid contents were analyzed using an automatic amino acid analyzer (e.g., Hitachi L-8900, Tokyo, Japan) [21]. Based on the results, amino acids were classified as essential amino acids (EAAs) or non-essential amino acids (NEAAs). EAAs—including lysine, methionine, threonine, valine, isoleucine, leucine, phenylalanine, and histidine—must be supplied through the diet, as they cannot be synthesized de novo. NEAAs, such as glycine, glutamic acid, aspartic acid, alanine, and proline, can be synthesized endogenously. Arginine was considered a conditionally essential amino acid in aquatic animals, particularly during early developmental stages or under certain stress conditions.

2.7. Histological Staining of the Body Wall

Body wall tissues were fixed in 4% paraformaldehyde at room temperature for 24 h, dehydrated through a graded ethanol series, cleared with xylene, embedded in paraffin, and sectioned using a rotary microtome (Leica RM2016, Leica Microsystems GmbH, Wetzlar, Germany). Hematoxylin–eosin (HE) staining was conducted according to standard protocols, and sections were mounted with neutral balsam. Van Gieson and Sirius Red staining followed identical sample preparation steps. Tissue morphology was examined under an upright light microscope (ECLIPSE E100, Nikon, Japan), and collagen distribution after Sirius Red staining was visualized using a polarizing microscope (ECLIPSE Ci, Nikon, Japan).

2.8. Characterization of Water Absorption of the Body Wall

Water absorption, also referred to as the soaking ratio, represents the percentage increase in body wall weight following soaking under standardized conditions and serves as an indicator of the water-holding capacity of the sea cucumber body wall. The soaking ratio of sea cucumbers is a crucial parameter reflecting the extent of expansion of dried sea cucumbers upon rehydration, and it directly indicates their quality and economic value. Sea cucumbers with higher soaking ratios generally exhibit plump flesh and abundant collagen, making them more desirable in the market, while also affecting processing costs and sensory attributes [22]. Fresh body wall samples were boiled in tap water at a ratio of 1:4 (*w*:*w*) for 30 min, cooled, and weighed. They were then immersed in deionized water at a ratio of 1:10 (*w*:*w*) and stored at 4 °C for 48 h. Final weights after soaking were recorded, and water absorption capacity was calculated accordingly.

2.9. Texture Profile Analysis (TPA)

TPA was performed using a texture analyzer (TMS-Pro, Beijing Yingsheng Hengtai Technology Co., Ltd., Beijing, China) equipped with a P20 cylindrical probe. Tests were conducted at room temperature with pre-test, test, and post-test speeds of 0.50 mm/s, a compression ratio of 75%, and a trigger force of 0.5 g.

2.10. Real-Time PCR (RT-PCR) of Collagen Synthesis-Related Genes

The expression pattern of collagen synthesis-related genes in the body wall was detected with RT-PCR using the specific primers. The sequences of specific primers are present in Table 2. The detailed procedures are described in the study by Zuo et al [23].. Briefly, total RNA was extracted and examined for concentration and integrity of RNA. Then, Total RNA was reverse transcribed using a Commercial kit (TaKaRa, Beijing, China). LightCycler® 96 real-time PCR system was used, and the relative gene expression was determined by using the $2^{-\Delta\Delta CT}$ [24].

Table 2. Real-time PCR primers used in the present study.

Genes	Gene ID	Annealing Temperature (°C)	Primer Sequences (5′-3′)
Col1A2 [1]	c54738.graph_c0	56	F:5′-CGGACTTTTACTTTGGCGTTAT-3′ R:5′-TTTCTGGCGGTCTGCCTAT-3′
TGF-β [2]	BSL78_20284	59	F:5′-ACCGCTCCTCACCCTTTAACAC-3′ R:5′-CCACTAGCACTAAGCAGCATATCAG-3′
SMAD1 [3]	BSL78_15508	58	F:5′-GGCATACTCGCAGCAGTCTAAAG-3′ R:5′-AGGTGTCTCATCGGAAAGGTCTAC-3′
SMAD2/3 [4]	BSL78_11878	59	F:5′-AGAACCACCACGAACTCAAACATG-3′ R:5′-GCAGACAGCAGCAGGGATAAAC-3′

Note: [1] Col1A2, collagen type I alpha 2 chain. [2] TGF-β, transforming growth factor beta. [3] SMAD1, Sma- and Mad-related protein 1. [4] SMAD2/3, sp-smad2/3.

2.11. Calculations and Statistical Analysis

$$\text{Survival rate (SR, \%)} = N_f/N_i \times 100$$

$$\text{Weight growth rate (WGR, \%)} = \left(W_f - W_i\right)/W_i \times 100$$

$$\text{Body wall index (BWI, \%)} = P_W/W \times 100$$

$$\text{Expansion multiple (EM)} = W_r/W_0$$

where N_i and N_f represent the initial and final numbers of *A. japonicus* in each tank, respectively; W_i and W_f denote the initial average weight and final average weight; P_W is the wet weight of the body wall, and W is the total wet body weight; W_0 and W_r correspond to the wet mass of the body wall before and after soaking in water, respectively.

Statistical analyses were performed using SPSS 22.0 (IBM, Armonk, NY, USA). Data were subjected to one-way analysis of variance (ANOVA), and significant differences among dietary treatments ($p < 0.05$) were identified using Tukey's post hoc test.

3. Results

3.1. Growth Performance

There were no significant differences in survival rates among the different dietary groups ($p > 0.05$). As the dietary proline level increased, the WGR and body wall weight significantly increased ($p < 0.05$). When the dietary proline supplementation level increased to 3%, the WGR of *A. japonicus* reached 115.30%, which was significantly higher than those of the other two groups ($p < 0.05$). As the dietary proline level increased, the body length of *A. japonicus* significantly decreased ($p < 0.05$), while the weight of the body wall significantly increased ($p < 0.05$) (Table 3).

Table 3. Effects of proline addition levels on the growth performance of sea cucumber (*Apostichopus japonicus*).

	Dietary Proline Levels (% Dry Diet)		
	0	**1.5**	**3**
Survival Rate (%)	100.00	100.00	100.00
Initial Body Weight (g)	30.40 ± 2.03	30.02 ± 2.06	30.35 ± 3.49
Final Body Weight (g)	55.54 ± 12.46	60.92 ± 11.08	66.39 ± 19.85
Weight Growth Rate (%)	82.20 ± 34.57 [a]	101.76 ± 25.09 [ab]	115.30 ± 42.34 [b]
Body Wall Weight (g)	34.53 ± 9.16 [a]	40.34 ± 8.85 [ab]	42.16 ± 10.89 [b]
Body Wall Index (%)	69.36 ± 6.37	69.80 ± 5.94	67.93 ± 6.59

Note: Data with different lowercase letters in the same horizontal column are significantly different between the various diet groups ($p < 0.05$).

3.2. Collagen Content and Structure

The supplementation of proline in diets had a significant effect on collagen deposition in the body wall of *A. japonicus*. As the dietary proline levels increased, the collagen content in the body wall of sea cucumbers initially increased and then decreased. However, the collagen content in the two experimental groups was significantly higher than that in the control group ($p < 0.05$) (Figure 1).

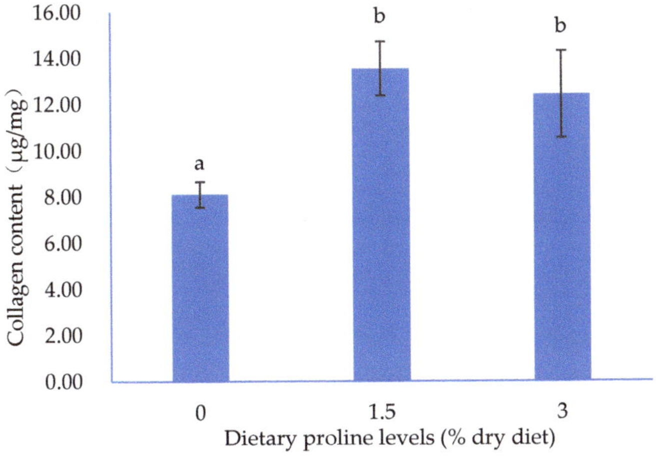

Figure 1. Effects of proline on collagen content in the body wall of sea cucumber (*Apostichopus japonicus*). Note: Bars bearing different lowercase letters are significantly different between the dietary groups ($p < 0.05$).

Following staining, the body wall tissues of *A. japonicus* were observed under a polarized light microscope. Type I collagen fibers exhibited strong birefringence, appearing as thick red or yellow fibers, whereas Type III collagen fibers displayed weaker birefringence, appearing as thin green fibers. In the proline-supplemented groups, the collagen fibers in the body wall of sea cucumbers were markedly thicker, with a significant increase in the number of Type I collagen fibers (Figure 2).

After staining, collagen fibers appear pink, while muscle fibers are displayed in yellow. Proline supplementation in diets induced notable structural changes in the body wall tissue of *A. japonicus*. Compared to the control group, the collagen fiber layer in the group with 1.5% proline was thicker and more densely arranged. In contrast, the *A. japonicus* in the group with 3% proline showed a reduction in the thickness of both the cuticle and epidermis layers compared to those in the other two groups (Figure 3).

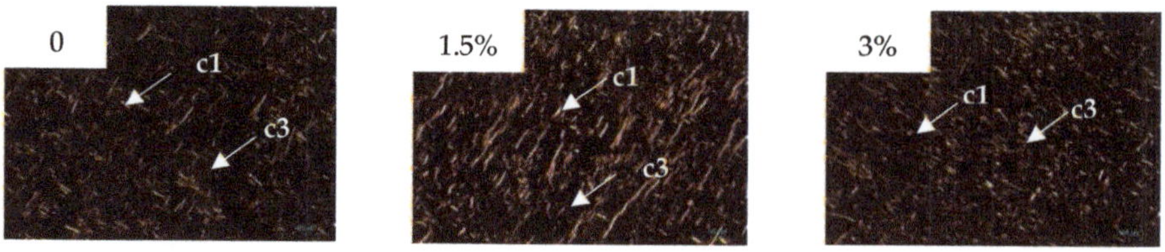

Figure 2. Effects of proline on collagen fiber morphology in the body wall of sea cucumber (*Apostichopus japonicus*). Notes: Sirius red staining (200×), c1: Type I collagen fibers, c3: Type III collagen fibers.

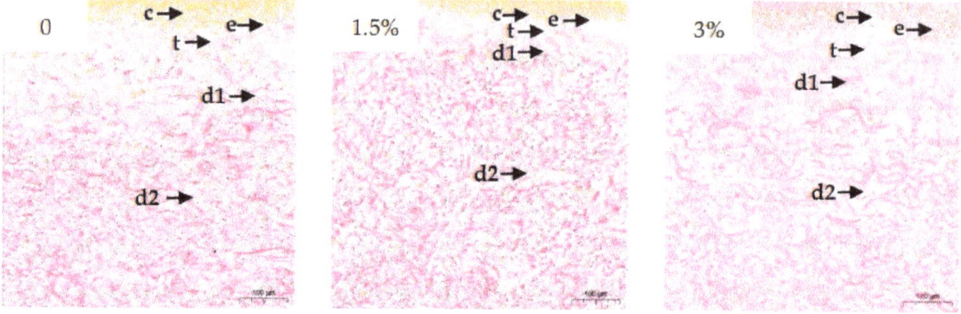

Figure 3. Effects of proline on collagen fiber morphology in the body wall of sea cucumber (*Apostichopus japonicus*). Note: Van Gieson staining (100×), c: stratum corneum; e: epidermis; t: transition layer; d1: loose collagen fiber layer; d2: dense fiber layer.

3.3. Amino Acid Profile

After appropriate supplementation of proline in the diet, the content of proline, a characteristic amino acid of collagen, in the body wall of *A. japonicus* showed significant changes. When the dietary proline supplementation level reached 3%, the contents of proline, glycine, and alanine in the body wall of *A. japonicus* significantly increased ($p < 0.05$), reaching 9.67%, 13.96%, and 6.90%, respectively. Overall, glutamic acid was the most abundant amino acid in the body, followed by glycine (Table 4).

Table 4. Effects of proline addition levels on amino acid contents of sea cucumber (*Apostichopus japonicus)* (% dry matter basis).

Amino Acid	Dietary Proline Levels (% Dry Diet)		
	0	1.5	3
Aspartic	8.17 ± 0.47	8.52 ± 0.26	8.15 ± 0.50
Glutamic	15.47 ± 0.44 ^{b}	15.63 ± 0.75 ^{b}	14.35 ± 0.62 ^{a}
Serine	4.90 ± 0.08	4.83 ± 0.31	4.69 ± 0.17
Glycine	11.49 ± 1.66 ^{a}	9.92 ± 0.93 ^{a}	13.96 ± 1.87 ^{b}
Histidine	1.99 ± 0.17 ^{b}	2.07 ± 0.14 ^{b}	1.71 ± 0.05 ^{a}
Argnine	10.14 ± 0.45	9.81 ± 0.53	9.57 ± 0.34
Threonine	5.23 ± 0.19	5.15 ± 0.37	5.21 ± 0.12
Alanine	6.76 ± 0.30 ^{a}	6.37 ± 0.18 ^{a}	6.90 ± 0.44 ^{b}
Proline	8.67 ± 0.37 ^{a}	8.40 ± 0.50 ^{a}	9.67 ± 0.54 ^{b}
Tyrosine	3.40 ± 0.16 ^{b}	3.53 ± 0.12 ^{ab}	3.25 ± 0.09 ^{a}
Valine	4.26 ± 0.07	4.41 ± 0.04	4.36 ± 0.19
Methionine	1.49 ± 0.44 ^{a}	2.12 ± 0.41 ^{ab}	1.76 ± 0.56 ^{b}
Isoleucine	3.76 ± 0.12 ^{b}	3.96 ± 0.10 ^{c}	3.57 ± 0.18 ^{a}

Table 4. *Cont.*

Amino Acid	Dietary Proline Levels (% Dry Diet)		
	0	1.5	3
Leucine	0.30 ± 0.12 [b]	0.29 ± 0.12 [b]	0.44 ± 0.18 [a]
Phenylalanine	3.09 ± 0.12 [ab]	3.21 ± 0.08 [b]	2.98 ± 0.19 [a]
Lysine	5.18 ± 0.44 [b]	5.74 ± 0.52 [b]	4.40 ± 0.64 [a]

Note: Data with different lowercase letters in the same horizontal column are significantly different between the various diet groups ($p < 0.05$).

With the increasing dietary proline levels, both the content of EAAs in the body wall of *A. japonicus* and the ratio of EAAs to NEAAs (EAA/NEAA) initially increased and then decreased. Compared with the other two groups, the P3 group exhibited a significant reduction in both EAA content and the EAA/NEAA (Figure 4). The EAA/NEAA ratio first slightly increased at proline levels below 1.5%, likely due to adequate support for collagen synthesis without altering other amino acids. At higher levels (>1.5%), excess proline was primarily used for collagen production, causing accumulation of non-essential amino acids and a relative decrease in essential amino acids, thereby significantly lowering the EAA/NEAA ratio [25].

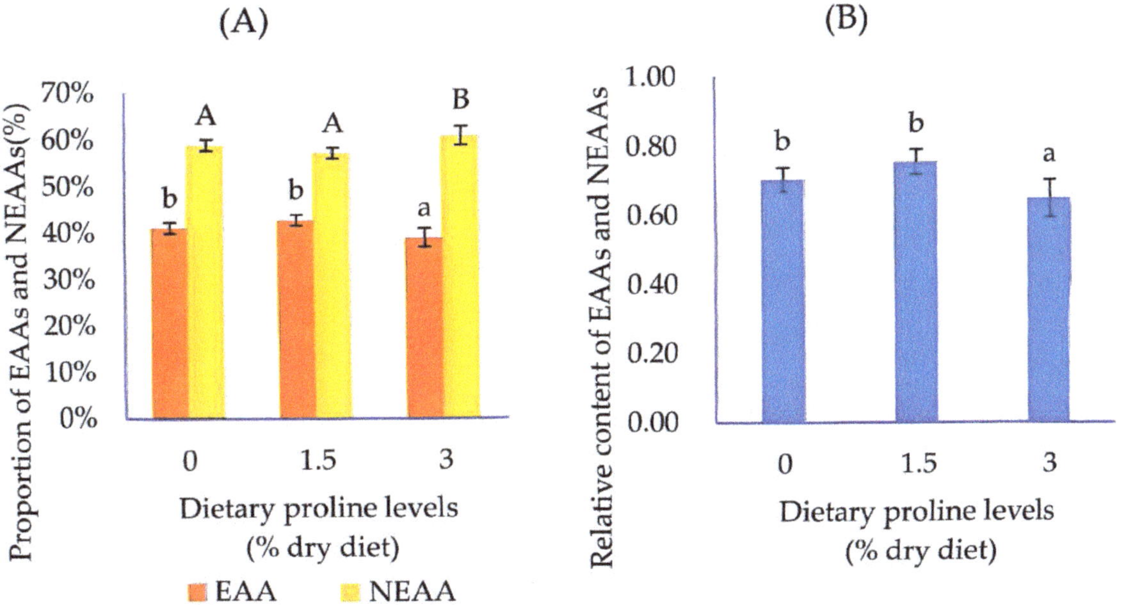

Figure 4. Effects of proline addition levels on the proportion of essential and non-essential amino acids in sea cucumber (*Apostichopus japonicus*) (% dry matter basis). Note: (**A**) The proportion of EAA and NEAA; (**B**) Relative content of EAA and NEAA. Bars bearing different lowercase letters are significantly different between the dietary groups ($p < 0.05$).

3.4. Organoleptic Quality

Dietary proline supplementation had a significant effect on the water adsorption of *A. japonicus* after boiling. As the level of proline addition in the feed increased, the water adsorption of the sea cucumber body wall initially increased and then decreased. The water adsorption of the body wall in the experimental groups was significantly higher than that in the control group ($p < 0.05$) (Figure 5).

Dietary proline supplementation had a significant effect on the textural properties of *A. japonicus*. With increasing levels of proline in the diet, the hardness, cohesiveness, chewiness, and adhesiveness of the sea cucumber body wall were all significantly enhanced ($p < 0.05$), reaching maximum values at a 3% proline supplementation level. In addition, no significant differences were observed in elasticity among the three dietary groups ($p > 0.05$) (Table 5).

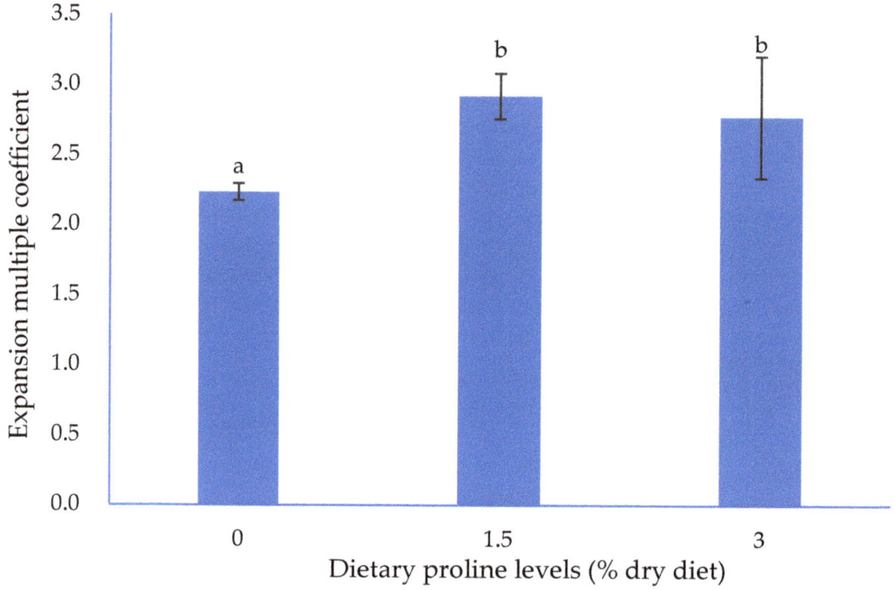

Figure 5. Effects of proline addition levels on the expansion of multiple coefficients of sea cucumber (*Apostichopus japonicus*) after boiling. Note: Bars bearing different lowercase letters are significantly different between the dietary groups ($p < 0.05$).

Table 5. Effects of proline addition levels on the texture profile analysis of sea cucumber (*Apostichopus japonicus*).

	Dietary Proline Levels (% Dry Diet)		
	0	**1.5**	**3**
Hardness [1]	2.80 ± 0.08^{a}	3.32 ± 0.73^{ab}	3.76 ± 0.55^{b}
Springiness [2]	2.64 ± 0.33	2.84 ± 0.29	2.78 ± 0.27
Cohesiveness [3]	0.80 ± 0.05^{a}	0.83 ± 0.03^{ab}	0.85 ± 0.03^{b}
Chewiness [4]	5.86 ± 1.03^{a}	7.79 ± 2.44^{ab}	8.90 ± 1.88^{b}
Gumminess [5]	2.21 ± 0.16^{a}	2.72 ± 0.70^{ab}	3.18 ± 0.46^{b}

Note: [1] Hardness (N) was defined as the maximum peak value when compressing the sample for the first time. [2] Springiness (mm) was defined as the degree to which the sample can be recovered after the first compression. [3] Cohesiveness (Ratio)was defined as the adhesion within the sample. [4] Chewiness (mJ) was defined as the work required to chew a solid sample (springiness × gumminess). [5] Gumminess (N) was defined as the viscosity characteristics of semi-solid samples (hardness × cohesiveness). Data with different lowercase letters in the same horizontal column are significantly different between the various diet groups ($p < 0.05$).

3.5. Expression of Collagen Synthesis-Related Genes

Compared to the P0 group, the relative expression levels of collagen type I alpha 2 chain (*COL1A1*), Sma- and Mad-related protein 1(*SMAD1*), and sp-smad2/3 (*SMAD2/3*) in the body wall of *A. japonicus* were significantly upregulated in both the P1.5 and P3 groups ($p < 0.05$). However, compared to the P1.5 group, the relative expression levels of *COL1A1*, transforming growth factor beta (*TGF-β*), *SMAD1*, and *SMAD2/3* were significantly reduced in the P3 group ($p < 0.05$) (Figure 6).

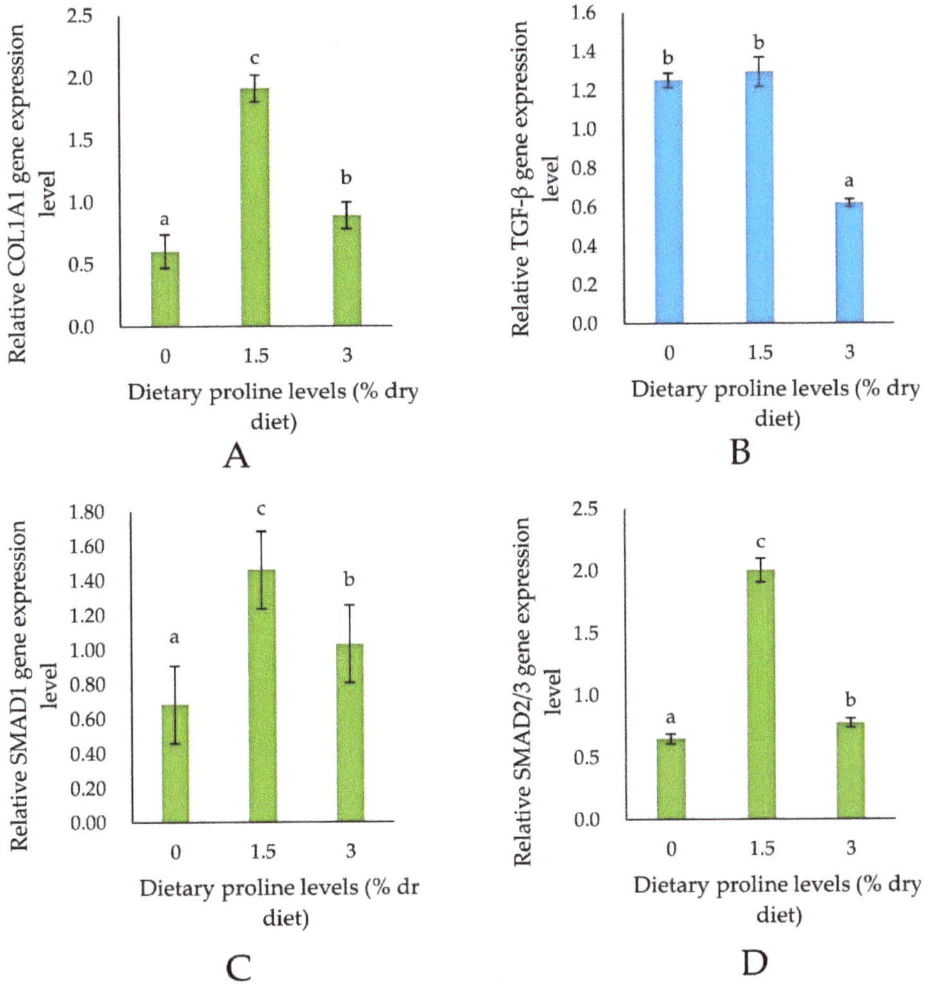

Figure 6. Effects of proline addition levels on collagen deposition gene expression in the body walls of sea cucumber (*Apostichopus japonicus*). Note: Bars bearing different lowercase letters are significantly different between the dietary groups ($p < 0.05$). (**A**) COL1A1; (**B**) TGF-β; (**C**) SMAD1; (**D**) SMAD2/3.

4. Discussion

Proline, as a non-polar, neutral amino acid with a cyclic structure, is widely present in both plant and animal organisms and serves as an important component of proteins. Traditionally, proline has been classified as a non-essential amino acid in most animals, including aquatic species. However, recent studies have indicated that endogenously synthesized proline could not sufficiently meet the physiological demands of animals, especially during rapid growth, tissue repair, or under the state of environmental stress [26,27]. Therefore, proline is now considered a member of "conditionally essential amino acids". Increasing evidence suggests that dietary proline supplementation can significantly promote growth in aquatic animals.

Research has shown that proline supplementation significantly promotes growth performance in various aquatic species. Taziki et al. [28] demonstrated that dietary supplementation with 5 g/kg of L-proline, in combination with an equal amount of L-alanine, significantly enhanced the growth performance and survival rate of juvenile common carp (*Cyprinus carpio*). Additionally, supplementation of 1.0% proline in the diets significantly increased the weight gain rate of large yellow croaker (*Larimichthys crocea*) [28]. In *L.vannamei*, dietary supplementation with 2.29–2.34% proline has been found to significantly enhance antioxidant capacity, immune function, and stress resistance, thereby providing a robust foundation for subsequent growth [27]. In addition to proline, hydroxyproline—the major metabolite of proline—has also been shown to exert similar growth-promoting effects in aquatic species, further supporting the functional importance of this amino acid family in aquaculture nutrition. Aksnes et al. (2008) found that the addition of 2.9 g/kg crystalline hydroxyproline to high plant protein-based diets significantly promoted growth and vertebral development of Atlantic salmon (*Salmo salar*) [29]. In the present study, dietary supplementation with 1.5–3% proline significantly improved the weight gain rate of *A. japonicus*, which differs markedly from previously reported optimal levels in other aquatic species. This discrepancy may be attributed to the fact that collagen synthesis in the body wall of sea cucumbers relies heavily on proline and hydroxyproline as precursors, resulting in a higher dependence on exogenous proline. Additionally, the basal diet used in this experiment was primarily composed of plant-based protein sources, which may have contained lower intrinsic levels of proline, thereby increasing the requirement for dietary supplementation. Moreover, variations in proline utilization efficiency, differences among species, and the growth stage of the animals may also contribute to the observed differences.

Collagen is one of the most abundant structural proteins in living organisms, widely distributed in the skin, bones, blood vessels, and connective tissues [30]. Proline is not only an essential amino acid component of collagen molecules but also plays a critical role in the hydroxylation of collagen precursors [16,31]. Numerous studies have demonstrated that appropriate dietary supplementation of proline can significantly promote collagen deposition in aquatic animals, thereby improving muscle quality and tissue structure. In yellow drum (*Nibea coibor*), dietary proline supplementation significantly enhanced collagen synthesis and deposition, and muscle elasticity [32]. Similarly, in triploid crucian carp (*Carassius auratus*), proline supplementation not only significantly promoted body weight gain and improved feed utilization but also notably increased collagen content in the dorsal muscle [33]. In olive flounder (*Paralichthys olivaceus*), although proline supplementation did not significantly enhance growth performance, it markedly increased collagen content in the muscle tissue [34]. These findings confirm the promoting effect of proline on collagen synthesis in aquatic animals. In shrimp, dietary proline has been shown to significantly increase collagen content in both muscle tissue and the exoskeleton [35]. The body wall of sea cucumbers is primarily composed of collagen and proline, which as a key substrate in collagen biosynthesis, can promote collagen deposition, thereby enhancing body wall hardness and elasticity. In the present study, appropriate proline supplementation in diets significantly increased the collagen content in the body wall of *A. japonicus*, particularly in the 1.5% proline supplementation group, where collagen deposition was approximately 14% higher than that of the control group. This suggests that proline supplementation can directly promote the synthesis and deposition of collagen in the body wall of *A. japonicus*.

Proline is one of the primary amino acids constituting collagen and plays a critical role in the process of collagen synthesis [36,37]. As a structural protein, collagen is widely distributed in the skin, bones, muscles, and connective tissues, and its contents directly influence muscle hardness, elasticity, and ultimately, market quality of aquatic animals [38]. Textural properties are essential indicators for evaluating food quality, as texture analysis provides insights into attributes such as mouthfeel, chewiness, and overall sensory experience, which are crucial for quality control and consumer satisfaction [39,40]. Proline not only serves as a necessary precursor for collagen synthesis but also contributes to increasing the degree of intermolecular cross-linking within collagen, thereby enhancing its stability and resistance to degradation. This, in turn, improves muscle hardness and elasticity in aquatic animals, ultimately enhancing flesh firmness and mouthfeel [41]. In the present study, the elasticity of the body wall of *A. japonicus* in the proline-supplemented groups improved to some extent, although there was no significant difference compared to the control group. Meanwhile, the hardness, cohesiveness, chewiness, and adhesiveness of the body wall increased significantly with the increase of dietary proline. This was consistent with the results of Wang et al. [42] who found that appropriate myo-inositol(MI) levels increased flesh texture quality by increasing muscle hardness, adhesive force, springiness, resilience and chewiness. The proportion of type I and type III collagen fibers could influence the texture characteristics of the body wall of *A. japonicus*. The hardness of the body wall of *A. japonicus* is positively correlated with its type I collagen content [43]. In the periodontal ligament, the tissue's resistance to deformation is positively correlated with its collagen fiber content [44]. Similarly, in bone tissue, mechanical strength and deformation resistance are enhanced when collagen fibers are organized in a denser and more orderly manner [45]. In this study, Sirius Red staining, a technique specific for labeling collagen fibers, demonstrated that the collagen fibers in the experimental groups were markedly thicker than those observed in the control group, accompanied by a significant increase in the proportion of type I collagen fibers. Furthermore, Van Gieson staining corroborated these findings, revealing a denser arrangement of collagen fibers in the experimental groups. Collagen fibers are formed by the self-assembly of collagen triple-helical molecules into microfibrils, which subsequently aggregate into thicker fibers [46]. Therefore, when the concentration of collagen molecules increases or the fibers are more tightly aligned, lateral associations between microfibrils may occur, leading to the formation of thicker fibers. Alternatively, an increase in intra- and intermolecular cross-linking can stabilize the interactions between microfibrils, thereby increasing fiber diameter. In addition, elevated levels of extracellular matrix proteins may promote tighter alignment and aggregation of collagen fibers.

In the present study, dietary supplementation with 1.5% proline significantly upregulated the expression of *COL1A1*, *SMAD1*, and *SMAD2/3* genes in the body wall of *A. japonicus*, thereby promoting collagen synthesis. *COL1A1* encodes the α1 chain of type I collagen, a fundamental structural component of collagen fibers that plays a critical role in maintaining tissue integrity and elasticity. Proline contributes to collagen synthesis through dual mechanisms, functioning both as a structural amino acid in procollagen and as a co-substrate for prolyl-4-hydroxylase (P4H) that catalyzes the formation of hydroxyproline, a modification essential for stabilizing the collagen triple-helix structure [19,47]. Previous studies have demonstrated that supplementation with proline or its derivatives can significantly enhance *COL1A1* transcription. For instance, in spotted drum, dietary proline has been shown to upregulate *COL1A1* expression and facilitate collagen deposition [48]. In the present study, *COL1A1* expression in the P1.5 group was significantly higher than that in the control group, supporting the conclusion that proline promotes the transcription and synthesis of procollagen.

It is widely acknowledged that the *TGF-β/Smad* signaling pathway serves as a central regulatory network governing collagen synthesis [49]. Upon TGF-β activation, receptor-regulated *SMAD2/3* are phosphorylated and subsequently form a complex with co-Smad (*SMAD4*), which translocates into the nucleus to activate target genes such as *COL1A1* [18,50]. In fish studies, dietary proline has been reported to activate *SMAD2/3* expression and enhance collagen deposition via the *TGF-β* signaling pathway. For example, in spotted drum, increased *SMAD2* expression was associated with elevated collagen content in tendons and the swim bladder. Furthermore, studies suggest that proline not only upregulates *SMAD2/3* transcription but may also enhance their activation, thereby facilitating *COL1A1* transcription [48]. In the present study, the expression levels of *SMAD2/3* in the P1.5 group were significantly higher than those in the control group, accompanied by elevated *SMAD1* expression. *SMAD1*, a receptor-regulated Smad in the BMP signaling pathway, may also participate in collagen gene regulation through crosstalk with the *TGF-β* pathway. Studies in dermal fibroblasts have shown that *SMAD1* and *SMAD2/3* can synergistically regulate collagen gene expression downstream of *TGF-β* receptor activation [51–53]. Our findings suggest that the upregulation of *SMAD1* and *SMAD2/3* in the body wall of sea cucumber under 1.5% proline supplementation may enhance Smad signal transduction, thereby promoting the transcription of collagen-related genes.

In this study, TPA parameters and tissue amino acid contents generally increased with dietary proline levels up to 3.0%, whereas the most pronounced improvements in histological features and gene expression were observed in the 1.5% proline group. Notably, collagen content and gene expression were measured in fresh body wall samples, while TPA parameters were assessed in rehydrated samples after processing. This methodological difference may partly explain the observed divergence, as rehydration can alter water content and the spatial structure of collagen, thereby affecting TPA parameters such as hardness and chewiness [54–56]. Furthermore, the disparity between indicators can also be attributed to their underlying physiological response mechanisms. TPA parameters and amino acid contents primarily reflect the accumulation of terminal metabolic products, showing a linear or near-linear response to increasing proline levels [57]. In contrast, tissue morphology and gene expression are subject to physiological regulation and feedback control, reaching optimal activation at specific proline levels, with higher levels potentially causing saturation or negative feedback inhibition [58]. Temporal differences in response also play a role, improvements in TPA parameters and amino acid contents reflect long-term cumulative effects, whereas changes in gene expression and tissue morphology represent more immediate physiological responses.

The findings of this study can provide important insights for the optimization of formulated feeds to produce high quality sea cucumbers characterized by more collagen deposition and better texture quality.

5. Conclusions

In conclusion, dietary supplementation with 1.5% proline in *A. japonicus* feed upregulated the expression of collagen synthesis-related genes, including *COL1A1*, *SMAD1*, and *SMAD2/3*, thereby enhancing growth performance and promoting hydroxyproline synthesis. Furthermore, the 1.5% proline supplementation facilitated collagen deposition, improved muscle structure, increased collagen fiber density, and enhanced the rehydration ratio of the body wall, ultimately leading to an improvement in the overall quality of *A. japonicus*. Based on the outcomes related to growth performance, collagen content in the body wall, rehydration ratio, textural properties, and the relative expression levels of collagen-related genes, a dietary proline supplementation level of 1.5% was found to be more beneficial for the growth of *A. japonicus*.

Author Contributions: R.X.: writing—original draft, investigation, data curation, formal analysis. Z.W.: review and editing, data curation. H.L.: investigation, data curation. R.L.: investigation, data curation. X.W.: investigation, data curation. H.Y.: investigation, data curation. J.D.: methodology, resources. Y.C.: supervision, funding acquisition. R.Z.: data curation, writing—review and editing, supervision, funding acquisition. All authors have read and agreed to the published version of the manuscript.

Funding: This research was funded by the Xingliao Talent Plan for Top Young Talents (XLYC2203036).

Institutional Review Board Statement: Following the animal experiment guidelines for the care and use of laboratory animals at Dalian Ocean University, the studies on animals were reviewed by the Committee for the Welfare and Ethics of Laboratory Animals at Dalian Ocean University. (Approval code 202351; date 8 June 2023).

Informed Consent Statement: Not applicable.

Data Availability Statement: Data are contained within the article.

Acknowledgments: The authors thank all the students and staff who contributed to and supported the entire study.

Conflicts of Interest: The authors declare no conflicts of interest.

References

1. Wang, G.; Dong, S.; Tian, X.; Gao, Q.; Wang, F.; Xu, K. Life Cycle Assessment of Different Sea Cucumber (*Apostichopus japonicus Selenka*) Farming Systems. *J. Ocean Univ. China* **2015**, *14*, 1068–1074. [CrossRef]
2. Wang, G.-D.; Li, C.-Y.; Hu, M.; Xu, T.; Xu, K.-F. Current Situation Analysis and Development Suggestions of the *Apostichopus japonicus* Culture Industry in Shandong Province. *Isr. J. Aquac.-Bamidgeh* **2022**, *74*, 1–9. [CrossRef]
3. Ru, X.; Feng, Q.; Zhang, S.; Liu, S.; Zhang, L.; Yang, H. Eco-Friendly Method for Rearing Sea Cucumber (*Apostichopus japonicus*) Larvae. *Aquac. Res.* **2022**, *53*, 3759–3766. [CrossRef]
4. Xia, S.; Wang, X. Chapter 19—Nutritional and Medicinal Value. In *Developments in Aquaculture and Fisheries Science*; Yang, H., Hamel, J.-F., Mercier, A., Eds.; Elsevier: Amsterdam, The Netherlands, 2015; Volume 39, pp. 353–365.
5. Guo, Z.; Meng, X.; Zhang, Y.; Wu, D.; Zhang, R.; Zhao, X.; Ren, T.; Han, Y. Replacing Sea Mud with Attachment of Suspension Cage Can Improve Growth and Gut Health for Sea Cucumber *Apostichopus Japonicus*. *Front. Mar. Sci.* **2024**, *11*. [CrossRef]
6. Kobayashi, M.; Msangi, S.; Batka, M.; Vannuccini, S.; Dey, M.M.; Anderson, J.L. Fish to 2030: The Role and Opportunity for Aquaculture. *Aquac. Econ. Manag.* **2015**, *19*, 282–300. [CrossRef]
7. Saito, M.; Kunisaki, N.; Urano, N.; Kimura, S. Collagen as the Major Edible Component of Sea Cucumber (*Stichopus japonicus*). *J. Food Sci.* **2002**, *67*, 1319–1322. [CrossRef]
8. Frantz, C.; Stewart, K.M.; Weaver, V.M. The Extracellular Matrix at a Glance. *J. Cell Sci.* **2010**, *123*, 4195–4200. [CrossRef]
9. To, V.-A.; Liou, C.-H. Taurine Supplementation Enhances the Replacement Level of Fishmeal by Soybean Concentrate in Diets of Juvenile Pacific White Shrimp (*Litopenaeus vannamei* Boone, 1931). *Aquac. Res.* **2021**, *52*, 3771–3784. [CrossRef]
10. Rong, H.; Zhang, Y.; Hao, M.; Lin, F.; Zou, W.; Zhang, H.; Yu, C.; Yu, J.; Shi, Q.; Aweya, J.J.; et al. Effect of Hydroxyproline Supplementation on Growth Performance, Body Composition, Amino Acid Profiles, Blood-Biochemistry and Collagen Synthesis of Juvenile Chu's Croaker (*Nibea coibor*). *Aquac. Res.* **2020**, *51*, 1264–1275. [CrossRef]
11. Wu, G.; Bazer, F.W.; Burghardt, R.C.; Johnson, G.A.; Kim, S.W.; Knabe, D.A.; Li, P.; Li, X.; McKnight, J.R.; Satterfield, M.C.; et al. Proline and Hydroxyproline Metabolism: Implications for Animal and Human Nutrition. *Amino Acids* **2011**, *40*, 1053–1063. [CrossRef]
12. Albaugh, V.L.; Mukherjee, K.; Barbul, A. Proline Precursors and Collagen Synthesis: Biochemical Challenges of Nutrient Supplementation and Wound Healing. *J. Nutr.* **2017**, *147*, 2011–2017. [CrossRef]
13. Rong, H.; Xia, Y.; Wang, X.; Hu, Q.; Jia, D.; Wu, X.; Kong, L.; Bi, B. Effects of dietary proline on growth, body composition and antioxidant capacity of Chu' s croaker (*Nibea coibor*). *J. Shanghai Ocean Univ.* **2023**, *32*, 89–97.
14. Rong, H.; Yang, Y.; Wang, X.; Yan, H.; Gao, Y.; Bi, B.; Shi, Q.; Kong, L. Effects of Proline on Collagen Deposition and Related Gene Expression in Swim Bladder of Chu' s Croaker (*Nibea coibor*). *Chin. J. Anim. Nutr.* **2022**, *34*, 3184–3193.
15. Deng, Z.; Song, W.; Rui, Q.; Shi, C.; Zhang, J.; Li, C. Effects of Proline on Growth Performance, Antioxidant Capacity and Resistance to Aeromonas hydrophila Infection of *Pelodiscus sinensis*. *Chin. J. Anim. Nutr.* **2023**, *35*, 3213–3227.
16. Salo, A.M.; Myllyharju, J. Prolyl and Lysyl Hydroxylases in Collagen Synthesis. *Exp. Dermatol.* **2021**, *30*, 38–49. [CrossRef]
17. Leask, A.; Abraham, D.J. TGF-Beta Signaling and the Fibrotic Response. *FASEB J.* **2004**, *18*, 816–827. [CrossRef]
18. Hillege, M.M.G.; Galli Caro, R.A.; Offringa, C.; de Wit, G.M.J.; Jaspers, R.T.; Hoogaars, W.M.H. TGF-β Regulates Collagen Type I Expression in Myoblasts and Myotubes via Transient Ctgf and Fgf-2 Expression. *Cells* **2020**, *9*, 375. [CrossRef]
19. Wang, Z.; Xu, R.; Yang, H.; Li, R.; Ding, J.; Chang, Y.; Zuo, R. Vitamin E Regulates the Collagen Contents in the Body Wall of Sea Cucumber (*Apostichopus japonicus*) via Its Antioxidant Effects and the TGF-β/Smads Pathway. *Antioxidants* **2024**, *13*, 847. [CrossRef]
20. AOAC International. *Official Methods of Analysis (OMA)*, 16th ed.; AOAC International: Arlington, VA, USA, 1995.
21. AOAC International. *Official Methods of Analysis (OMA)*, 17th ed.; AOAC International: Arlington, VA, USA, 2002.

22. Dong, X.; Liu, W.; Song, X.; Lin, X.; Yu, D.; Yu, C.; Zhu, B. Characterization of Heat-Induced Water Adsorption of Sea Cucumber Body Wall. *J. Food Sci.* **2019**, *84*, 92–100. [CrossRef]
23. Zuo, R.; Li, M.; Ding, J.; Chang, Y. Higher Dietary Arachidonic Acid Levels Improved the Growth Performance, Gonad Development, Nutritional Value, and Antioxidant Enzyme Activities of Adult Sea Urchin (*Strongylocentrotus intermedius*). *J. Ocean Univ. China* **2018**, *17*, 932–940. [CrossRef]
24. Livak, K.J.; Schmittgen, T.D. Analysis of Relative Gene Expression Data Using Real-Time Quantitative PCR and the $2^{-\Delta\Delta CT}$ Method. *Methods* **2001**, *25*, 402–408. [CrossRef]
25. Church, D.D.; Hirsch, K.R.; Park, S.; Kim, I.-Y.; Gwin, J.A.; Pasiakos, S.M.; Wolfe, R.R.; Ferrando, A.A. Essential Amino Acids and Protein Synthesis: Insights into Maximizing the Muscle and Whole-Body Response to Feeding. *Nutrients* **2020**, *12*, 3717. [CrossRef]
26. Wu, G. Functional Amino Acids in Growth, Reproduction, and Health. *Adv. Nutr.* **2010**, *1*, 31–37. [CrossRef]
27. Xie, S.-W.; Tian, L.-X.; Li, Y.-M.; Zhou, W.; Zeng, S.-L.; Yang, H.-J.; Liu, Y.-J. Effect of Proline Supplementation on Anti-Oxidative Capacity, Immune Response and Stress Tolerance of Juvenile Pacific White Shrimp, *Litopenaeus vannamei*. *Aquaculture* **2015**, *448*, 105–111. [CrossRef]
28. Taziki, T.; Jafari, V.; Mazandarani, M.; Hoseinifar, S.H. Effects of Dietary L-Proline and L-Alanine on Growth Performance, and Flesh Quality of Common Carp (*Cyprinus carpio*) Juveniles. *Ann. Anim. Sci.* **2023**, *23*, 195–204. [CrossRef]
29. Aksnes, A.; Mundheim, H.; Toppe, J.; Albrektsen, S. The Effect of Dietary Hydroxyproline Supplementation on Salmon (*Salmo salar* L.) Fed High Plant Protein Diets. *Aquaculture* **2008**, *275*, 242–249. [CrossRef]
30. Nagaoka, I.; Tsuruta, A.; Yoshimura, M. Chondroprotective Action of Glucosamine, a Chitosan Monomer, on the Joint Health of Athletes. *Int. J. Biol. Macromol.* **2019**, *132*, 795–800. [CrossRef]
31. Sipilä, K.H.; Drushinin, K.; Rappu, P.; Jokinen, J.; Salminen, T.A.; Salo, A.M.; Käpylä, J.; Myllyharju, J.; Heino, J. Proline Hydroxylation in Collagen Supports Integrin Binding by Two Distinct Mechanisms. *J. Biol. Chem.* **2018**, *293*, 7645–7658. [CrossRef]
32. Lin, F.; Rong, H.; Lin, J.; Yuan, Y.; Yu, J.; Yu, C.; You, C.; Wang, S.; Sun, Z.; Wen, X. Enhancement of Collagen Deposition in Swim Bladder of Chu's Croaker (*Nibea coibor*) by Proline: View from *In-Vitro* and *In-Vivo* Study. *Aquaculture* **2020**, *523*, 735175. [CrossRef]
33. Cao, S.; Xiao, Y.; Huang, R.; Zhao, D.; Xu, W.; Li, S.; Tang, J.; Qu, F.; Jin, J.; Xie, S.; et al. Dietary Supplementation with Hydroxyproline Enhances Growth Performance, Collagen Synthesis and Muscle Quality of *Carassius auratus* Triploid. *Front. Physiol.* **2022**, *13*. [CrossRef]
34. Zhang, K.; Ai, Q.; Mai, K.; Xu, W.; Liufu, Z.; Zhang, Y.; Peng, M. Effects of Dietary Hydroxyproline on Growth Performance, Body Composition, Hydroxyproline and Collagen Concentrations in Tissues in Relation to Prolyl 4-Hydroxylase α(I) Gene Expression of Juvenile Turbot, *Scophthalmus maximus* L. Fed High Plant Protein Diets. *Aquaculture* **2013**, *404–405*, 77–84. [CrossRef]
35. Shi, M.; Li, H.; Chen, T.; Huang, B.; Li, X.; Dong, X.; Chi, S.; Yang, Q.; Liu, H.; Deng, J.; et al. Effects of Hydroxyproline Supplementation in Low Fish Meal Diets on Collagen Synthesis, Myofiber Development and Muscular Texture of Juvenile Pacific White Shrimp (*Litopenaeus vannamei*). *Anim. Nutr.* **2024**, *17*, 428–437. [CrossRef]
36. Barbul, A. Proline Precursors to Sustain Mammalian Collagen Synthesis12. *J. Nutr.* **2008**, *138*, 2021S–2024S. [CrossRef]
37. Li, P.; Wu, G. Roles of Dietary Glycine, Proline, and Hydroxyproline in Collagen Synthesis and Animal Growth. *Amino Acids* **2017**, *50*, 29–38. [CrossRef]
38. Tong, F.; Bai, J.; Tang, Z.; Li, C.; Liu, S.; Wei, Z. Replacing Fish Meal with Hydrolyzed Collagen Derived from Fish By-Products Improved Muscle Quality and Glycolipid Metabolism of Triploid Crucian Carp. *Foods* **2023**, *12*, 1235. [CrossRef]
39. Wolinska-Kennard, K.; Schönberger, C.; Fenton, A.; Sahin, A.W. Mouthfeel of Food and Beverages: A Comprehensive Review of Physiology, Biochemistry, and Key Sensory Compounds. *Compr. Rev. Food Sci. Food Saf.* **2025**, *24*, e70223. [CrossRef]
40. Zhang, Y.; Pandiselvam, R.; Zhu, H.; Su, D.; Wang, H.; Ai, Z.; Kothakota, A.; Khaneghah, A.M.; Liu, Y. Impact of Radio Frequency Treatment on Textural Properties of Food Products: An Updated Review. *Trends Food Sci. Technol.* **2022**, *124*, 154–166. [CrossRef]
41. Li, X.; Bickerdike, R.; Lindsay, E.; Campbell, P.; Nickell, D.; Dingwall, A.; Johnston, I.A. Hydroxylysyl Pyridinoline Cross-Link Concentration Affects the Textural Properties of Fresh and Smoked Atlantic Salmon (*Salmo salar L.*) Flesh. *J. Agric. Food Chem.* **2005**, *53*, 6844–6850. [CrossRef]
42. Wang, M.; Feng, L.; Wu, P.; Liu, Y.; Ren, H.; Jin, X.; Zhou, X.; Jiang, W. Regulation of Muscle Springiness and Hardness: The Role of *Myo*-Inositol in Enhancing Fish Flesh Texture. *Food Chem. X* **2025**, *29*, 102788. [CrossRef]
43. Li, S.; Zhou, Y.; Sun, L.; Wang, Y.; Song, S.; Ai, C.; Yang, J. Effects of Storage Method on the Quality of Processed Sea Cucumbers (*Apostichopus japonicus*). *Foods* **2022**, *11*, 4098. [CrossRef]
44. Wu, B.; Li, N.; Liu, M.; Cheng, K.; Jiang, D.; Yi, Y.; Ma, S.; Yan, B.; Lu, Y. Construction of Human Periodontal Ligament Constitutive Model Based on Collagen Fiber Content. *Materials* **2023**, *16*, 6582. [CrossRef] [PubMed]
45. Niu, Y.; Du, T.; Liu, Y. Biomechanical Characteristics and Analysis Approaches of Bone and Bone Substitute Materials. *J. Funct. Biomater.* **2023**, *14*, 212. [CrossRef] [PubMed]
46. Orgel, J.P.R.O.; Irving, T.C.; Miller, A.; Wess, T.J. Microfibrillar Structure of Type I Collagen *In Situ*. *Proc. Natl. Acad. Sci. USA* **2006**, *103*, 9001–9005. [CrossRef] [PubMed]

47. Rong, H.; Zhang, H.; Ning, L.; Wu, K.; Limbu, S.M.; Shi, Q.; Qin, C.; Wen, X. The Transforming Growth Factor Beta (TGF-β/Smads) Pathway Regulates Collagen Synthesis and Deposition in Swim Bladder of Chu's Croaker (*Nibea coibor*) Stimulated by Proline. *Aquaculture* **2022**, *558*, 738360. [CrossRef]
48. Rong, H.; Lin, F.; Limbu, S.M.; Lin, Z.; Bi, B.; Dou, T.; Zhao, L.; Wen, X. Effects of Dietary Proline on Swim Bladder Collagen Synthesis and Its Possible Regulation by the TGF-β/Smad Pathway in Spotted Drum, *Nibea diacanthus*. *Aquac. Nutr.* **2020**, *26*, 1792–1805. [CrossRef]
49. Meng, X.; Nikolic-Paterson, D.J.; Lan, H.Y. TGF-β: The Master Regulator of Fibrosis. *Nat. Rev. Nephrol.* **2016**, *12*, 325–338. [CrossRef]
50. Devos, H.; Zoidakis, J.; Roubelakis, M.G.; Latosinska, A.; Vlahou, A. Reviewing the Regulators of COL1A1. *Int. J. Mol. Sci.* **2023**, *24*, 10004. [CrossRef]
51. Guo, X.; Wang, X.-F. Signaling Cross-Talk between TGF-β/BMP and Other Pathways. *Cell Res.* **2009**, *19*, 71–88. [CrossRef]
52. Saadat, S.; Noureddini, M.; Mahjoubin-Tehran, M.; Nazemi, S.; Shojaie, L.; Aschner, M.; Maleki, B.; Abbasi-kolli, M.; Rajabi Moghadam, H.; Alani, B.; et al. Pivotal Role of TGF-β/Smad Signaling in Cardiac Fibrosis: Non-Coding RNAs as Effectual Players. *Front. Cardiovasc. Med.* **2021**, *7*, 588347. [CrossRef]
53. Wu, M.; Chen, G.; Li, Y.-P. TGF-β and BMP Signaling in Osteoblast, Skeletal Development, and Bone Formation, Homeostasis and Disease. *Bone Res.* **2016**, *4*, 16009. [CrossRef]
54. Li, Y.; Li, Y.; Lai, Q.; Yao, Z.; Zhou, K.; Niu, D. Proline Metabolism and Its Osmotic Regulatory Role in Enhancing Hypersalinity Tolerance of the Razor Clam (*Sinonovacula constricta*). *Aquaculture* **2024**, *589*, 741019. [CrossRef]
55. Wang, S.; Lin, R.; Cheng, S.; Tan, M. Water Dynamics Changes and Protein Denaturation in Surf Clam Evaluated by Two-Dimensional LF-NMR T1-T2 Relaxation Technique during Heating Process. *Food Chem.* **2020**, *320*, 126622. [CrossRef]
56. Ge, X.; Wang, H.; Yin, M.; Wang, X. Effect of Different Thawing Methods on the Physicochemical Properties and Microstructure of Frozen Instant Sea Cucumber. *Foods* **2022**, *11*, 2616. [CrossRef]
57. Meena, M.; Divyanshu, K.; Kumar, S.; Swapnil, P.; Zehra, A.; Shukla, V.; Yadav, M.; Upadhyay, R.S. Regulation of L-Proline Biosynthesis, Signal Transduction, Transport, Accumulation and Its Vital Role in Plants during Variable Environmental Conditions. *Heliyon* **2019**, *5*. [CrossRef]
58. Geng, S.; Wang, H.; Wang, X.; Ma, X.; Xiao, S.; Wang, J.; Tan, M. A Non-Invasive NMR and MRI Method to Analyze the Rehydration of Dried Sea Cucumber. *Anal. Methods* **2015**, *7*, 2413–2419. [CrossRef]

fishes

MDPI

Article

Four Organic Protein Source Alternatives to Fish Meal for Pacific White Shrimp (*Penaeus vannamei*) Feeding

Yosu Candela-Maldonado [1], Imane Megder [1], Eslam Tefal [2], David S. Peñaranda [1], Silvia Martínez-Llorens [1], Ana Tomás-Vidal [1], Miguel Jover-Cerdá [1] and Ignacio Jauralde [1,*]

[1] Research Group of Aquaculture and Biodiversity, Institute of Animal Science and Technology, Universitat Politècnica de València, 46022 Valencia, Spain; yocanmal@etsiamn.upv.es (Y.C.-M.); dasncpea@upvnet.upv.es (D.S.P.); silmarll@dca.upv.es (S.M.-L.); atomasv@dca.upv.es (A.T.-V.); mjover@dca.upv.es (M.J.-C.)

[2] Department of Animal and Poultry Production, Faculty of Agriculture, Damanhour University, Damanhour 22516, Egypt; etefal@doctor.upv.es

[*] Correspondence: igjaugar@upv.edu.es

Abstract: The use of eco-organic ingredients as a source of protein in aquaculture diets needs important attention due to the growing demand for organic seafood products. The present study evaluated the effects of fish meal substitution by different organic ingredients on the growth, body composition, retention efficiency, enzyme activity, and nutrient digestibility of white shrimp *Penaeus vannamei*. The four dietary formulations tested were formulated with organic ingredients and the fish meal was replaced by the following organic protein meals: Iberian pig viscera meal (PIG), trout by-product meal (TRO), insect meal (FLY), and organic vegetable meal (WHT), in addition to a control diet (CON) that included 15% fish meal. A growth trial was carried out for 83 days, raising 1 g shrimp to commercial size (20 g). Shrimp were stocked at 167 shrimp/m^3 (15 individuals per 90 L tank). The results showed that the growth obtained by shrimp fed with TRO (19.27 g) and PIG (19.35 g) were similar in weight gain to the control diet (20.76 g), while FLY (16.04 g) and WHT (16.73 g) meals resulted in a significant lower final weight. The FLY diet showed significantly lower protein digestibility (68.89%) compared to the CON, PIG, TRO, and WHT diets, and significantly higher trypsin activity (0.17 mU/g) compared to shrimp fed with the PIG, TRO, and WHT diets. Shrimp fed with WHT have a significantly lower body weight percentage of protein (19.69%) than shrimp fed with the WHT and TRO diets, and some significant differences in dietary aminoacidic levels affecting amino acid body composition. These results indicate that Iberian pig viscera and trout by-product meal can successfully replace fish meal in Pacific white shrimp aquaculture.

Keywords: organic; shrimp; digestibility; fish meal; diets; protein

Key Contribution: The substitution of fish meal in shrimp feed formulations with organic ingredients such as rainbow trout by-products, Iberian pig viscera, and insect meal has shown potential to enhance nutrient absorption, digestibility, and growth rates in shrimp. Evaluating the impact of these alternative diets on digestive efficiency and enzymatic activity offers promising strategies to support environmentally friendly shrimp farming and promote more sustainable and health-conscious aquaculture systems.

1. Introduction

Organic aquaculture is a term usually understood as synonymous with ecological aquaculture and is a comprehensive method of farming fish and other marine species

 https://doi.org/10.3390/fishes10080384

that adheres to organic principles [1]. These principles can be summarized as sustainable resource management, environmental protection, principles of social responsibility and equity, and careful location selection to protect ecosystems, among others including the use of certified organic feed. Shrimp is the aquaculture product with the highest economic value around the world (USD 32 million), with a total production of almost 7 million tonnes in 2024, and although its organic production growth is very recent, with only 421 tonnes in 2018 [2], some companies are producing organic shrimp —both *P. vannamei* and *P. monodon*—as an alternative strategy for diversification to serve a growing niche of consumers demanding organic products.

Compared to organic land-based farming, organic aquaculture is a very recent form of production. However, it is considered a rapidly growing sector due to the increasing demand for low-environmental-impact products [1]. Different countries, regions, and organizations have developed different requirements for labeling shrimp as organic or eco-friendly, including concepts like density, welfare, soil practices, water quality, input limitation for reproduction, health management, and feeding (hormones, antibiotics, and feed ingredients) [1]. The European Union regulates organic aquaculture through Regulation (EU) 2018/848, ensuring environmentally sustainable and high-welfare aquaculture practices [3]. In 2020, the EU's total organic aquaculture production reached 74,032 tonnes, accounting for 6.4% of the region's total aquaculture output [4]. Organic shrimp production is subject to strict guidelines, including organic feed requirements and sustainable stocking densities [3]. The leading shrimp organic aquaculture producer within the EU is Italy. The sector has significant growth potential, supported by the EU's Farm to Fork Strategy, which aims to expand organic aquaculture by 2030, ensuring economic viability and environmental sustainability [4].

Organic ingredients are defined as raw materials produced in accordance with organic farming standards, excluding the use of synthetic chemicals, genetically modified organisms, and artificial additives. As outlined by Lembo and Mente and the European Commission, ingredients must be certified by authorized bodies, fully traceable, and sourced through environmentally sustainable practices that prioritize animal welfare and ecological integrity [1,3].

The market value of shrimp presents the highest commercial value in the global aquaculture market (USD 31.241 million in 2022) and continues to expand due to increasing global demand, making shrimp aquaculture one of the most profitable sectors in the seafood industry [5]. Intensive shrimp farming has demonstrated high economic feasibility, with projected global market values reaching billions of dollars annually [5]. This growth is driven by rising consumer preference for protein-rich seafood and advancements in sustainable farming techniques. The industry presents significant growth opportunities, particularly in Asia and Latin America, where good climatic conditions and innovative technological systems have improved production efficiency [5]. With an increasing focus on sustainability and value-added products, shrimp aquaculture remains a highly lucrative market. Strategic investment in feed optimization, certified organic products, and efficient supply chains will be crucial in sustaining long-term profitability while meeting global demand for high-quality shrimp products [5].

The feed used to raise animals is probably the main issue for developing organic aquaculture. European Union (EU) legislation protects the terms organic or ecological with a label that can only be used if the production is certified, and to obtain that label the feed used to raise organic shrimp must be certified as organic and conform to the minimum standards considered in the regulations, usually prohibiting GMOs, limiting pesticides and antibiotics, banning the use of synthetic amino acids, and many other requirements [3]. Other certificates have similar criteria, typically allowing the use of sustainable raw materials due to the scarcity of organic raw materials on the market, mainly as a protein source, and particularly fish meal [1].

Fish meal from sustainable sources can be used in organic shrimp diets, but price and scarcity make the use of other protein sources preferable. Several studies have been conducted with organic plant ingredients [6–8]. A study by Browdy [8] was the first to use an organic vegetable diet to feed *P. vannamei* shrimp in ponds, and no significant differences were found between the experimental diet and a commercial diet containing fish meal in terms of growth or feed conversion ratio (FCR). Several studies on substitution by non-organic sources, such as those by Oujifard and colleagues [9], have completely replaced fish meal with rice protein concentrate, but included 15% shrimp meal to ensure an aminoacidic balanced source.

Regarding animal sources not labeled as eco-friendly, several non-eco-friendly ingredients, both animal and vegetal, have been used in the past as an alternative to fish meal in shrimp diets [10–12]. World rainbow trout aquaculture production in 2020 was 82 million tons [13], therefore its by-products represent a valuable resource for circular economy practices in sustainable aquaculture feed.

Amaya [12] reported good results without fish meal, using poultry by-products or vegetal mix. Some studies have been conducted with insects that, although not certified as organic, are likely to be labeled as organic in the near future [10,14–16]. Defatted insect meals are especially interesting due to their amino acid composition and high protein content, similar to that of fish meal. However, from an ecological perspective, whole, undefatted insect meals are more sustainable, especially compared to those defatted using a solvent [17].

Another approach has been the use of used fermented corn protein concentrate, such as in a study by Galkanda-Arachchige [18], but other studies showed the poorest results with several vegetal concentrates or foods fermented without fish meal [19–21]. Good results were only obtained when including small quantities of fish meal [22–24], but in other cases the inclusion of fish meal, chicken, and blood meal resulted in poorer growth [25].

The development of organic shrimp aquaculture requires certified organic feed, and the requirements to obtain certification will become more demanding in the future. The scarcity of certified organic animal ingredients, particularly fish meal, poses a major challenge. Although plant-based eco-organic alternatives exist, their effectiveness in shrimp diets remains uncertain [8,21]. This study is essential to evaluate the limits of replacing fish meal with organic ingredients and ensuring sustainability, growth performance, and nutritional adequacy in organic shrimp farming.

According to this context, the main challenge to developing organic shrimp production is in obtaining organic diets based on organic ingredients, considering the lack of alternative organic animal ingredients available on the market, as Tefal [26–28] has shown regarding fish.

To this end, four alternative ingredients with low ecological impact have been considered: Iberian pork by-products, due to their high availability, consistency, and quality of animal origin. Animal species generally contain good quality protein that contains all the essential amino acids. Organic trout by-products are currently less available on the market, but, a priori, present a very interesting profile, as we replace fish meal with organically produced fish by-products, thus contributing to the circular economy. Organic vegetable meals, the most available, represent the best non-by-product option, as they have lower trophic values, but also have amino acid profiles that could be less suitable. Some vegetables such as soya beans also contain good quality protein, but some vegetables do not and are deficient in one or more essential amino acids. Wheat is an example of a vegetable that is deficient in lysine and other essential amino acids. To use vegetables to provide all the essential amino acids there are two strategies: (a) the use of a vegetable that contains all of the essential amino acids, or (b) use a combination of vegetables that singly are deficient in one or more essential amino acids, but in combination supply all the amino acids. There is a third strategy that involves supplementation with synthetic amino acids, which is not permitted in the case of organic production, or insect protein, emerging meals with a still very high price but with a very promising future in aquaculture.

The aim of this research is to study the total replacement of fish meal with organic ingredients, both animal and vegetable, in shrimp feed for ecological and sustainable aquaculture, studying growth, nutritional parameters and digestibility.

2. Materials and Methods

2.1. Formulation and Manufacture of Experimental Diets

Five experimental diets were formulated and manufactured with raw materials from both plant and animal organic origin. The total fish meal in the control diet (CON) was replaced by Iberian pig viscera by-products (PIG), organic insect meal from *Hemetia illucens* larvae (FLY), organic trout by-products (TRO), or organic wheat gluten (WHT), which all maintained isoprotein and isolipid dietary content. As different raw materials have different amounts of protein and fat, the values of oils and some ingredients were varied to maintain isoprotein and isolipid content. The ingredients and nutritional compositions of experimental diets are shown in Table 1. Amino acid and fatty acid compositions are shown in Table 2 and Table 3, respectively.

Table 1. Ingredients and nutritional composition in dry matter of the experimental feeds.

Ingredients (g kg^{-1})	CON	PIG	FLY	TRO	WHT
Raw materials (g kg^{-1})					
Fish meal [a]	150				
Organic trout by-product [b]				205	
Organic insect meal [c]			286		
Organic Iberian pig meal [d]		173			
Organic wheat [e]	420	435	331	396	432
Organic wheat gluten [f]	90	88	104	94	214
Organic soybean meal [g]	225	225	225	225	225
Organic fish oil	32.5	29	4	15	42
Organic soybean oil	32.5			15	37
Calcium phosphate	25	25	25	25	25
Lecithin	15	15	15	15	15
Vitamins [h]	10	10	10	10	10
Nutritional composition (% Dry Matter)					
Dry matter	86.88	86.82	88.57	88.02	89.16
Crude protein	36.02	38.3	38.56	35.87	35.84
Crude lipid	10.07	10.61	9.45	11.29	10.34
Ash	6.24	4.71	7.16	6.46	4.49

[a] Fish meal (91.9% DM, 73.9% CP, 11.2% CL, 14.3% Ash); CORPESCA S.A. [b] Organic trout by-product (97.5% DM, 34.5% CP, 41.7% CL, 3% CHO, 20.80% Ash) (Naturix, Spain). [c] Organic insect meal (92.6% DM, 37.6% CP, 28.5% CL, 20% CHO, 13.9% Ash) (Entomoch, Spain). [d] Organic Iberian pig by-products (92.6% DM, 53.0% CP, 28.6% CL, 14.6% CHO, 3.8% Ash) (Jamón y Salud, Spain)). [e] Organic wheat (92.3% DM, 12.7% CP, 1.3% CL, 1.7% Ash, 19.7 kJ^{-1} Energy); (PIENSOS ecoLUCAT, Barrax, Albacete, Spain). [f] Organic wheat gluten (93.7% DM, 87.6% CP, 2.2% CL, 10.2% CHO, 0.05% Ash); (PIENSOS ecoLUCAT, Barrax, Albacete, Spain). [g] Organic soybean meal (94.6% DM, 43.1% CP, 9.3% CL, 6.3% Ash, 19.7 kJ^{-1} Energy); (PIENSOS ecoLUCAT, Barrax, Albacete, Spain). [h] Vitamin and mineral mix (per kg): colloidal silica, 176.7 g; sepiolitic clay, 357.3 g; butylhydroxytoluene, 20 g; vitamin B12, 0.010 g; niacinamide, 20 g; folic acid, 1.5 g; vitamin D3, 2 × 105 UI; vitamin A, 2 × 106 UI; vitamin E, 10 g; vitamin K3, 2.5 g; vitamin B1, 3 g; vitamin B2, 3 g; vitamin B6, 2 g; calcium d-pantothenate, 10 g; biotin, 0.3 g; inositol, 50 g; betaine anhydrous, 50 g; iron (II) sulfate, monohydrate, 0.6 g; potassium iodide 0.05; copper (II) sulfate pentahydrate, 0.9 g; manganese (II) oxide, 0.96 g; zinc sulfate monohydrate, 0.75 g; sodium selenite, 0.001 g; medium: calcium carbonate, sodium chloride, and potassium chloride. CON—control diet; PIG—organic Iberian pig diet; FLY—organic insect diet; TRO—organic trout by-product diet; WHT—organic vegetal diet.

Table 2. Amino acid composition of experimental diets.

Diets	CON	PIG	FLY	TRO	WHT
Essential Amino Acids (g/100 g dry sample)					
Arginine	2.07	1.87	1.72	2.26	1.48
Histidine	0.90	0.64	0.68	0.68	0.65
Isoleucine	1.38	1.19	1.32	1.54	1.27
Leucine	2.51	2.32	2.54	2.63	2.40
Lysine	1.72	1.44	1.72	1.83	1.02
Methionine	0.59	0.59	0.48	0.68	0.40
Phenylalanine	1.41	1.35	1.51	1.47	1.49
Threonine	1.23	1.13	1.25	1.29	0.96
Valine	1.61	1.61	1.85	1.72	1.46
Non-Essential Amino Acids (g/100 g dry sample)					
Alanine	1.46	1.38	1.51	1.62	1.03
Aspartate	2.76	2.62	3.01	3.04	2.07
Cystine	0.42	0.51	0.53	0.42	0.46
Glutamine	8.01	7.90	8.22	9.89	10.64
Glycine	1.62	1.74	1.35	2.52	1.20
Proline	2.32	2.43	2.50	2.64	3.23
Serine	1.56	1.69	1.7	1.74	1.58
Tyrosine	0.94	0.91	1.17	0.97	0.88

CON—control diet; PIG—organic Iberian pig diet; FLY—organic insect diet; TRO—organic trout by-product diet; WHT—organic vegetal diet.

Table 3. Fatty acid composition (g/100 g of dry matter) of experimental diets.

	CON	PIG	FLY	TRO	WHT
Saturated fatty acids					
(C13:0)	0.266	0.374	0.391	0.393	0.222
(C14:0)	0.170	0.162	0.393	0.147	0.414
(C15:0)	0.019	0.017	0.021	0.014	0.014
(C16:0)	1.538	1.696	1.508	2.075	1.300
(C17:0)	0.022	0.018	0.021	0.030	0.015
(C18:0)	0.431	0.446	0.434	1.015	0.240
(C20:0)	0.033	0.037	0.030	0.031	0.016
MUFA [1]					
(C16:1)	0.195	0.179	0.284	0.197	0.126
(C17:1)	0.010	0.009	0.014	0.017	0.008
(C18:1n9c)	2.112	2.600	2.449	3.126	1.266
(18:1(n-7))	0.206	0.229	0.220	0.271	0.103
(C20:1)	0.115	0.141	0.185	0.136	0.041
(C24:1)	0.021	0.019	0.022	0.016	0.006
PUFA [2]					
(C18:2n6c) LA [3]	4.020	4.670	2.420	3.336	2.690
(C18:3n3) LNA [4]	0.513	0.622	0.339	0.361	0.355
(C20:2)	0.076	0.081	0.100	0.083	0.021
(C20:3n6)	0.015	0.021	0.026	0.016	0.005
(C20:4n6)	0.025	0.030	0.033	0.021	0.008
(22:4n-6)	0.004	0.004	0.006	0.018	0.001
(22:5n-3)	0.395	0.265	0.339	0.195	0.072
EPA [5]	0.293	0.251	0.248	0.169	0.062
DHA [6]	0.395	0.265	0.339	0.195	0.072

[1] MUFA—monounsaturated fatty acids. [2] PUFA—polyunsaturated Fatty acids. [3] LA—linoleic acid. [4] LNA—linolenic acid. [5] EPA—eicosapentaenoic acid. [6] DHA—docosahexaenoic acid. CON—control diet; PIG—organic Iberian pig diet; FLY—organic insect diet; TRO—organic trout by-product diet; WHT—organic vegetal diet.

All the organic ingredients came from organically certified producers with the EU organic label. Organic trout was self-processed to extract meat slices and the remaining parts were carefully cut, oven-dried, and ground into a suitable form for use in the feed. The remaining parts used for the Iberian pig meal were the liver, intestines, and heart, which were cut into small pieces. After oven-drying, they were ground into a suitable form for incorporation into the feed. Dry organic insect larvae were utilized after they were ground for inclusion in the diets. Dietary formulation and processing used organic raw components labeled and approved following Regulation (EU) 2018/848 [29]. The feed was extruded with two nozzle sizes: 2 mm for shrimp weighing between 2 and 12 g and 3 mm for those weighing more than 12 g. For shrimp weighing less than 2 g, the 2 mm feed was ground and sieved to obtain 1 mm crumbs. Once completed, they were packaged and stored in a commercial refrigerator at 4 °C.

The diets used in the trial were manufactured in the feed factory facilities of the Universitat Politècnica de València (UPV) by a cooking extrusion process, using a semi-industrial double-screw extruder (CLEXTRAL BC 45, St. Etienne, France) from the Animal Technology and Science Institute. The following processing parameters were used: a screw speed of 100 rpm, a pressure range of 20–30 atm, and a temperature of 110 °C.

The dry ingredients were ground in a mill with a 1 mm mesh screen. For the mixture, we mixed the macro-ingredients on one side and the micro-ingredients (calcium phosphate, lecithin, and vitamins) on the other. The micro-ingredient mixture was mixed with a small portion of the macro-ingredient mixture, increasing the quantity of the macro-ingredient mixture in successive proportions. Finally, the oils were added to the horizontal screw mixer.

2.2. Experimental Design

Shrimp growth trials were conducted in the facilities of the Wet Aquaculture Laboratory of the Universitat Politècnica de València. Shrimp used for this study came as larvae from the White Panther hatchery in Austria. Before the beginning of the experiment, larvae were kept in an adaptation period and fed with a commercial diet until they reached 1 g.

The shrimp were placed in 30 plastic rectangular tanks with a capacity of 90 L, with independent and constant aeration and temperature regulation. Tanks were included within an indoor RAS system, which allowed a water flow of 40% per hour. Water was purified using a mechanical filter, a foam fractionator, and a biofilter.

Fifteen shrimps, with an average weight of approximately 1 g, were placed in each 90 L tank, which equates to 167 shrimp/m^3. Shrimp from six tanks were fed with one of the 5 experimental diets (CON, PIG, FLY, TRO, and WHT). All shrimps were individually weighed every 14 days, and the daily rations were changed depending on liveweight and biomass. The growth was estimated using the specific growth rate parameter (SGR) in the following equation:

$$SGR = 100 \times (LN(BW_f) - LN(BW_i))/\text{days}$$

where the initial body weight is BWi, BW_f is the final body weight, and days are the number of days between both weights.

The daily feeding rate varied from 10% for average weights less than 4 g, 8% for weights between 4 and 7 g, and 5% after the average weight reached 7 g. Feeding was carried out manually and divided into three doses per day during the first period and into two doses per day in the second period. The experiment lasted a total of 83 days.

For feed conversion ratio (FCR), feed intake (FI), ingested protein retention (IPR), and ingested fat retention (IFR) estimates, the final biomass was corrected to consider the biomass of dead shrimp during the trial:

$$\text{Final biomass} = \text{Final number of shrimp} \times \text{Final weight} + \text{Estimated biomass of dead shrimp}.$$

The biomass of dead shrimp was estimated from the average weight between sampling periods and the number of dead shrimps.

$$\text{FI (g 100 g fish} - 1 \text{ day} - 1) = 100 \times \text{feed consumption (g)}/(((\text{Final biomass} + \text{Initial biomass})/2)\ (\text{g}) \times \text{days}).$$

$$\text{FCR} = \text{feed consumption (g)}/(\text{Final biomass} - \text{Initial biomass})\ (\text{g}).$$

$$\text{IPR (\%)} = 100 \times ((\text{Final Biomass (g)} \times \text{\%final protein}) - (\text{Initial Biomass (g)} \times \text{\%Initial protein}))/(\text{feed consumption (g)} \times \text{\%crude protein}).$$

$$\text{IFR (\%)} = 100 \times ((\text{Final Biomass (g)} \times \text{\%final fat}) - (\text{Initial Biomass (g)} \times \text{\%Initial fat}))/(\text{feed consumption (g)} \times \text{\%gross fat}).$$

2.3. Chemical and Biochemical Analysis

Water quality parameters, including temperature, oxygen levels, and N compounds (ammonia, nitrites, and nitrates), were monitored daily. These parameters remained within the optimal range for the species' development: temperature 28 °C ± 1, pH 7.92, O_2 6.36 mg/L, salinity 30 ± 1 psu, and NO_3^- 26.3 mg/L. Temperature, salinity, pH, and O_2 were measured using a multiparametric device (HANNA, HI19829 Model, Woonsocket, RI, USA). Ammonia and nitrites were measured using the Baumgarten protocol [30] and remained lower than the detection limit of the method throughout the growth experiment ($N\text{-}NH_4^+ < 0.001$ and $N\text{-}NO_2^- < 0.001$ mg/L).

All biochemical analyses were developed in the laboratories of the Animal Technology and Science Institute, where samples of ingredients, diets, shrimp, and their feces were analyzed.

The diet's approximate composition (Table 1), as well as the whole shrimp, were examined using the methods described in AOAC [31] and analyzed for the following metrics: dry matter (105 °C to constant weight); ash (incinerated at 550 °C for five hours); crude protein (determined by the direct combustion method DUMAS using LECO CN628, Geleen, The Netherlands); and crude lipid (extracted with methyl-ether using ANKOMXT10 Extractor (Macedon, NY, USA)). Each analysis was carried out in triplicate.

A Waters HPLC system (Waters 474, Waters, Milford, MA, USA) composed of two pumps (Model 515, Waters), an autosampler (Model 717, Waters), a fluorescence detector (Model 474, Waters), and a temperature-control module was used to analyze the levels of amino acids (AA) in the diets and in the shrimp using a procedure previously described by Bosch [32].

Before hydrolysis, aminobutyric acid was introduced as an internal standard. AQC was used to derivatize AA (6-aminoquinolyl-N-hydroxysuccinimidyl carbamate). After oxidation with performic acid, methionine and cysteine were identified individually as methionine sulphone and cystic acid. AA was converted to methionine and cystine after it was separated with a reverse-phase C-18 column by Waters Acc—Tag (150 mm 3.9 mm). Table 3 shows the essential amino acid (EAA) content of experimental diets. All amino acid analyses were carried out in duplicate.

For the determination of fatty acids, a direct method of synthesis of fatty acid methyl esters (FAME) was performed. Fatty acids were analyzed in a Focus Gas Chromatograph (Thermo, Milan, Italy), equipped with a split/splitless injector and an ionization detector of flame. The separation was performed on an SPTM 2560 capillary column (Supelco, PA, USA). Shrimp meat fat analysis was carried out by extracting 10–30 mg of crude fat from each sample.

First, 1 mL of tridecanoic acid (C13:0) was used. Next, 0.7 mL of 10 N KOH and 5.3 mL of methanol were added. The tubes were incubated at 55 °C in a thermoblock for an hour and a half, and were subjected to vigorous shaking for 5 s every 20 min. After cooling to room temperature in a water bath, 1.5 mL of HPLC-grade hexane was added to the reaction tubes, vortexed, and centrifuged at 1006× g for 5 min. The hexane layer, containing the FAMEs, was then placed in vials for analysis by gas chromatography. Fatty acids were identified by comparing their retention times with those of a standard of fatty acid methyl esters (47885-U) from Supelco® (Bellefonte, PA, USA) and quantified using C13:0 as an internal standard [33]. Initial and final shrimp samples were analyzed to evaluate the nutritional value of each diet and to estimate the nutrient retention in the shrimp.

2.4. Enzymatic Activity Analysis

Regarding the evaluation of the activity of digestive enzymes, this study was carried out through the analysis of the enzymes trypsin and chymotrypsin at the level of the hepatopancreas. For sampling, the shrimp were slaughtered 3 h after sampling, and the cephalothorax part was cut from 12 shrimps per treatment (3 shrimps per bucket and 4 replicates/treatment). The weight of the cephalothorax was recorded, and it was stored at a temperature of −20 °C. Digestive enzyme analyses and trypsin and chymotrypsin activities were measured using 50 mM BApNA (N-α-benzoyl-L-arginine-p-nitroanilide hydrochloride, Merck KGaA, Darmstadt, Germany) and 50 mM GApNA (N-glutaryl-L-phenylalanine-p-nitroanilide, Merck KGaA, Darmstadt, Germany) as substrates, respectively. The absorbance at 405 nm was measured using an Ex multiscan spectrophotometer (Thermolab Systems, Helsinki, Finland), where a unit of activity was defined as 1 μg of p-nitroaniline released per minute.

2.5. Digestibility Study

After the feeding trial, the shrimp were individually placed in 80 L glass aquariums to continue with the digestibility trial and 3 replicates were made for each feed to obtain feces. Changes of water in each aquarium were carried out to maintain good water quality, twice per week, with a continuous filtration system during the test period. For the technique, two aquariums were used for each batch of shrimp. In one, the shrimp were fed and approximately 30 min later they were moved to a second aquarium, clean of food remains and molting, to ensure that only feces were collected. Feces were collected daily by manually siphoning the aquariums after feeding (at 8:30 a.m. and 2:00 p.m.). This process requires a lot of staff time but ensures proper collection of feces. The collected fecal material was stored at −20 °C, dried weekly in an oven at 55 °C with crucibles, and stored in hermetic plastic jars until reaching 4 to 5 g to carry out the analyses of nutritional components and markers.

Apparent digestibility of the diet was estimated indirectly using chromic oxide (Cr_2O_3) as an inert and indigestible marker, which was added to the mixtures shown in Table 1 at 5 g/kg. Additionally, its concentration was measured in the diet and feces to analyze dry matter, crude protein, and energy in both. An atomic absorption spectrometer was used to determine the amount of chromium oxide in diets and feces after acid digestion (Perkin Elmer 3300, Perkin Elmer, Boston, MA, USA). The apparent digestibility coefficients (ADCs) of the diets were calculated according to Cho [34].

2.6. Economic Indexes

For the economic study, two economic efficiency indicators were estimated. Economic conversion (EC) was calculated using the following formula:

$$\text{EC} = \text{FCR} \times \text{Price of diet (€/Kg)}$$

And Profit Index (PI), which takes into account increase in profit, was calculated using the formula

$$\text{PI} = \text{Organic shrimp value } (12\text{ €/Kg}) - \text{Cost of feeding } (\text{€/Kg shrimp})$$

2.7. Statistical Analyses

Growth data, nutrient utilization, biometric parameters, body composition, amino acid retention, protein and fat retention, and digestibility data were treated using analysis of variance (ANOVA). The Student–Newman–Keuls test was used to evaluate specific differences between diets. Levene's test was used to check for homogeneity of variance, and the Chi-square and Kolmogorov–Smirnov tests were used to check for normality. If the variables did not pass the normality test, data were transformed using ASIN or SQRT functions before performing the ANOVA test. Initial bodyweight was used as a covariate when analyzing Final Weight, Weight Gain, or SGR. Data were considered statistically significant when $p < 0.05$, and data are shown as the mean with standard error of the mean (±SEM). Statistical analyses of the data were performed with Statgraphics Statistical Graphics System, version Centurion 18—X64.

3. Results

Growth of *Penaeus vannamei* fed with the different organic experimental diets (CON; PIG; TRO; FLY; WHT) can be observed in Figure 1. Differences among treatments began on day 70, where shrimp fed with the FLY and WHT diets grew less than those on the rest of the treatments. On the final day 83, significant differences in weight appeared between shrimp fed with the CON diet and shrimp fed with the FLY and WHT diets, but not with PIG and TRO.

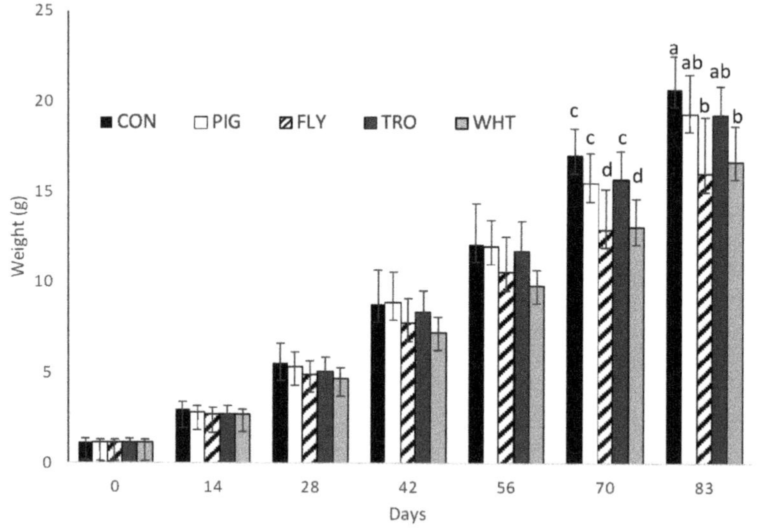

Figure 1. Average weight evolution of shrimps fed with experimental diets for 83 days. The data shown in the figure represents the mean ± standard deviation of 6 replicates per treatment. Different letters indicate significant differences ($p < 0.05$) for day 83 (p-value = 0.018) and for day 70 (p-value = 0.043). The means comparison test used was Student–Newman–Keuls. CON—control diet; PIG—organic Iberian pig diet; FLY—organic insect diet; TRO—organic trout by-product diet; WHT—organic vegetal diet.

The results of shrimp growth and nutritional parameters are shown in Table 4. No significant differences were seen among treatments in terms of survival, specific growth rate (SGR), feed intake (FI), and feed conversion ratio (FCR). However, there were statistical differences in the final weight; shrimp fed with the control diet (CON) had the highest final weight (20.7 g), with no differences between those fed with Iberian pig by-product (PIG) and organic trout by-product (TRO) diets (both 19.3 g), while the lowest final weights (16.7 g and 16.0 g) were in shrimp fed with the organic vegetal mixture (WHT) and organic insect meal (FLY) diets, respectively. Likewise, weekly gains in shrimp fed CON (1.65 g/week) were higher than in shrimp fed WHT (1.32 g/week) and FLY (1.26 g/week), but similar to shrimp fed PIG (1.54 g/week) and TRO (1.53 g/week).

Table 4. Growth and nutritional parameters of the shrimp fed with the organic experimental diets for 83 days.

Diets	CON	PIG	FLY	TRO	WHT	*p* Value
Initial Weight (g)	1.3 ±0.02	1.12 ±0.07	1.13 ±0.08	1.11 ±0.10	1.11 ±0.09	0.76
Final Weight (g)	22.4 ^{a} ±1.77	19.4 ^{ab} ±0.98	16.0 ^{b} ±0.98	19.1 ^{ab} ±1.07	16.9 ^{b} ±1.07	0.018
Survival Rate (%)	70.0 ±9.5	70.0 ±5.5	71.1 ±5.5	69.3 ±6.0	58.6 ±6.0	0.3
SGR (%/day)	3.62 ^{a} ±0.12	3.43 ^{ab} ±0.07	3.19 ^{ab} ±0.07	3.41 ^{ab} ±0.08	3.25 ^{b} ±0.08	0.036
Weekly Gain (g/week)	1.80 ^{a} ±0.14	1.54 ^{ab} ±0.08	1.26 ^{b} ±0.08	1.51 ^{ab} ±0.08	1.32 ^{b} ±0.08	0.017
FI (g/100 g shrimp day)	4.66 ±0.43	4.46 ±0.25	4.82 ±0.25	3.9 ±0.28	4.26 ±0.28	0.24
FCR	2.27 ±0.28	2.13 ±0.14	2.34 ±0.16	1.89 ±0.16	2.08 ±0.16	0.33

Values are represented as mean ± standard error of the mean (n = 6). For each treatment, the values that do not share the same letter are significantly different ($p < 0.05$). The means comparison test used was Student–Newman–Keuls. Specific Growth Rate (SGR) = 100 × ln (final weight/initial weight)/days. Week Gain (WG) = (final weight − initial weight)/weeks. Feed intake (FI) (g 100 g fish − 1 day − 1) = 100 × feed consumption (g)/(average biomass (g) × days). Feed conversion ratio (FCR) = feed consumption (g)/biomass gain (g). CON—control diet; PIG—organic Iberian pig diet; FLY—organic insect diet; TRO—organic trout by-product diet; WHT—organic vegetal diet.

Body composition and protein and fat retention efficiencies are presented in Table 5. All shrimp showed increased dry matter and protein from the beginning, but the percentage of fat was similar. Shrimp fed with the FLY diet had significantly higher dry matter and ash (25.75 and 2.90%, respectively), followed by shrimp fed with the TRO diet (25.39 and 2.69%, respectively). No differences in body fat appeared, but there were significant differences in body protein, with the highest values in shrimp fed FLY and TRO (both 20.5%) and lowest in shrimp fed WHT (19.7%).

Regarding retention efficiencies, no significant differences were found between all the organic diets and the ranges were 29–33% in IPR and 6–7% in IFR.

Table 6 shows amino acid composition of whole shrimp expressed as g per 100 g wet weight. For the essential amino acids, there were no significant differences in histidine, lysine, phenylalanine, or threonine in non-essential amino acids, except for alanine and proline, which were higher (1.27% and 1.28%, respectively) in shrimp fed with the FLY diet. Shrimp fed with TRO presented the highest level of methionine (0.52%) and those fed the FLY diet had the highest content of arginine, isoleucine, leucine, and valine (2.04, 0.85, 1.46, and 1.00%, respectively).

Table 5. Corporal composition, fat and protein retention, after feeding with organic experimental diets for 83 days.

	Initial	CON	PIG	WHT	TRO	WHT	*p* Value
Dry matter (%)	21.88	24.90 ^{c} ±0.13	24.54 ^{d} ±0.09	25.75 ^{a} ±0.13	25.39 ^{b} ±0.13	24.93 ^{c} ±0.12	<0.0001
Crude protein (%)	15.13	20.20 ^{ab} ±0.09	20.09 ^{ab} ±0.32	20.55 ^{a} ±0.12	20.52 ^{a} ±0.12	19.69 ^{b} ±0.39	0.0022
Crude lipid (%)	1.05	1.26 ±0.17	1.15 ±0.11	1.06 ±0.02	1.27 ±0.11	1.16 ±0.03	0.15
Ash	2.99	2.40 ^{b} ±0.01	2.37 ^{b} ±0.19	2.90 ^{a} ±0.17	2.69 ^{ab} ±0.37	2.64 ^{ab} ±0.26	0.0231
IPR (%)		29.0 ±7.7	29.5 ±2.6	29.2 ±4.2	33.5 ±8.7	30.8 ±4.2	0.67
IFR (%)		6.4 ±1.7	5.9 ±0.5	6.7 ±1.0	6.5 ±1.7	6.2 ±0.9	0.85

Values are represented as mean ± standard error of the mean (n = 3). For each treatment, the values that do not share the same letter are significantly different ($p < 0.05$). The means comparison test used was Student–Newman–Keuls. Ingested protein retention (%) IPR = 100 × ((Final Biomass (g) × %final protein) − (Initial Biomass (g) × %Initial protein))/(feed consumption (g) × %crude protein). Ingested fat retention (%) IFR = 100 × ((Final Biomass (g) × %final fat) − (Initial Biomass (g) × %Initial fat))/(feed consumption (g) × %gross fat).

Table 6. Composition of whole shrimp in essential and non-essential amino acids at the beginning and after feeding with organic experimental diets for 83 days.

Amino Acids g/100 g	Initial	CON	PIG	FLY	TRO	WHT	*p* Value
EAA [1]							
Histidine	0.39	0.44 ±0.01	0.47 ±0.03	0.46 ±0.02	0.47 ±0.08	0.47 ±0.01	0.23
Arginine	1.71	1.86 ^{ab} ±0.05	1.86 ^{ab} ±0.04	2.04 ^{a} ±0.06	1.77 ^{bc} ±0.14	1.63 ^{c} ±0.13	0.02
Valine	0.82	0.91 ^{c} ±0.04	0.98 ^{ab} ±0.01	1.00 ^{a} ±0.03	0.95 ^{abc} ±0.03	0.92 ^{bc} ±0.03	0.023
Methionine	0.34	0.45 ^{ab} ±0.02	0.43 ^{b} ±0.05	0.44 ^{ab} ±0.03	0.52 ^{a} ±0.06	0.43 ^{b} ±0.04	0.022
Lysine	1.15	1.39 ±0.02	1.43 ±0.05	1.43 ±0.03	1.40 ±0.06	1.32 ±0.04	0.83
Isoleucine	0.68	0.76 ^{b} ±0.04	0.82 ^{a} ±0.01	0.85 ^{a} ±0.03	0.79 ^{ab} ±0.03	0.75 ^{b} ±0.04	0.0065
Leucine	1.15	1.38 ^{ab} ±0.06	1.45 ^{ab} ±0.03	1.46 ^{a} ±0.06	1.41 ^{ab} ±0.07	1.33 ^{b} ±0.09	0.032
Phenylalanine	0.70	0.81 ±0.03	0.85 ±0.01	0.86 ±0.02	0.84 ±0.03	0.81 ±0.04	0.82
Threonine	0.66	0.75 ±0.03	0.76 ±0.01	0.80 ±0.02	0.75 ±0.04	0.73 ±0.003	0.65
NEAA [2]							
Aspartate	1.51	1.81 ±0.09	1.90 ±0.03	2.00 ±0.13	1.89 ±0.07	1.84 ±0.11	0.6
Serine	0.69	0.77 ±0.04	0.82 ±0.05	0.82 ±0.02	0.87 ±0.24	0.94 ±0.22	0.66
Glutamine	2.40	2.80 ±0.14	2.94 ±0.09	2.96 ±0.01	2.86 ±0.18	2.70 ±0.16	0.75
Glysine	1.32	1.56 ±0.09	1.60 ±0.03	1.58 ±0.09	1.64 ±0.06	1.58 ±0.02	0.67
Alanine	1.02	1.16 ^{b} ±0.04	1.23 ^{ab} ±0.03	1.27 ^{a} ±0.05	1.22 ^{ab} ±0.02	1.16 ^{b} ±0.04	0.032
Proline	0.92	1.08 ^{c} ±0.05	1.17 ^{b} ±0.03	1.28 ^{a} ±0.01	1.14 ^{bc} ±0.03	1.13 ^{bc} ±0.06	0.0095
Cystine	0.17	0.20 ±0.02	0.21 ±0.02	0.18 ±0.02	0.20 ±0.04	0.22 ±0.03	0.5
Tyrosine	0.62	0.74 ±0.03	0.76 ±0.01	0.75 ±0.02	0.76 ±0.02	0.74 ±0.02	0.94

Values are represented as mean ± standard error of the mean (n = 6). For each treatment, the values that do not share the same letter are significantly different ($p < 0.05$). [1]: Essential amino acids, [2]: Non-essential amino acids. The means comparison test used was Student–Newman–Keuls.

In amino acid retention efficiency, significant differences were found between shrimp fed with different diets for most essential amino acids (Figure 2). Shrimp fed with the WHT diet had the highest retention efficiency of lysine (64.4%), methionine (53.6%), and threonine (37.8%), and shrimp fed CON presented the lowest values in some amino acids.

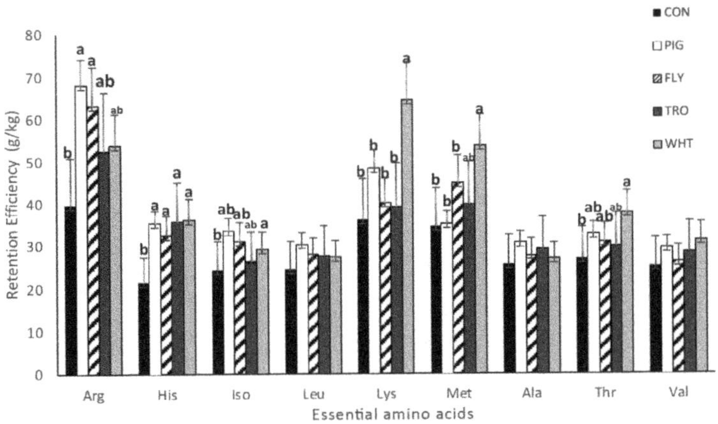

Figure 2. Retention Efficiency (Gained AA/Ingested AA) of essential amino acids in experimental diets. The data shown in the figure represents the mean ± standard deviation (N = 6). Different letters indicate significant differences ($p < 0.05$). The means comparison test used was Student–Newman–Keuls.

In the case of Lysine, shrimp fed with the WHT diet presented significant differences across all treatments. Regarding methionine values, shrimp fed the WHT diet presented significant differences to the shrimp fed with all treatments, except the shrimp fed with the TRO diet. Finally, retention efficiency of arginine showed significant differences; shrimp fed with the CON diet had lower values (39.8%) than shrimp fed with the PIG (68.0%) and FLY (63.0%) diets.

As shown in Table 7, significant differences in enzymatic activity were observed. Shrimp fed with the FLY diet showed higher trypsin activity (0.17 U/g) compared to those fed with WHT, TRO and PIG diets. Similarly, shrimp fed with the FLY diet also showed higher chymotrypsin activity (0.62 U/g) than those fed with the PIG diet.

Table 7. Apparent digestibility coefficients of dry matter, protein, and energy, and trypsin and chymotrypsin concentration of shrimp fed with organic experimental diets.

	DIETS					
ADC (%)	**CON**	**PIG**	**FLY**	**TRO**	**WHT**	***p* Value**
Dry matter (%)	56.72 [a] ±4.7	58.58 [a] ±3.2	33.50 [c] ±2.0	49.44 [b] ±1.1	46.33 [b] ±1.8	<0.0001
Crude protein (%)	80.43 [a] ±4.2	82.43 [a] ±0.5	68.89 [b] ±4.5	79.72 [a] ±1.4	79.91 [a] ±0.8	0.0022
Gross energy (%)	73.68 [ab] ±6.0	77.50 [a] ±1.8	64.95 [b] ±6.1	75.40 [ab] ±1.9	71.99 [ab] ±3.1	0.0478
Trypsin (mU/g)	0.12 [ab] ±0.02	0.10 [b] ±0.01	0.17 [a] ±0.03	0.12 [b] ±0.01	0.11 [b] ±0.01	0.0042
Chymotrypsin (mU/g)	0.51 [ab] ±0.05	0.36 [b] 0.05	0.62 [a] ±0.07	0.49 [ab] ±0.07	0.55 [ab] ±0.05	<0.0001

Values are represented as mean ± standard error of the mean (n = 3). For each treatment, the values that do not share the same letter are significantly different ($p < 0.05$). The means comparison test used was Student–Newman–Keuls. ADC (%) = 100 − 100 × [(marker in diet / marker in feces) × (AA in feces/AA in diet)]. Gross Energy (%) = [51.8 × (%C/100)) − (19.4 × (%N/100)].

Apparent digestibility coefficients (Table 7) revealed some differences, the FLY diet presented significantly lower values for dry matter, crude protein and gross energy (33.5, 68.9 and 64.9%, respectively) compared to the other diets. The ADC of dry matter registered higher values in the PIG and CON diets (58.6 and 56.7%, respectively), and intermediate values in TRO and WHT diets (49.4 and 46.3%, respectively). No differences can be observed on protein and energy ADC for the rest of diets.

The cost of organic experimental diets (Table 8) was different in the function of the price of ingredients. The cost was highest for FLY due to it consisting of very expensive organic insect meal, followed by WHT and CON because of the high prices of organic wheat gluten and soybean, and fish meal, respectively, making the PIG and TRO diets cheaper due to the low prices of by-products. Economic conversion and profit index were similar in all diets except in FLY, which presented the highest value of EC and lowest of PI. When comparisons were made without FLY, new differences appeared and PIG and TRO presented better EC and PI values than CON and WHT.

Table 8. Economic analysis of shrimp fed with organic experimental diets.

	DIETS				
	CON	**PIG**	**FLY**	**TRO**	**WHT**
Price of diet (€/Kg)	1.50	1.27	2.89	1.26	1.51
Economic conversion (€/Kg)	3.10 [b] ±0.3	2.70 [b] ±0.3	6.86 [a] ±1.9	2.36 [b] ±0.3	3.10 [b] ±0.6
Profit index (€/Kg)	8.90 [a] ±0.3	9.30 [a] ±0.3	5.14 [b] ±1.9	9.64 [a] ±0.3	8.90 [a] ±0.6

Values are represented as mean ± standard error of the mean (n = 6). For each treatment, the values that do not share the same letter are significantly different ($p < 0.05$). The means comparison test used was Student–Newman–Keuls. Price of ingredients (March 2022): fish meal—EUR 1933/t; organic trout by-product—EUR 750/t; organic insect meal—EUR 6500/t; Iberian pig viscera—EUR 750/t; organic wheat—EUR 515/t; organic wheat gluten—EUR 2450/t; organic soybean—EUR 1355/t; fish oil—EUR 3671/t; organic soybean oil—EUR 1650/t. Economic conversion (EUR/Kg shrimp), EC = FCR × Price of diet (EUR/Kg) Profit Index, PI = Organic shrimp value (12 EUR/Kg) − Cost of feeding (EUR/Kg shrimp).

4. Discussion

The exploration of alternative raw materials to protein derived from fish meal is complicated when they need to be certified as organic. Organic certification regulations promote the use of organic vegetable meals, animal by-products, which are preferable when they come from organic production, and insect production, which is seen as a good alternative to animal protein.

Shrimp survival can vary depending on many factors: initial and final weight, management, diet, and the use of probiotics, among others [35]. In the present test, it ranged between 58.6% and 71%, typical values for this species considering the entire growth range [35]. The estimated biomass data from mortality was considered in the calculations so as not to distort the FI and FCR results.

The growth, measured as SGR, of the organic diets was between 88 and 95% of the SGR obtained with the CON diet (Table 4), with similar SGR values (3.19–3.43%/day) to the values obtained by Liao [19] and Hlordzi [23] with a control diet, but these authors grew shrimp to lower weights (5.7–7.5 g). When weekly gain is considered, the growth obtained with the best diets (CON, PIG, and TRO) ranged from 1.53 to 1.65 g/w, which was in the same interval as values cited by other authors [12,20–22] and with a similar final weight. Although some tendencies were observed on days 42 and 56, growth differences between diets only appeared on day 70 (Figure 1).

Previously, only one study has researched the growth of shrimp on organic plant-based diets [8]. Comparison of results is limited by several factors. The feed composition used by Browdy et al. [8] includes 581 g/kg of soybean meal, 100 g/kg of pea meal, and 20 g/kg of squid and liquid fish meal. In the present study, the VEG diet has 214 g kg^{-1} of wheat gluten and 225 g kg^{-1} of soybean meal, which is responsible for the low lysine content (Table 2). In contrast, soybean and pea meals are richer in lysine, and it could be assumed that the diet used by Browdy et al. [8] was not deficient in lysine. Browdy et al. [8] developed their trial in earth ponds and reported a lower weekly gain (1.43 g/w) than the current trial but a better FCR (1.33), perhaps due to natural productivity and diet. Likewise, the organic plant diet included a small amount of squid meal and liquid fish solubles, and the good results may have originated from natural nutrients.

Currently, the growth of shrimp fed diets containing Iberian pig viscera (PIG) and organic trout by-products (TRO) was good, with no differences in relation to the control diet. These results agree with Amaya [12], who obtained similar shrimp growth with diets in which the protein source was a mix of soybean and poultry by-product or even only soybean. However, in the present study the vegetal diet (WHT) initiated lower growth due to the insufficient lysine content of the diet. Likewise, Galkanda-Arachchige [18] reported good results with diets without fish meal and fermented corn protein, but the duration of the trial was only 8 weeks and the shrimps' final weights were lower (3.8–4.7 g), so it is possible that differences did not appear. On the other hand, some authors obtained poor results with a total substitution of fish meal by soybean [21], by *Rhodobacter* fermented protein [19], or by fermented soybean [20]. Although the substitution was not total (5–10% fish meal), Manikandan and Felix [22] reported better growth with single-cell protein than the control diet, which is a promising result. Likewise, organic insect meal showed a bad result (FLY), in agreement with Cummins Jr [36] who cited lower growth (final weight 8 g) with a complete substitution of fish meal with *Hermetia* meal, and even with partial substitution (9–12 g), in relation to control diets (16 g).

Concerning FCR, the results were not as expected. The lowest value (1.83) was obtained with the diet TRO; however, no significant differences appeared between experimental diets, and FCR values were in general similar to those obtained by other authors, even with lower final weights [18–23], but clearly worse than those cited by Amaya [12] (around 1.1–1.2).

In general, similar values of FCR have been found when fish meal was replaced by alternative ingredients, but some differences can be cited. For example, Liao [19] obtained lower FCR values of 1.71 and 1.58 with 12 and 16% fish meal level, respectively, than diets including *Rhodobacter sphaeroides* protein, which showed an FCR of 2.24. Hlordzi [23] tried a 37.5% fish meal substitution (15% fish meal) by hydrolyzed fish protein meal (4%), obtaining a promising FCR of 1.5 in relation to 1.8–2.2 in other diets. Other alternative vegetable sources [18], such as fermented corn protein concentrate [21], soy protein concentrate [20], and fermented soybean meal, did not show differences in FCR between diets, and values were similar or higher (2.1–2.9) than for the shrimps fed with WHT diets of the present study (2.1). In the case of insect meal, FCR values were worse with high inclusion (4.0–4.5), in comparison with the control diet (FCR around 2.0), and much higher than in the current trial with organic insect meal (2.4).

When alternative protein sources, primarily plant-based, are recommended, the amino acid profile can be limiting. Raw materials such as soy flour provide more complete profiles than other plant-based sources. Wheat gluten is limiting in this case, due to its lysine content. In non-organic feeds, supplementation with synthetic amino acids is generally required. However, in organic aquaculture, this is difficult because organically certified synthetic amino acids are not usually available [26]. Nevertheless, Nunes [37] did not find global differences in the performance of shrimp fed diets containing several levels of fish meal (0, 6, 12, and 18%) supplemented with methionine (0.58, 0.69 and 0.82%), but considering only diets without fish meal, a methionine level of 0.58% gave a lower growth in comparison with 0.69%. In the current trial, diets of FLY and WHT, containing 0.48% and 0.40% met, gave a lower growth than the rest of the diets containing 0.59–0.68% met.

The diets were designed to have 35% crude protein and 10% lipids, using only organic raw materials and without supplementation with amino acids. But the amino acid needs were not considered; consequently, this WHT diet has some amino acid deficiencies. According to recommendations for meeting nutritional needs, lysine should be above 1.6% (Table 2) [38,39]. This is almost achieved in the case of the PIG diet (1.44), but was deficient in the case of the WHT diet (1.07). Assuming a lysine content in wheat gluten of 0.5% and with a lysine content of 2.6% in organic soybean meal (47% protein), it can be hypothesized that by varying the WHT diet by increasing the soybean meal up to 450 g/kg, decreasing the wheat gluten to 89 g/kg, and decreasing the organic wheat to 332 g/kg, we would obtain a WHT diet formulation with enough lysine to not be limiting, reaching the required 1.6% and maintaining the 35% of crude protein required. But with the current composition of the WHT diet, the main reason for the lower growth is the lysine content.

On the other hand, experimental organic diets presented different levels of essential fatty acids (Table 3), mainly EPA and DHA, which could have some effect on growth. The National Research Council [39] recommends minimum levels of LC-PUFA of 0.25–0.5%, which is provided by LNA in diets. But previous studies have demonstrated the inability of shrimp to elongate and desaturate PUFA into HUFA [40], and in some studies it seems that this does not affect growth performance [41]. Other studies have shown that EPA and DHA fatty acid levels above 1.7% and 2.7% appear advisable to optimize growth [42]. CON, PIG, and FLY presented similar values, and TRO slightly lower, but WHT had the lowest values, which could also partially explain its worse growth results.

Protein composition is usually a stable parameter, in the present study it varies slightly between 19.7% dm and 20.5% dm of body composition (Table 5), as expected and in agreement with other studies with alternative proteins, where the protein composition was very similar and between 20.0% and 21.5% in shrimp weighing 10 g [24], similar in small shrimp of 6–9 g with 17.5–19.1% [23] but higher than in big shrimp of 23.3 g, with 16.1–16.9% protein [22]. The significant differences in body protein found between shrimps fed with FLY treatment and those fed with WHT do not have enough importance to offer a clear result; for example, in the experiments of Cummins Jr., fish meal was replaced by insect meal in different proportions until total substitution, and a variation of between 18 and 20% protein was obtained without any significant differences or trends being noted. In the present study, all values are very close to 20.

IPR, a ratio that offers information on protein quality without distinguishing between amino acid profile or digestibility, ranged between 29% (CON) and 33.5% (TRO), but no significant differences among groups were found, which partially agrees with the only results obtained by Hlordzi [18] with a hydrolyzed fish protein, which report a good amino acid profile and good digestibility, obtaining an IPR that improved with dietary inclusion from 23 to 35%.

If experimental organic diets were not supplemented with amino acids then essential AAs were not balanced (Table 2) and, for example, arginine, lysine, methionine, threonine, and valine presented at lower levels in the WHT diet. Nevertheless, the final composition of shrimp did not reflect this (Table 6) because no differences were found in the same Aas, such as lysine and threonine, and in others, such as methionine, the differences were small. Despite the lower lysine content in the diet already mentioned, amino acid retention (Figure 2) is a good way to approach this matter. Shrimp fed the WHT diet showed the highest values of lysine, methionine, and threonine retention as a response to lower amounts of these amino acids in its composition, which means more effective retention in shrimp fed the WHT diet to cover requirements. This could mean that higher levels of methionine and threonine in the WHT diet could be desirable to maximize growth, in addition to the already known deficiency of lysine in the WHT diet. This effect has also been observed in several studies [26–28] in organic diets for seabream and seabass, containing lower amounts of a particular amino acid, methionine, in most cases.

Conversely, when the amino acid profile of essentials is well balanced, as expected in shrimp fed with the CON diet, retention of essential amino acids could be lower and the retention of essential amino acids may decrease due to reduced catabolic recycling. In such cases, amino acids are more efficiently utilized for protein synthesis, minimizing the need for excess retention and resulting in improved growth or feed conversion efficiency in shrimp. The poor digestibility of the FLY diet seems to have some influence on the bad performance in the shrimp fed it, which would partially agree with Tefal et al. [26], because seabass organic diets containing insect meal had a lower digestibility, but not with Tefal et al. [28], because seabream organic diets with insect meal presented the highest digestibility. Organic animal ingredients (PIG and TRO) had good digestibility, such as the organic vegetal diet (WHT), the results for which are in agreement with findings of high digestibility with fermented soybean [20] and single-cell protein [22]. In this study, the influence of the different diets in digestibility could be partially explained by enzyme activity.

Furthermore, chymotrypsin activity was higher than trypsin activity in shrimp, which agrees with other studies [43–45]. Although in the past trypsin was considered the main peptidase, chymotrypsin is nowadays recognized as the one with the highest activity [46]. Their activities have been shown to be a sensitive key in the protein digestion process, strongly modulated by the quality and quantity of food proteins, because trypsin shows a preference for amide substrates with arginine and lysine, and chymotrypsin for carbonyl groups of tyrosine, phenylalanine, tryptophan and leucine [44].

In this study, trypsin activity was significantly higher in shrimp fed with the FLY diet than the other shrimp fed with alternative protein ingredients, which could indicate greater difficulty in digesting the insect meal protein. This is probably caused by the probable chitin content of insect-based meals, which may impact digestive enzyme reactions. Increased chitin has been shown to have a negative influence on protein digestibility, at least in vitro [47].

Furthermore, in *P. monodon*, an increase in chitin has also been shown to decrease the digestibility of protein and lipids, although to a lesser extent [48]. A higher trypsin activity was found in low-protein diets, but no differences appeared in relation animal-to-plant protein ratio [49], which partially corroborates results for similar trypsin activity in diets of CON, PIG, TRO, and WHT.

Omont [45] cited higher trypsin activity in shrimp fed with marine algae in comparison with other plant ingredients (wheat), and although no data for amino acid composition was presented, the authors commented that it could be due to the amount of lysine and arginine. However, in the present study several experimental diets contained different levels of arginine and lysine and similar trypsin activity was obtained, except in the shrimp fed the WHT diet, which had intermediate levels of these amino acids.

In the case of chymotrypsin activity, differences only appeared between the shrimp fed with the FLY and PIG diets, which showed lower values, indicating that shrimp fed with the PIG diet required less digestive effort than the shrimp fed with the FLY diet. Other authors [45] have not found differences in chymotrypsin activity using several plant sources with similar levels of fish meal. For a more complete understanding, the interaction of different raw materials with digestibility and enzymes should also be studied with studies on intestinal health and immunology.

Alternative protein ingredients hold interest for aquaculture in improving sustainability, both environmental and economic. The environmental advantages of reducing the use of fish meal and reutilizing by-products are obvious, but economic benefits depend on the price of raw materials. In the case of organic feeding some ingredients can be more expensive, which makes an economic evaluation necessary. The economic productivity parameters shown in Table 9, in terms of kilograms per hectare, include some extrapolations that should be interpreted with caution. These experiments were conducted in RAS systems with small tanks, and caution should be taken in extrapolating to production sizes. Furthermore, ingredient prices can vary on the market; by-products do not have an established price, unlike other, more common raw materials. Insect meal also has a high current price, which could decrease in the future with the development of the industry.

Table 9. Productivity projection in the RAS system and in semi-intensive ponds.

	DIETS				
	CON	**PIG**	**FLY**	**TRO**	**WHT**
RAS productivity kg/m^2	0.58	0.50	0.42	0.49	0.36
RAS profit €/ha	51,432	46,546	21,550	47,026	32,484
Semi-intensive productivity	0.41	0.36	0.30	0.35	0.24
Semi-intensive profit €/ha	36,737	33,247	15,488	33,454	21,397

For projection, the trial density (36.86 shrimp/m^2), survival, and final weight (FW) from Table 4 was used to estimated RAS Productivity with the formula Productivity = Density × Survival(%) × FW (kg) and for profit, PI from Table 8 was used in the formula Profit = Productivity × PI × 10,000. For semi-intensive systems, survival is derived from Table 4 minus 20%.

Any different mechanism can explain the good results using organic ingredients because these products do not contain different nutrients to conventional ingredients. Iberian and trout by-products provided similar results to fish meal diets, but the low content of some EAAs, such as arginine and methionine, or the lower lysine content in the WHT diet, could explain the poor results for the wheat organic diet and partially for the INS diet. Furthermore, protein digestibility may also influence the INS diet.

5. Conclusions

In conclusion, the feeding of white shrimp *Penaeus vannamei* with organic protein ingredients can eliminate the use of fish meal from traditional fisheries without negative effects on growth performance, feed efficiency, retention efficiency, and shrimp body composition, obtaining economic advantages with animal by-products from Iberian pig and organic trout.

For future studies, it would be interesting to use new fermented organic plants, hydrolyzed organic animal by-products, and some vegetal protein concentrates to improve the availability and quantity of protein.

Author Contributions: Conceptualization, I.J. and M.J.-C.; data curation, Y.C.-M., A.T.-V., and S.M.-L.; formal analysis, A.T.-V. and I.J.; funding acquisition, M.J.-C.; investigation, Y.C.-M. and I.J.; methodology, S.M.-L., I.M., I.J., and M.J.-C.; supervision, A.T.-V. and D.S.P.; validation, A.T.-V. and D.S.P.; Writing—original draft, Y.C.-M., I.M., and E.T.; Writing—review and editing, Y.C.-M., A.T.-V., and I.J. All authors have read and agreed to the published version of the manuscript.

Funding: The research project had been developed with the collaboration of the Biodiversity Foundation (Spanish Ministry for Ecological Transition and the Demographic Challenge), through the Pleamar Program, co-financed by the European Maritime and Fisheries Fund (EMFF).

Institutional Review Board Statement: In accordance with Royal Decree 53/2013 and European Directive 2010/63/EU concerning the protection of animals used for scientific research, in which invertebrates are not included, official approval was not required. All the experiments were carried out at the Animal Experimentation Centre (code: ES462500001091).

Data Availability Statement: All data are contained within the article.

Conflicts of Interest: The authors declare no conflicts of interest.

References

1. Lembo, G.; Mente, E. *Organic Aquaculture*; Lembo, G., Mente, E., Eds.; Springer International Publishing: Cham, Switzerland, 2019; ISBN 978-3-030-05602-5.
2. FAO. *The State of World Fisheries and Aquaculture 2024*; FAO: Rome, Italy, 2024. [CrossRef]
3. European Union Regulation (EU) 2018/848 of the European Parliament and of the Council of 30 May 2018 on Organic Production and Labelling of Organic Products and Repealing Council Regulation (EC) No 834/2007. Available online: https://eur-lex.europa.eu/legal-content/EN/TXT/PDF/?uri=CELEX:02018R0848-20250325 (accessed on 29 July 2025).
4. European Commission. *Organic Aquaculture in the EU*; Publications Office of the European Union: Luxembourg, 2022.
5. Wafi, A.; Ariadi, H. The Dynamic Model Analysis of Production Feasibility and Market Valuation of Intensive Shrimp Culture Business. *Aquac. Aquar. Conserv. Legis.* **2024**, *17*, 173–179.
6. Ray, A.J.; Lewis, B.L.; Browdy, C.L.; Leffler, J.W. Suspended Solids Removal to Improve Shrimp (*Litopenaeus vannamei*) Production and an Evaluation of a Plant-Based Feed in Minimal-Exchange, Superintensive Culture Systems. *Aquaculture* **2010**, *299*, 89–98. [CrossRef]
7. Davis, D.A.; Samocha, T.M.; Bullis, R.A.; Patnaik, S.; Browdy, C.L.; Stokes, A.D.; Atwood, H.L. Practical Diets for Litopenaeus vannamei (Boone, 1931): Working Towards Organic and/or all Plant Production Diets. 2004. Available online: https://nutricionacuicola.uanl.mx/index.php/acu/article/view/196/195 (accessed on 1 June 2025).
8. Browdy, C.; Seaborn, G.; Atwood, H.; Davis, D.A.; Bullis, R.A.; Samocha, T.M.; Wirth, E.; Leffler, J.W. Comparison of Pond Production Efficiency, Fatty Acid Profiles, and Contaminants in *Litopenaeus vannamei* Fed Organic Plant-based and Fish-meal-based Diets. *J. World Aquac. Soc.* **2006**, *37*, 437–451. [CrossRef]
9. Oujifard, A.; Seyfabadi, J.; Kenari, A.A.; Rezaei, M. Fish Meal Replacement with Rice Protein Concentrate in a Practical Diet for the Pacific White Shrimp, *Litopenaeus vannamei* Boone, 1931. *Aquac. Int.* **2012**, *20*, 117–129. [CrossRef]
10. Sánchez-Muros, M.J.; Renteria, P.; Vizcaino, A.; Barroso, F.G. Innovative Protein Sources in Shrimp (*Litopenaeus vannamei*) Feeding. *Rev. Aquac.* **2020**, *12*, 186–203. [CrossRef]
11. Chen, Y.; Mitra, A.; Rahimnejad, S.; Chi, S.; Kumar, V.; Tan, B.; Niu, J.; Xie, S. Retrospect of Fish Meal Substitution in Pacific White Shrimp (*Litopenaeus vannamei*) Feed: Alternatives, Limitations and Future Prospects. *Rev. Aquac.* **2024**, *16*, 382–409. [CrossRef]
12. Amaya, E.A.; Davis, D.A.; Rouse, D.B. Replacement of Fish Meal in Practical Diets for the Pacific White Shrimp (*Litopenaeus vannamei*) Reared Under Pond Conditions. *Aquaculture* **2007**, *262*, 393–401. [CrossRef]
13. D'Agaro, E.; Gibertoni, P.; Esposito, S. Recent Trends and Economic Aspects in the Rainbow Trout (*Oncorhynchus mykiss*) Sector. *Appl. Sci.* **2022**, *12*, 8773. [CrossRef]
14. Li, S.; Ji, H.; Zhang, B.; Zhou, J.; Yu, H. Defatted Black Soldier Fly (*Hermetia illucens*) Larvae Meal in Diets for Juvenile Jian Carp (Cyprinus Carpio Var. Jian): Growth Performance, Antioxidant Enzyme Activities, Digestive Enzyme Activities, Intestine and Hepatopancreas Histological Structure. *Aquaculture* **2017**, *477*, 62–70. [CrossRef]
15. Chen, Y.; Chi, S.; Zhang, S.; Dong, X.; Yang, Q.; Liu, H.; Tan, B.; Xie, S. Evaluation of the Dietary Black Soldier Fly Larvae Meal (*Hermetia illucens*) on Growth Performance, Intestinal Health, and Disease Resistance to Vibrio Parahaemolyticus of the Pacific White Shrimp (*Litopenaeus vannamei*). *Front. Mar. Sci.* **2021**, *8*, 706463. [CrossRef]
16. Rahimnejad, S.; Hu, S.; Song, K.; Wang, L.; Lu, K.; Wu, R.; Zhang, C. Replacement of Fish Meal with Defatted Silkworm (*Bombyx mori* L.) Pupae Meal in Diets for Pacific White Shrimp (*Litopenaeus vannamei*). *Aquaculture* **2019**, *510*, 150–159. [CrossRef]
17. FAO. *Edible Insects: Future Prospects for Food and Feed Security*; FAO: Rome, Italy, 2013; ISBN 978-92-5-107595-1.
18. Galkanda-Arachchige, H.S.C.; Hussain, A.S.; Davis, D.A. Fermented Corn Protein Concentrate to Replace Fishmeal in Practical Diets for Pacific White Shrimp *Litopenaeus vannamei*. *Aquac. Nutr.* **2021**, *27*, 1640–1649. [CrossRef]
19. Liao, Z.; Gong, Y.; Zhao, W.; He, X.; Wei, D.; Niu, J. Comparison Effect of Rhodobacter Sphaeroides Protein Replace Fishmeal on Growth Performance, Intestinal Morphology, Hepatic Antioxidant Capacity and Immune Gene Expression of *Litopenaeus vannamei* Under Low Salt Stress. *Aquaculture* **2022**, *547*, 737488. [CrossRef]
20. Lin, Y.-H.; Chen, Y.-T. Lactobacillus Spp. Fermented Soybean Meal Partially Substitution to Fish Meal Enhances Innate Immune Responses and Nutrient Digestibility of White Shrimp (*Litopenaeus vannamei*) Fed Diet with Low Fish Meal. *Aquaculture* **2022**, *548*, 737634. [CrossRef]

21. Jatobá, A.; Vieira, F.d.N.; Silva, B.C.d.; Soares, M.; Mouriño, J.L.P.; Seiffert, W.Q. Replacement of Fishmeal for Soy Protein Concentrate in Diets for Juvenile *Litopenaeus vannamei* in Biofloc-Based Rearing System. *Rev. Bras. De Zootec.* **2017**, *46*, 705–713. [CrossRef]
22. Manikandan, K.; Felix, N. Growth Performance, Apparent Digestibility Coefficients and Digestive Enzyme Activities of Pacific White Shrimp, *Penaeus vannamei* Fed Low Fishmeal Diets Supplemented with DL-methionine and Phytase. *Aquac. Nutr.* **2021**, *27*, 2240–2250. [CrossRef]
23. Hlordzi, V.; Wang, J.; Kuebutornye, F.K.A.; Yang, X.; Tan, B.; Li, T.; Cui, Z.; Lv, S.; Lao, T.; Chi, S. Hydrolysed Fish Protein Powder is Better at the Growth Performance, Hepatopancreas and Intestinal Development of Pacific White Shrimp (*Litopenaeus vannamei*). *Aquac. Rep.* **2022**, *23*, 101025. [CrossRef]
24. Novriadi, R.; Wahyudi, A.E.; Fadhilah, R.; Seiler, B.; Balk, D.; Jolly-Breithaupt, M. Evaluation of Dietary Corn Fermented Protein on Growth Performance and Haemato-Immunological Parameters of the Pacific White Shrimp *Litopenaeus vannamei*. *Aquac. Res.* **2022**, *53*, 851–859. [CrossRef]
25. Xie, B.; Qin, J.; Yang, H.; Wang, X.; Wang, Y.-H.; Li, T.-Y. Organic Aquaculture in China: A Review from a Global Perspective. *Aquaculture* **2013**, *414–415*, 243–253. [CrossRef]
26. Tefal, E.; Jauralde, I.; Tomás-Vidal, A.; Martínez-Llorens, S.; Peñaranda, D.S.; Jover-Cerdá, M. New Organic Raw Materials for *Gilthead seabream* (*Sparus aurata*) Feeding and the Effects on Growth, Nutritive Parameters, Digestibility, and Histology. *Fishes* **2023**, *8*, 330. [CrossRef]
27. Tefal, E.; Tomás-Vidal, A.; Martínez-Llorens, S.; Jauralde, I.; Sánchez-Peñaranda, D.; Jover-Cerdá, M. Effects of Eco-Organic Feed on Growth Performance, Biometric Indices, and Nutrient Retention of *Gilthead seabream* (*Sparus aurata*). *Sustainability* **2023**, *15*, 10750. [CrossRef]
28. Tefal, E.; Jauralde, I.; Martínez-Llorens, S.; Tomás-Vidal, A.; Milián-Sorribes, M.C.; Moyano, F.J.; Peñaranda, D.S.; Jover-Cerdá, M. Organic Ingredients as Alternative Protein Sources in the Diet of Juvenile Organic Seabass (*Dicentrarchus labrax*). *Animals* **2023**, *13*, 3816. [CrossRef]
29. Regulation—2018/848—EN—EUR-Lex. Available online: https://eur-lex.europa.eu/eli/reg/2018/848/oj/eng (accessed on 6 March 2025).
30. Baumgarten, M.G.Z.; Wallner-Kersanach, M.; Niencheski, L.F.H. *Manual de Análises Em Oceanografia Química*; FURG: Rio Grande, Brazil, 2010; p. 172.
31. AOAC. *Official Methods of Analysis*, 15th ed.; AOAC: Arlington, VA, USA, 1990.
32. Bosch, L.; Alegría, A.; Farré, R. Application of the 6-Aminoquinolyl-N-Hydroxysccinimidyl Carbamate (AQC) Reagent to the RP-HPLC Determination of Amino Acids in Infant Foods. *J. Chromatogr. B* **2006**, *831*, 176–183. [CrossRef]
33. O'Fallon, J.V.; Busboom, J.R.; Nelson, M.L.; Gaskins, C.T. A Direct Method for Fatty Acid Methyl Ester Synthesis: Application to Wet Meat Tissues, Oils, and Feedstuffs. *J. Anim. Sci.* **2007**, *85*, 1511–1521. [CrossRef]
34. Cho, C.Y.; Slinger, S.J.; Bayley, H.S. Bioenergetics of Salmonid Fishes: Energy Intake, Expenditure and Productivity. *Comp. Biochem. Physiol. Part B: Comp. Biochem.* **1982**, *73*, 25–41. [CrossRef]
35. McKay, H.; McAuliffe, W. Pre-Slaughter Mortality of Farmed Shrimp. *Rethink. Priorities* **2024**. Available online: https://rethinkpriorities.org/wp-content/uploads/2024/03/Pre-slaughter-mortality-of-farmed-shrimp.pdf (accessed on 1 June 2025). [CrossRef]
36. Cummins, V.C., Jr.; Rawles, S.D.; Thompson, K.R.; Velasquez, A.; Kobayashi, Y.; Hager, J.; Webster, C.D. Evaluation of Black Soldier Fly (*Hermetia illucens*) Larvae Meal as Partial or Total Replacement of Marine Fish Meal in Practical Diets for Pacific White Shrimp (*Litopenaeus vannamei*). *Aquaculture* **2017**, *473*, 337–344. [CrossRef]
37. Nunes, A.J.P.; Sabry-Neto, H.; da Silva, F.H.P.; de Oliveira-Neto, A.R.; Masagounder, K. Multiple Feedings Enhance the Growth Performance and Feed Efficiency of Juvenile *Litopenaeus vannamei* When Fed a Low-Fish Meal Amino Acid-Supplemented Diet. *Aquac. Int.* **2019**, *27*, 337–347. [CrossRef]
38. Fox, J.M.; Lawrence, A.L.; Li-Chan, E. Dietary Requirement for Lysine by Juvenile *Penaeus vannamei* Using Intact and Free Amino Acid Sources. *Aquaculture* **1995**, *13*, 279–290. [CrossRef]
39. National Research Council. *Nutrient Requirements of Fish and Shrimp*; National Academies Press: Washington, DC, USA, 2011; ISBN 978-0-309-16338-5.
40. Monroig, Ó.; Shu-Chien, A.C.; Kabeya, N.; Tocher, D.R.; Castro, L.F.C. Desaturases and Elongases Involved in Long-Chain Polyunsaturated Fatty Acid Biosynthesis in Aquatic Animals: From Genes to Functions. *Prog. Lipid Res.* **2022**, *86*, 101157. [CrossRef]
41. González-Félix, M.L.; da Silva, F.S.D.; Davis, D.A.; Samocha, T.M.; Morris, T.C.; Wilkenfeld, J.S.; Perez-Velazquez, M. Replacement of Fish Oil in Plant Based Diets for Pacific White Shrimp (*Litopenaeus vannamei*). *Aquaculture* **2010**, *309*, 152–158. [CrossRef]
42. Feng, Z.F.; Dong, W.X.; Bu, F.; Zhu, W. Influence of Dietary N-3 Polyunsaturated Fatty Acids (PUFAs) on Growth, PUFA Composition and the Expression of FAD6 and ELOV5 in Pacific White Shrimp (*Litopenaeus vannamei*). *Aquac. Res.* **2021**, *52*, 1666–1676. [CrossRef]

43. Gora, A.; Jayasankar, V.; Rehman, S.; Kizhakudan, J.K.; Laxmilatha, P.; Vijayagopal, P. Biochemical Responses of Juvenile Rock Spiny Lobster Panulirus Homarus Under Different Feeding Regimes. *J. Appl. Anim. Res.* **2018**, *46*, 1462–1468. [CrossRef]
44. Navarrete del Toro, M.A.; García-Carreño, F. The Toolbox for Protein Digestion in Decapod Crustaceans: A Review. *Rev. Aquac.* **2019**, *11*, 1005–1021. [CrossRef]
45. Omont, A.; Elizondo-González, R.; Escobedo-Fregoso, C.; Tovar-Ramírez, D.; Hinojosa-Baltazar, P.; Peña-Rodríguez, A. Bacterial Communities and Digestive Enzymatic Activities of *Litopenaeus vannamei* Shrimp Fed Pre-Digested Seaweeds as a Functional Ingredient. *J. Appl. Phycol.* **2021**, *33*, 1239–1251. [CrossRef]
46. Castellanos-Ochoa, C.; Torres-Ochoa, E.; Pacheco-Vega, J.; Cortés-Sánchez, A.; Espinosa-Chaurand, D. Crustacean Chymotrypsin: Estate of the Art. *Abanico Vet.* **2022**, *12*, 1–18. [CrossRef]
47. Marono, S.; Piccolo, G.; Loponte, R.; Meo, C.D.; Attia, Y.A.; Nizza, A.; Bovera, F. In Vitro Crude Protein Digestibility of Tenebrio Molitor and *Hermetia illucens* Insect Meals and Its Correlation with Chemical Composition Traits. *Ital. J. Anim. Sci.* **2015**, *14*, 338–343. [CrossRef]
48. Shiau, S.-Y.; Yu, Y.-P. Chitin but Not Chitosan Supplementation Enhances Growth of Grass Shrimp, *Penaeus monodon*. *J. Nutr.* **1998**, *128*, 908–912. [CrossRef]
49. Hernández, J.C.S.; Murueta, J.H.C. Activity of Trypsin from *Litopenaeus vannamei*. *Aquaculture* **2009**, *290*, 190–195. [CrossRef]

fishes

MDPI

Article

Asparagopsis taxiformis Feed Supplementation as a Tool to Improve the Resilience of Farmed *Diplodus sargus* to Marine Heatwave Events—A Metabolomics Approach

Marta Dias [1,*], Isa Marmelo [1,2,3], Carla António [4,5], Ana M. Rodrigues [4,5,6], António Marques [1,3], Mário S. Diniz [2,7] and Ana Luísa Maulvault [2,3,7]

1 CIIMAR—Interdisciplinary Centre of Marine and Environmental Research, University of Porto, Terminal de Cruzeiros do Porto de Leixões, Avenida General Norton de Matos S/N, 4450-208 Matosinhos, Portugal; isa.marmelo@ipma.pt (I.M.); amarques@ipma.pt (A.M.)
2 UCIBIO—Applied Molecular Biosciences Unit, Department of Chemistry, NOVA School of Science and Technology, NOVA University of Lisbon, 2829-516 Caparica, Portugal; mesd@fct.unl.pt (M.S.D.); aluisa@ipma.pt (A.L.M.)
3 IPMA DivAV—Division of Aquaculture, Upgrading and Bioprospection, Portuguese Institute for the Sea and Atmosphere, Avenida Doutor Alfredo Magalhães Ramalho, 6, 1495-165 Algés, Portugal
4 Plant Metabolomics Lab Portugal, 1349-017 Lisboa, Portugal; plantmetabolomicslabpt@gmail.com (C.A.); amargaridacrodrigues@gmail.com (A.M.R.)
5 Forest Research Centre, Associate Laboratory TERRA, School of Agriculture, University of Lisbon, Tapada da Ajuda, 1349-017 Lisboa, Portugal
6 iBET—Instituto de Biologia Experimental e Tecnológica, Apartado 12, 2781-901 Oeiras, Portugal
7 Associate Laboratory i4HB—Institute for Health and Bioeconomy, NOVA School of Science and Technology, NOVA University of Lisbon, 2829-516 Caparica, Portugal
* Correspondence: mdias@ciimar.up.pt

Abstract: The need to maximize aquaculture production while addressing environmental and food security challenges posed by climate change has driven research towards the development of functional aquafeeds that enhance performance and immunity in farmed species. However, exposure to dietary and environmental stressors affects marine organisms, altering key metabolic pathways best understood through high-throughput "omics" tools. This study assessed the effects of *Asparagopsis taxiformis* supplementation on central metabolic pathways by analyzing changes in primary metabolite levels in the liver of farmed *Diplodus sargus* under optimal and suboptimal temperature conditions. Results showed that seaweed supplementation had a beneficial effect on the fish's primary metabolome; however, inclusion levels and rearing conditions played a crucial role in determining outcomes. While 1.5% supplementation maintained a balanced primary metabolome under optimal temperature conditions, 3.0% supplementation most effectively mitigated the adverse effects of acute thermal stress during a marine heatwave. These findings highlight the nutritive and functional potential of *A. taxiformis* supplementation in aquafeeds for marine omnivorous fish species and emphasize the importance of evaluating functional aquafeeds under suboptimal rearing conditions. Overall, our results demonstrate the value of metabolomics in elucidating the molecular basis underlying biological pathways in farmed marine fish and optimizing production through climate-smart dietary strategies.

Keywords: climate change; functional feeds; GC-TOF-MS; white seabream; primary metabolome; red macroalga; suboptimal rearing conditions

 https://doi.org/10.3390/fishes10070350

Key Contribution: Seaweed supplementation and metabolic reprogramming are crucial for the survival of farmed fish during heat stress. Inclusion percentages and rearing conditions play a crucial role in the development of climate-smart feeds.

1. Introduction

Aquaculture's fast expansion and contribution to global seafood production in the last decades are widely recognized. Currently, more than 300 different finfish species are farmed worldwide [1], each with distinct ecological traits, feeding patterns, and nutritional requirements that must be carefully addressed to maximize productivity. Hence, over the last years, fish nutrition has become one of the most investigated areas in aquaculture, with one of the primary focus being the nutritional value, safety (to both farmed species and consumers) and functionality (in terms of animal growth, performance, immunocompetence and resilience to stress) of aquafeed ingredients [2–4]. In parallel, the limited availability and recent surge in prices of raw materials traditionally used as ingredients in fish feed, either from animal (i.e., fishmeals and oils extracted from wild pelagic species) or plant (cereals and vegetable oils) origins, has rushed the aquaculture industry to find alternative ingredients from sustainable sources in a sort of "hunt for gold". In this sense, marine macroalgae have been recently placed in the spotlight, as this group is a rich source of essential nutrients and bioactive compounds [5,6]. Seaweeds' reported beneficial effects on farmed fish include the improvement of growth performance and physiological state (e.g., lipid metabolism) [7–9], modulation of metabolic rates [10,11], immunostimulatory effects [10,12], genoprotection [13], and enhancement of antimicrobial [14–17] and antioxidant [10,12,18] responses. Still, many questions remain unanswered regarding the use of seaweeds in aquafeeds. Key uncertainties include the metabolic and nutrient conversion implications, optimal inclusion levels to achieve beneficial outcomes, and the impact of dietary modulation on product nutritional value and safety [9,19,20], all of which hamper the acceptance of these plant-based ingredients by aquafeed and fish producers.

In addition, nowadays, it has also become imperative to validate the effectiveness of any dietary modulation strategy under suboptimal animal rearing conditions, as farmed marine fish are increasingly exposed to sudden and severe environmental shifts caused by extreme weather events linked to climate change, such as marine heatwave (MHW) events [21]. Hobday et al. [22] defined an MHW event as "a prolonged discrete anomalously warm water event that can be described by its duration, intensity, rate of evolution, and spatial extent". MHW events are increasing in number, intensity, and duration worldwide, particularly in the Mediterranean region, affecting inshore and offshore aquaculture facilities, and thus, strongly defying farmed species' physiological tolerance [23,24]. Despite the various uncertainties regarding the effects of MHWs in marine organisms/ecosystems, it is fairly well known that these events have devastating impacts on farmed species, making them particularly vulnerable to the various stress factors related to captivity (e.g., stocking density, disease outbreaks, malnutrition) that, in turn, often culminates in massive animal mortality and significant economic losses [25]. Hence, this urgently calls for climate-smart adaptation strategies that can build resilience against the present and future environmental challenges faced by the aquaculture sector. As such, eco-friendly nutritionally based approaches that maximize farmed animals' growth and welfare under suboptimal rearing conditions represent a valuable opportunity for the sector.

Either to validate the use of a certain ingredient/formula or to assess the metabolic mechanisms underpinning farmed species responses to environmental stressors, growth trials (e.g., performance, feed conversion [26]), biochemical (e.g., body composition, physiological biomarkers [12]) and aerobic scope [27] data are typically the common approaches. However, when a given nutritional strategy is unable to meet fish metabolic requirements, a physiological unbalance occurs, triggering a myriad of cascading effects (at developmental, growth, reproductive, and immune levels, and including gene regulation and the production of numerous metabolites [28–30]) that cannot be fully understood through the analysis of conventional zootechnical parameters. As such, metabolomics is a "systems wide" molecular approach that has recently gained attention in aquaculture research [31]. Metabolomics is as a well-established "omics" tool concerned with the study of the organism's physiological state at the metabolite level by offering means to obtain a comprehensive view of the changes in the levels of numerous low-molecular-weight endogenous metabolites simultaneously (e.g., lipids, sugars, amino acids, organic acids) as well as an holistic view of organisms' physiological state taking into account the interactions between genetic traits and environmental factors (e.g., dietary changes and/or stressors exposure [32,33]). Often described as the "molecular phenotype" of living organisms [34,35], metabolites are products of cellular processes, and in comparative experiments, metabolomics analyses define significance by the relative changes in metabolite levels that can reveal the integrated response of a given organism. However, metabolomics applications in modern aquaculture research remain relatively recent and are still largely confined to studies in livestock [36–41].

In metabolomics analysis, gas chromatography time-of-flight mass spectrometry (GC-TOF-MS) primary metabolite profiling platform can be used for a broad coverage of the primary metabolome, being considered the "gold standard" for the analysis of primary metabolites. GC-TOF-MS offers high chromatographic resolution, reproducibility, and informative fragmentation patterns, making it especially suitable for the analysis of small and polar primary metabolites [42].

In this setting, the current study aims to assess, for the first time, whether metabolomics by means of GC-TOF-MS primary metabolite profiling provides additional functional insights into explaining the effects of dietary supplementation with the red macroalga *Asparagopsis taxiformis* on farmed juvenile *Diplodus sargus* primary metabolome, and to explore its effectiveness as a nutritional modulation strategy to overcome the thermal stress prompted by MHWs by highlighting functional central metabolic pathways and relative changes in primary metabolite levels, enhancing a more comprehensive view of these phenomena.

2. Materials and Methods

2.1. Ethical Statement

The present study was performed by researchers with certification in animal experimentation (EU functions A and B). All animal handling and sampling procedures were strictly in accordance with the ARRIVE (Animal Research: Reporting of In Vivo Experiments) guidelines and ethical standards for the care and use of animals were consistently followed. These practices were in line with the recommendations of the Federation of European Laboratory Animal Science Associations (FELASA) and complied with the Portuguese legislation on laboratory animal science (EU Directive 2010/63; Decree-Law No. 113/2013). The research was approved by IPMA's Animal Welfare and Ethics Body (ORBEA; authorization number 001/2023, 2 November 2023), under the oversight of the National Authority for the use of live animals, also recognized as the Directorate-General for Food and Veterinary (DGAV).

2.2. Seaweed Collection and Experimental Diets

The seaExpert company harvested the gametophyte life stage of the invasive seaweed *Asparagopsis taxiformis* from rocky substrates at depths between 3 and 5 m through scuba diving in Angústias dock, Faial Island, Azores, Portugal (38°31′45.0″ N 28°37′09.0″ W). Subsequently, seaweeds were transported to the processing facility in a controlled environment, ensuring cool and dark conditions. Specially designed boards facilitated water drainage during transportation. To prepare seaweeds for further use, a Black Block solar dryer (BBKW, Lisboa, Portugal) was used, employing a dark setting, and maintaining at a maximum temperature of 40 °C for 2 days. Following the drying process, seaweeds were securely stored in pre-labeled dark plastic bags and sent to SPAROS Lda, a specialized company in Olhão, Portugal, focused on fish feed production. At SPAROS Lda, four distinct diets were meticulously formulated, considering the nutritional requirements of juvenile white seabream (*Diplodus sargus*) (detailed feed composition is reported in Marmelo et al. [43] and in Table 1). The diets shared similar nutritional compositions, with the key variation being the percentage of powdered dry seaweed incorporated at the expense of wheat meal. The red macroalga *A. taxiformis* was chosen as feed supplement as it has (i) relatively high protein content and structurally diverse bioactive compounds than the other taxonomic groups of macroalgae (i.e., brown and green macroalgae [44]); and (ii) a cosmopolitan distribution being, in some locations of the Southwestern Mediterranean Sea, undoubtedly invasive and outcompeting indigenous benthic organisms [45]. Thus, its incorporation in feeds may constitute a window of opportunity to mitigate its destructive effects on the ecosystem while being sustainable due to its high availability in the Mediterranean region. The formulated diets were as follows: (i) commercial control diet without seaweed supplementation (0%, CTR); (ii) diet supplemented with 1.5% *A. taxiformis* (1.5-AT); (iii) diet supplemented with 3.0% *A. taxiformis* (3.0-AT); and (iv) diet supplemented with 6.0% *A. taxiformis* (6.0-AT).

Table 1. Formulation and proximate chemical composition (% dry matter, DM) of the four diets fed to *Diplodus sargus* juveniles during the experimental trial. Abbreviations: CTR-fish fed a non-supplemented/commercial diet; 1.5-AT-fish fed a supplemented diet containing 1.5% of dried *Asparagopsis taxiformis*; 3.0-AT-fish fed a supplemented diet containing 3.0% of dried *A. taxiformis*; 6.0-AT-fish fed with supplemented diet containing 6.0% of dried *A. taxiformis*.

Ingredients (%)	**CTR**	**1.5-AT**	**3.0-AT**	**6.0-AT**
Fishmeal super prime [1]	25.0	25.0	25.0	25.0
Fish protein concentrate [2]	2.0	2.0	2.0	2.0
Soy protein concentrate [3]	10.0	10.0	10.0	10.0
Pea protein concentrate [4]	3.0	3.0	3.0	3.0
Wheat gluten [5]	6.5	6.5	6.5	6.5
Corn gluten meal [6]	10.0	10.0	10.0	10.0
Soybean meal 44 [7]	6.0	6.0	6.0	6.0
Rapeseed meal [8]	6.0	6.0	6.0	6.0
Wheat meal [9]	10.8	9.3	7.8	4.8
Faba beans (low tannins) [10]	6.0	6.0	6.0	6.0
Vitamin and mineral premix [11]	1.0	1.0	1.0	1.0
Choline chloride 50% [12]	0.2	0.2	0.2	0.2
Monoammonium phosphate [13]	1.2	1.2	1.2	1.2
Fish oil [14]	5.0	5.0	5.0	5.0
Soybean oil [15]	7.3	7.3	7.3	7.3
Macroalga *Asparagopsis taxiformis* [16]	0	1.5	3.0	6.0

Table 1. *Cont.*

Ingredients (%)	CTR	1.5-AT	3.0-AT	6.0-AT
Dry matter, DM (%)	94.2	94.0	93.9	94.1
Crude protein, %DM	46.0	46.0	45.9	45.7
Crude fat, %DM	16.0	16.0	16.1	16.1
Fiber, %DM	1.8	1.9	2.0	2.1
Starch, %DM	13.7	12.8	11.8	9.9
Ash, %DM	6.8	7.1	7.4	8.0
Gross energy, MJ kg^{-1}	21.0	21.0	20.9	20.8

[1] Diamante: 66.3% crude protein (CP), 11.5% crude fat (CF), South America, Pesquera Diamante, Peru. [2] CPSP90: 82.6% CP, 9.6% CF, Sopropêche, France. [3] Soycomil P: 62.2% CP, 0.7% CF, ADM, The Netherlands. [4] Lysamine GPS: 78.1% CP, 8.3% CF, Roquette, France. [5] VITAL: 80.4% CP, 5.8% CF, Roquette, France. [6] Corn gluten meal: 61.2% CP, 5.2% CF, COPAM, Portugal. [7] Soybean meal 44: 43.8% CP, 3.5% CF, solvent extracted, Ribeiro & Sousa Lda., Portugal. [8] Rapeseed meal: 34.3% CP, 2.1% CF, solvent extracted, Ribeiro & Sousa Lda., Portugal. [9] Wheat meal: 11.7% CP, 1.6% CF, Molisur, Spain. [10] Faba beans (low tannins): 24.5% CP, 1.7% CF, Ribeiro & Sousa Lda., Portugal. [11] Vitamins (IU or mg kg^{-1} diet): DL-alphatocopherol acetate, 100 mg; sodium menadione bisulphate, 25 mg; retinyl acetate, 20,000 IU; DL-cholecalciferol, 2000 IU; thiamine, 30 mg; riboflavin, 30 mg; pyridoxine, 20 mg; cyanocobalamin, 0.1 mg; nicotidin acid, 200 mg; folic acid, 15 mg; ascorbic acid, 1000 mg; inositol, 500 mg; biotin, 3 mg; calcium panthotenate, 100 mg; choline chloride, 1000 mg, betaine, 500 mg. Minerals (g or mg kg^{-1} diet): cobalt carbonate, 0.65 mg; copper sulphate, 9 mg; ferric sulphate, 6 mg; potassium iodide, 0.5 mg; manganese oxide, 9.6 mg; sodium selenite, 0.01 mg; zinc sulphate, 7.5 mg; sodium chloride, 400 mg; calcium carbonate, 1.86 g; excipient wheat middlings. Premix Lda., Portugal. [12] Choline chloride 50%: ORFFA, The Netherlands. [13] Windmill AQUAPHOS: 26% phosphorus, ALIPHOS, The Netherlands. [14] Fish oil: 98.1% CF, 16% EPA, 12% DHA, Sopropêche, France. [15] Soybean oil: 98.6% CF, JC Coimbra, Portugal. [16] *Asparagopsis taxiformis*: SeaExpert Company, Angústias dock, Faial Island, Azores, Portugal.

2.3. Organisms and Acclimation Period

The white seabream, *Diplodus sargus* (Linnaeus, 1758), was selected as a biological model because it is an ecologically relevant coastal species [46,47] and an emerging species for aquaculture with high sensitivity to drastic environmental variations, as well as one of the most important species for commercial and recreational fisheries in southern European countries and in the Mediterranean [48–51]. Specimens of *D. sargus* (n = 150; for the purposes of the present study, only 30 animals were used in total; i.e., the remaining fish were used in several analyses whose results are reported in other manuscripts of the authors, [43,52]) were reared at the Aquaculture Research Station of the Portuguese Institute for the Sea and Atmosphere (EPPO-IPMA, Olhão, Portugal) until they reached the juvenile stage (total length = 12.2 ± 0.3 cm and weight = 28.5 ± 1.1 g; mean ± standard deviation) in recirculation aquaculture system (RAS), under the following abiotic conditions: (i) temperature: 24.0 ± 0.5 °C; (ii) dissolved oxygen: 7.2 ± 0.2 mg L^{-1}; (iii) salinity: 35.0 ± 0.5 psu; (iv) pH: 8.0 ± 0.1 units; and (v) photoperiod: 12 h light/12 h dark. Specimens were transported to IPMA's Live Marine Organisms Bioterium (LABVIVOS) in Algés, Portugal. Upon arrival, juvenile fish were distributed into two tanks (660 L total capacity each) for a 3-week quarantine period, during which they were maintained in similar abiotic conditions as those previously described. Throughout the quarantine period, juvenile fish were fed a high-quality commercial diet (control) twice a day, an amount of feed equivalent to 1.5% of their average body weight (bw).

2.4. Experimental Design and Fish Rearing Conditions

For the purposes of the marine heatwave (MHW) event simulation, the average conditions (i.e., peak temperature, ramp temperature rise rate, duration of "plateau" phase at peak temperature) of a MHW category II "Strong" were used as a reference in the experimental design, as this corresponds to the most typically observed scenario in the Mediterranean environment [23], where *D. sargus* is grown in aquaculture. The other

MHW categories include category I "Moderate"; category III "Severe"; and category IV "Extreme" [23].

The trial involved two phases (see Figure A1): Phase 1—Supplementation period for 30 days, during which fish were kept at the optimal temperature of 24 °C (T30); and Phase 2—Marine Heatwave simulation, during which fish were exposed to a category II Mediterranean heatwave, including a ramp temperature increase period of 8 days (+0.5 °C/day) up to 28 °C [53], followed by a 15-day period of exposure to the peak temperature (i.e., 28 °C). After the 3-week quarantine period, *D. sargus* specimens were randomly and equitably distributed in 15 rectangular experimental glass tanks (200 L total capacity each), with independent RAS. Information regarding the RAS system equipment and functioning is described in detail in Marmelo et al. [43]. Seawater abiotic parameters were adjusted as necessary, with dissolved oxygen maintained at 7.2 ± 0.2 mg L^{-1}, salinity at 35.0 ± 0.5 psu, pH at 8.0 ± 0.1 units, and photoperiod of 12 h light and 12 h dark (12 L:12 D). Ammonia (NH_3/NH_4^+), nitrites (NO_2^-) and nitrates (NO_3^-) levels were measured on a weekly basis using colorimetric test kits (Salifert, The Netherlands) and maintained below detectable levels (except for nitrates that were kept < 50 mg L^{-1}), by daily water renewals of 25% in each tank.

Five treatments were assigned (each composed by three replicate tanks, Figure A1): (i) Control (CTR)-fish fed with non-supplemented/commercial diet while being exposed to optimal temperature conditions; i.e., 24 °C during the entire trial (53 days in total); (ii) Control+Heatwave (CTR+HW)-fish fed with non-supplemented/commercial diet, and exposed to a category II Mediterranean MHW after a 30-day supplementation period; (iii) 1.5% of *A. taxiformis*+Heatwave (1.5-AT-HW)-fish fed with the commercial diet supplemented with 1.5% of dried *A. taxiformis*, and exposed to a category II Mediterranean MHW after a 30-day supplementation period; (iv) 3.0% of *A. taxiformis*+Heatwave (3.0-AT-HW)-fish fed with the commercial diet supplemented with 3.0% of dried *A. taxiformis*, and exposed to a category II Mediterranean MHW after a 30-day supplementation period; and (v) 6.0% of *A. taxiformis*+Heatwave (6.0-AT-HW)-fish fed with the commercial diet supplemented with 6.0% of dried *A. taxiformis*, and exposed to a category II Mediterranean MHW after a 30-day supplementation period.

On Phase 1 of the trial (Supplementation period), fish were hand-fed twice daily (2.0% bw) with the respective aquafeed in each treatment, while being exposed to 24 °C for 30 consecutive days (Figure A1). Then, on Phase 2 of the trial, a category II (strong) MHW was simulated in all treatments except for the CTR treatment. Seawater temperature was maintained at the peak level of 28 °C for 15 days in a row (corresponding to the heatwave "plateau"; Appendix A, Figure A1). No animal mortality was observed throughout both phases of the trial.

2.5. Fish Sampling

Samplings were conducted after each phase of the trial: Phase 1 (Supplementation period)-T30; i.e., after a 30-day feeding trial (and before the heatwave was simulated); and Phase 2 (Marine Heatwave simulation)-T53; i.e., after 15 days of exposure to the peak marine heatwave temperature (corresponding to 53 days of trial in total; Figure A1).

Fish were fasted for 24 h prior to sampling. Three fish per treatment (i.e., one fish from each replicate tank) were randomly collected and anesthetized by immersion in an overdosed tricaine methanesulfonate solution (2 g L^{-1} of MS-222, Acros Organics, Geel, Belgium), buffered with sodium bicarbonate ($NaHCO_3$, Sigma-Aldrich, St. Louis, MO, USA), using a ratio of 1:2 of MS-222/sodium bicarbonate to reduce fish stress. Once the anesthetic took effect, fish death was confirmed by cervical cut, and the animals were then dissected. Fish liver was collected, placed in 2 mL tubes, and immediately frozen in liquid nitrogen. Liver samples were freeze-dried at −40 °C and 0.06 mbar for 48 h, weighed, and stored at −80 °C until further analyses.

2.6. Extraction of Primary Metabolites from *D. sargus* Liver Tissues

Primary metabolites were extracted following a previously described method [54] from 25 to 28 mg dry weight (DW) of finely homogenized *D. sargus* liver tissues (three biological replicates per treatment) in 700 μL ice-cold methanol (HPLC-grade; Merck, Lisbon, Portugal) containing ribitol as internal standard (0.2 mg mL^{-1} ribitol in water; Sigma-Aldrich, St. Louis, MO, USA). Samples were vortex-mixed and incubated in a shaker (ThermoMixer, Eppendorf, Hamburg, Germany) for 15 min at 70 °C and 950 rpm and, subsequently, centrifuged at 12,000× *g* for 10 min. The supernatant was collected, mixed with 370 μL chloroform (Merck) and 750 μL water (HPLC-grade; Merck), and mixed with a vortex. After centrifugation at 2200× *g* for 15 min, an aliquot of 150 μL of the polar (aqueous/methanol) phase was evaporated to dryness for 3 h at 30 °C using a centrifugal concentrator (Concentrator Plus, Eppendorf, Hamburg, Germany), and stored at −80 °C prior to derivatization and GC-TOF-MS analysis.

2.7. GC-TOF-MS Primary Metabolite Profiling Analysis

Dried polar extracts were derivatized following the Lisec et al. [54] protocol with 40 μL of 20 mg mL^{-1} methoxyamine hydrochloride (Sigma-Aldrich, St. Louis, MO, USA) in pure pyridine (Merck S.A., Lisbon Portugal), vortex-mixed, and incubated in a shaker (ThermoMixer, Eppendorf, Hamburg, Germany) for 2 h at 37 °C and 950 rpm. Subsequently, 70 μL of N-methyl-N-trimethylsilyltrifluoroacetamide (MSTFA reagent for TMS derivatization; Macherey-Nagel) and 20 μL mL^{-1} of a mixture of fatty acid methyl esters (FAMES) were added. Sample mixtures were vortex-mixed and, subsequently, incubated in a shaker (ThermoMixer, Eppendorf, Hamburg, Germany) for 30 min at 37 °C and 950 rpm. Biological variations were controlled by analyzing quality control (QC) standards by FAMES internal standard markers and a QC standard solution of 41 pure reference compounds (i.e., the most detected and abundant metabolites) throughout the analysis. Primary metabolite profiling analysis of the derivatized samples (1 μL injection) was performed on an Agilent 6890 N gas chromatograph (Agilent Technologies, Böblingen, Germany) coupled to a LECO Pegasus III TOF-MS running in electron ionization (EI) mode (LECO Instrumente, Mönchengladbach, Germany). The chromatographic separation was performed on a VF-5MS column (Varian Inc., Palo Alto, CA, USA; 30 m length, 0.25 mm inner diameter, and 0.25 mm film thickness). The temperature program was set as follows: isothermal for 2 min at 85 °C, followed by a 15 °C min^{-1} ramp to 360 °C, and hold at this temperature for 6 min. The injector and transfer line temperatures were set to 230 °C and 250 °C, respectively, and the helium carrier gas flow was set to 2 mL min^{-1}. After a solvent delay of 180 s, mass spectra were scanned from m/z 70 to 600 with an acquisition rate of 20 spectra s^{-1} and a detector voltage between 1700 and 1850 V.

2.8. Metabolite Data Processing and Statistical Analysis

GC-TOF-MS data were evaluated using AMDIS (Automated Mass Spectral Deconvolution and Identification System) software (version 2.71). Primary metabolites were annotated using the TagFinder 4.0 software [55] and a reference library of ambient mass spectra and retention indices from the Golm Metabolome Database [56,57]. The relative abundance of primary metabolite levels was normalized to the signal intensity of the internal standard (ribitol) and the DW of the samples (see Table S1). Metabolomics statistical analyses were performed in R Studio software (version 1.3-3; MA, USA) [58] using the "agricolae" [59], "gplots" [60], and "mixOmics" [61] packages. One-way ANOVA at a 95% confidence level was used to assess differences between treatments. Subsequently, fold changes were determined and log10-transformed for heatmap plotting. Supervised partial least squares discriminant analysis (PLS-DA) was performed using the leave-one-out cross-validation embedded in the "mixOmics" package. Supplementary Tables S2 and S3 present the VIP scores for each PLS-DA analysis, indicating the relative importance of each metabolite in discriminating between the classes.

3. Results

3.1. GC-TOF-MS Primary Metabolite Profiling of Diplodus Sargus Liver Tissues When Supplemented with *A. taxiformis*

The analysis of fish liver metabolome through GC-TOF-MS allowed us to identify 46 primary metabolites under both optimal and suboptimal growth conditions, including 22 amino acids (AAs) and derivatives, 11 sugars and sugar alcohols, four organic acids, and nine other metabolites.

Primary metabolite profiling revealed that, at optimal growth conditions (Phase 1: 24 °C–T30), liver tissues of supplemented fish were mainly characterized by an increase in the levels of most AAs, sugars, and organic acids, particularly at 3.0-AT and 6.0-AT. Among these, a significant increase in the levels of both the AAs 4-Hydroxiproline (up to 2-fold) at 3.0-AT and tyrosine (up to 1.5-fold) at 6.0-AT was observed (Figure 1, Table S4).

Central metabolic pathways representing the changes in the relative levels of primary metabolites in the liver tissues of *D. sargus* grown under optimal conditions are represented in Figure S1. In Phase 1, PLS-DA analysis revealed that, while the sample groups corresponding to the 3.0-AT and 6.0-AT were clearly discriminated, CTR and the 1.5-AT sample groups were clustered together (Figure 2A). The correlation circle plot (Figure 2B) identifies the metabolites most responsible for the separation between sample groups. In more detail, the contribution plots (Figure 3) show that among those metabolites, 4-hydroxyproline (which significantly increased, up to 2-fold), alanine, glycine, myo-inositol-1-phosphate (component 1), and uracil, phosphoric acid and myo-inositol (component 2) contributed the most to the discrimination within the 3.0-AT sample group. Tyrosine (which significantly increased, up to 1.5-fold), adenosine-5-mono-phosphate, β-alanine, aspartate, erythritol, succinate, phenylalanine (component 1), and maltose, serine, succinate, malate, and fructose (component 2) contributed the most to the discrimination within the 6.0-AT sample group.

3.2. GC-TOF-MS Primary Metabolite Profiling of Diplodus Sargus Liver Tissues When Exposed to the MHW

During Phase 2 of the study, the effect of increased seawater temperature was assessed in non-supplemented and supplemented fish. Starting with the effect of seawater temperature conditions alone (i.e., supplementation comparisons between CTR and CTR+HW), a general increase in the levels of AAs (glutamine, ornithine, pyroglutamate, and serine, up to 2.6-fold), organic acids (threonate; 2-fold increase), in a sugar alcohol (polyol myo-inositol; 1.8-fold increase), and other metabolites (adenine and nicotinamide; 3-fold and 1.3-fold increase, respectively), as well as a decrease in the levels of some sugars (maltose and trehalose, down to 1.0-fold) was observed in fish exposed to the MHW in relation to those reared under optimal conditions (Figure 4A and Figure S2; Table S5A).

As for the effect of the MHW upon supplementation with *A. taxiformis* (i.e., comparing CTR+HW against 1.5-AT-HW, 3.0-AT-HW, and 6.0-AT-HW treatments), a higher number of significant changes in most primary metabolite levels was, overall, found in supplemented animals with central metabolic pathways of *D. sargus* revealing marked differences in the levels of AAs, as well as in carbohydrate metabolism (Figure S3). While the levels of most AAs significantly decreased at 1.5-AT-HW and 6.0-AT-HW, 6.0-AT-HW showed an increased in the levels of most sugars (Figure 4B), namely, fructose (up to 4-fold), maltose (up to 37-fold), mannose (up to 8-fold), trehalose (up to 28-fold), and the sugar alcohol mannitol (up to 3.5-fold) (Table S5B).

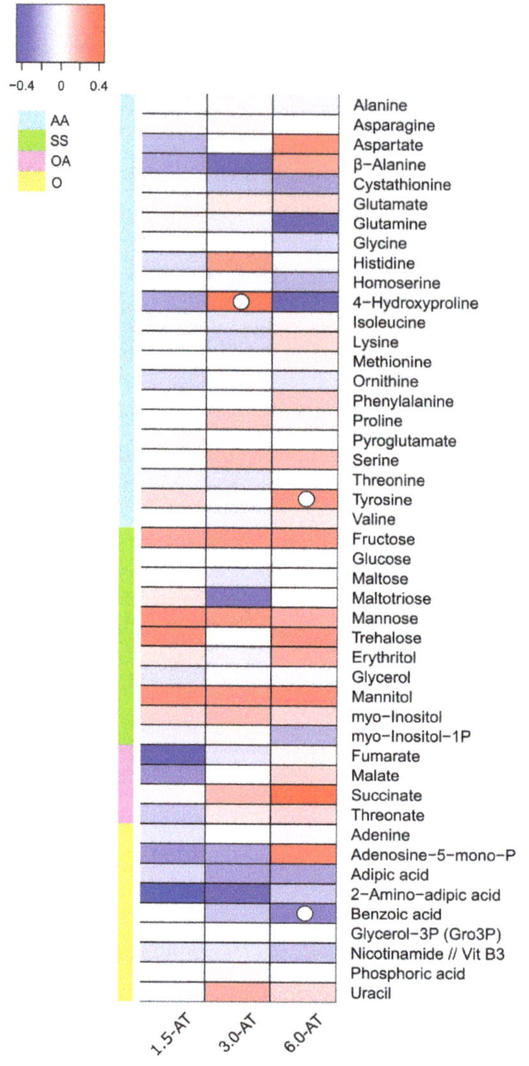

Figure 1. Heatmap visualization representing the changes in relative levels of primary metabolites in fish liver tissues at Phase 1—Supplementation period for 30 days, during which fish were kept at the optimal temperature of 24 °C (T30). Relative values (as means of 3 independent measurements) were normalized to the dry weight (DW) of the samples and IS ribitol. False color imaging was performed on log10-transformed GC-TOF-MS metabolite data. Fold changes were calculated in relation to the respective control (CTR) condition. Significant differences in relation to the respective control (CTR) condition using one-way ANOVA are indicated as ○ $p < 0.05$. Fishes were kept at the optimal temperature of 24 °C (CTR) and fed a supplemented diet with 1.5, 3, and 6.0% *Asparagopsis taxiformis* for 30 days (1.5-AT, 3.0-AT, and 6.0-AT, respectively). Metabolites grouped in amino acids and derivatives (AA), sugars and derivatives (sugar phosphate; sugar alcohol; sugar acid) (SS), organic acids (OA), and others (O).

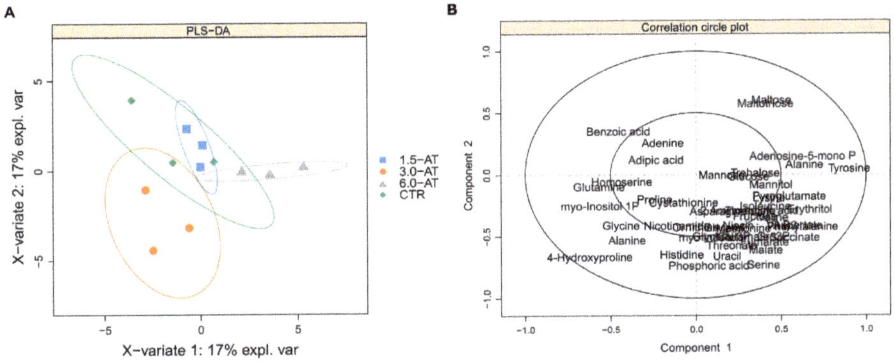

Figure 2. Partial least squares discriminant analysis (PLS-DA) score plot (**A**) and PLS-DA correlation circle plot (**B**) of the primary metabolite profile in fish liver tissues at Phase 1—Supplementation period for 30 days, during which fish were kept at the optimal temperature of 24 °C (T30). Abbreviations: CTR-fish fed a non-supplemented/commercial diet for 30 days at optimal temperature; 1.5-AT-fish fed a supplemented diet containing 1.5% of dried *Asparagopsis taxiformis* for 30 days at optimal temperature; 3.0-AT-fish fed a supplemented diet containing 3.0% of dried *A. taxiformis* for 30 days at optimal temperature; 6.0-AT-fish fed with supplemented diet containing 6.0% of dried *A. taxiformis* for 30 days at optimal temperature.

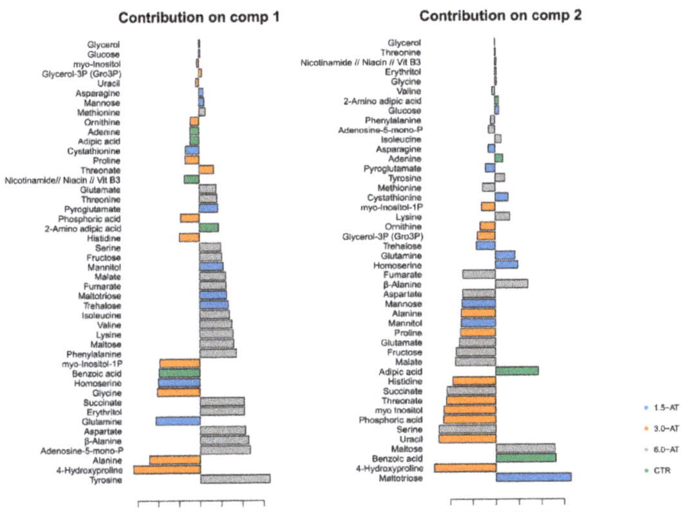

Figure 3. Partial least squares discriminant analysis (PLS-DA) contribution plot of the primary metabolite profile in fish liver tissues at Phase 1—Supplementation period for 30 days, during which fish were kept at the optimal temperature of 24 °C (T30). Abbreviations: CTR-fish fed a non-supplemented/commercial diet for 30 days at optimal temperature; 1.5-AT-fish fed a supplemented diet containing 1.5% of dried *Asparagopsis taxiformis* for 30 days at optimal temperature; 3.0-AT-fish fed a supplemented diet containing 3.0% of dried *A. taxiformis* for 30 days at optimal temperature; 6.0-AT-fish fed with supplemented diet containing 6.0% of dried *A. taxiformis* for 30 days at optimal temperature.

Interestingly, in 3.0-AT-HW treatment, only a few metabolites showed altered abundance under suboptimal conditions (Figure 4B); i.e., the AAs alanine, glutamine, and threonine significantly increased up to 1.5-fold (Table S5B), while the levels of other primary metabolites remained largely unchanged.

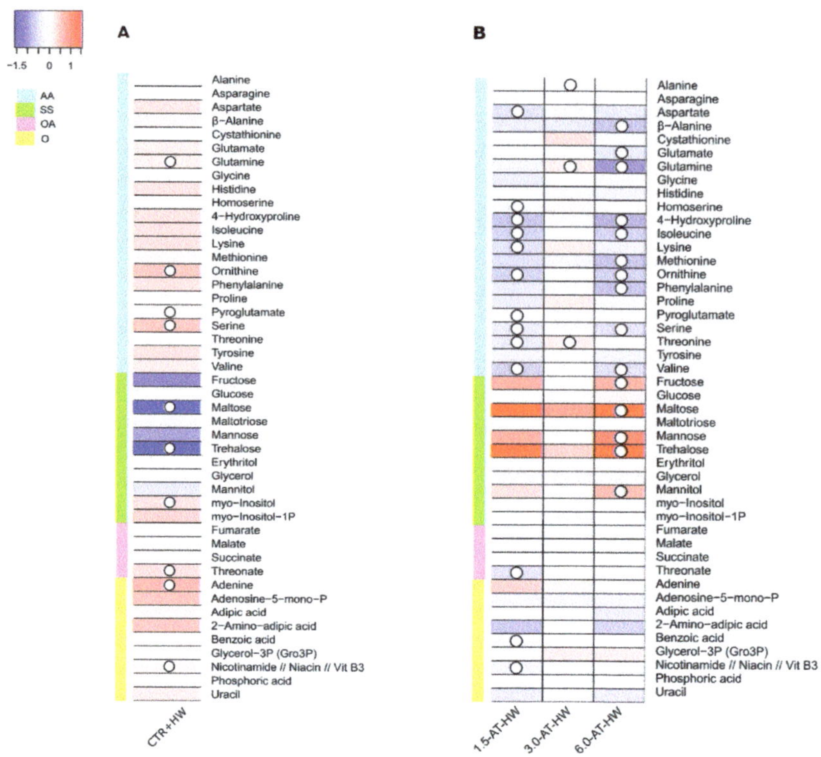

Figure 4. Heatmap visualization representing the changes in relative levels of primary metabolites in fish liver tissues at Phase 2—Marine Heatwave simulation, during which fish were exposed to a category II Mediterranean heatwave (CTR+HW), including a ramp temperature increase period of 8 days (+0.5 °C/day) up to 28 °C [53], followed by a 15-day period of exposure to the peak temperature (i.e., 28 °C, CTR+HW, T53). Relative values (as means of 3 independent measurements) were normalized to the dry weight (DW) of the samples and IS ribitol. False color imaging was performed on log10-transformed GC-TOF-MS metabolite data. Fold changes were calculated in relation to the respective control condition at 24 °C (CTR) (**A**) and the CTR+HW treatment; i.e., fish fed with non-supplemented/commercial diet and exposed to a category II Mediterranean marine heatwave after a 30-day supplementation period (**B**). Significant differences in relation to the respective control conditions (i.e., CTR in (**A**) and CTR+HW in (**B**)) using one-way ANOVA are indicated as ○ $p < 0.05$. Metabolites were grouped in amino acids and derivatives (AA), sugars and derivatives (sugar phosphate; sugar alcohol; sugar acid) (SS), organic acids (OA), and others (O). Abbreviations: 1.5-AT-HW-fish fed with supplemented diet containing 1.5% of dried *Asparagopsis taxiformis*, and exposed to a category II Mediterranean marine heatwave after a 30-day supplementation period; 3.0-AT-HW-fish fed with supplemented diet containing 3.0% of dried *A. taxiformis*, and exposed to a category II Mediterranean marine heatwave after a 30-day supplementation period; 6.0-AT-HW-fish fed with supplemented diet containing 6.0% of dried *A. taxiformis*, and exposed to a category II Mediterranean marine heatwave after a 30-day supplementation period.

In Phase 2, PLS-DA analysis revealed clear discrimination between all sample groups (Figure 5A). The correlation circle plot (Figure 5B) identifies the key metabolites most responsible for discriminating between the different sample groups. Among those metabolites, the contribution plots did not highlight key metabolites that contributed the most to the discrimination of the 1.5-AT-HW treatment (Figure 6). However, the contribution plots highlight glutamine, which significantly increased up to 1.5-fold, lysine, proline (component 1), threonine, which also significantly increased up to 1.5-fold, and glycerol-

3-phosphate (component 2) as the key metabolites that contributed the most for the discrimination of the 3.0-AT-HW treatment (Figure 6). The most relevant contribution to the cluster of the 6.0-AT-HW treatment was observed for those sugars whose levels were shown to significantly increase, namely, trehalose, fructose, maltose, mannose, and the sugar alcohol mannitol (component 1) (Table S5B), and phosphoric acid, mannitol, and succinate (component 2).

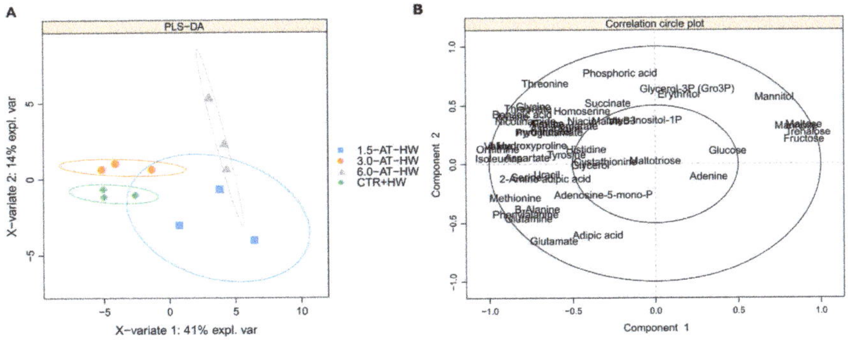

Figure 5. Partial least squares discriminant analysis (PLS-DA) score plot (**A**) and PLS-DA correlation circle plot (**B**) of the primary metabolite profile in fish liver tissues at Phase 2—Marine Heatwave simulation, during which fish were exposed to a category II Mediterranean heatwave (CTR+HW), including a ramp temperature increase period of 8 days (+0.5 °C/day) up to 28 °C [53], followed by a 15-day period of exposure to the peak temperature (i.e., 28 °C, CTR+HW, T53). Abbreviations: CTR+HW-fish fed with non-supplemented/commercial diet and exposed to a category II Mediterranean marine heatwave after a 30-day supplementation period; 1.5-AT-HW-fish fed with supplemented diet containing 1.5% of dried *Asparagopsis taxiformis*, and exposed to a category II Mediterranean marine heatwave after a 30-day supplementation period; 3.0-AT-HW-fish fed with supplemented diet containing 3.0% of dried *A. taxiformis*, and exposed to a category II Mediterranean marine heatwave after a 30-day supplementation period; 6.0-AT-HW-fish fed with supplemented diet containing 6.0% of dried *A. taxiformis*, and exposed to a category II Mediterranean marine heatwave after a 30-day supplementation period.

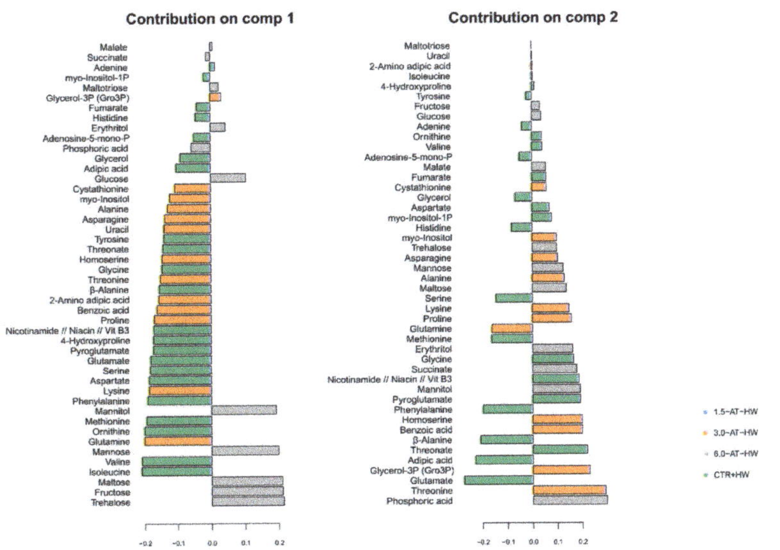

Figure 6. Partial least squares discriminant analysis PLS-DA Contribution plot of the primary metabolite profile in fish liver tissues at Phase 2—Marine Heatwave simulation, during which fish were

exposed to a category II Mediterranean heatwave (CTR+HW), including a ramp temperature increase period of 8 days (+0.5 °C/day) up to 28 °C [53], followed by a 15-day period of exposure to the peak temperature (i.e., 28 °C, CTR+HW, T53). Abbreviations: CTR+HW-fish fed with non-supplemented/commercial diet and exposed to a category II Mediterranean marine heatwave after a 30-day supplementation period; 1.5-AT-HW-fish fed with supplemented diet containing 1.5% of dried *Asparagopsis taxiformis*, and exposed to a category II Mediterranean marine heatwave after a 30-day supplementation period; 3.0-AT-HW-fish fed with supplemented diet containing 3.0% of dried *A. taxiformis*, and exposed to a category II Mediterranean marine heatwave after a 30-day supplementation period; 6.0-AT-HW-fish fed with supplemented diet containing 6.0% of dried *A. taxiformis*, and exposed to a category II Mediterranean marine heatwave after a 30-day supplementation period.

4. Discussion

4.1. Effects of *A. taxiformis* Supplementation on Fish Liver Primary Metabolome Under an Optimal Temperature Regime (CTR Versus 1.5-AT, 3.0-AT, and 6.0-AT)

Under optimal rearing conditions (i.e., 24 °C), supplementation with 1.5% of *A. taxiformis* did not significantly affect the primary metabolism of *D. sargus*. The absence of evident metabolic interferences, together with the maintenance of fish growth performance indicators (i.e., specific growth rate, SGR, and feed conversion ratio, FCR) and the improvement of antioxidant and immune responses reported in our previous studies [43,52] confirms that aquafeeds supplemented with low doses (i.e., 1.5%) of *A. taxiformis* provide a nutritionally balanced solution with beneficial functional outcomes.

Supplementation with higher levels of *A. taxiformis* (3.0% or 6.0%) led to notable changes in the primary metabolome of juvenile *D. sargus*, specifically increasing 4-hydroxyproline (4-Hyp; at 3.0%) and tyrosine (Tyr; at 6.0%). These AAs are not typically abundant in *Asparagopsis* spp. [62–64], and their presence may vary with the seaweed's harvest location and season [65]. Both 4-Hyp and Tyr have recognized benefits in fish nutrition: 4-Hyp supports collagen synthesis and growth, while Tyr is a precursor for hormones and neurotransmitters involved in metabolism and stress response [66–71]. Both play important roles in stress mitigation, redox balance, and cell protection in fish [66–75]. Therefore, the increased levels of these AAs may help explain the improved antioxidant activity and reduced cell damage observed in fish fed higher doses of *A. taxiformis*. Hence, the significant increase in 4-Hyp and Tyr might partially explain the increased antioxidant scavenging activity and decreased cell damage in fish supplemented with high doses of *A. taxiformis*, as previously reported in our studies [43,52].

In parallel, a decrease in the levels of benzoic acid was also observed in fish supplemented with 6.0% of *A. taxiformis*, most likely, due to the reduction in wheat meal (a major source of this phenolic compound [76]) in these feeds required to guarantee an isoproteic and isolipidic diet. Given the involvement of benzoic acid in various biological pathways (growth, digestion efficiency, immune function, and gut health [77–79]), as well as its anti-bacterial and antifungal properties [80], the significant reduction in this compound in fish supplemented with 6.0% of *A. taxiformis* can be considered an adverse outcome.

4.2. Effects of MHW on Fish Liver Primary Metabolome (CTR Versus CTR+HW)

Heat stress can lead to a wide range of metabolic disruptions in fish species, including increased glycolysis and lipid oxidation, accumulation of stress-related molecules (e.g., reactive oxygen species and heat shock proteins), and shifts in the metabolism of AAs and fatty acids [81–83]. The present findings highlight that metabolite profiling offers relevant insights for the early detection and management of stress signs in farmed marine fish, therefore, potentially contributing to the prevention of animal mortality and associated economic losses that often result from extreme weather events, such as MHWs. The acute exposure to increased temperatures prompted by a simulated MHW induced significant changes in the levels of AAs, sugars and sugar alcohols, organic acids, and other primary metabolites. As observed in other studies, the significant increase in AAs (i.e., glutamine—Gln, ornithine—Orn, pyroglutamate—Pyr, and serine—Ser) in fish exposed to the MHW might be related to an enhancement of protein and AA catabolic events to fuel the metabolic reprogramming associated with elevated temperatures [84,85]. In our study, the significant increase in Gln following exposure to MHW may suggest that (i) ammonia generated from AAs catabolism could have accumulated in the fish body, and consequently, the levels of Gln increased in order to reduce ammonia's toxicity, and/or (ii) a response to oxidative stress occurred, given that Gln is the precursor of glutathione (GSH [72]). Supporting our findings, Liu et al. [86] also observed a significant increase in hepatic Gln after juvenile lenok (*Brachymystax lenok*) exposure to thermal stress (ΔT = 8 °C for 7 days). Besides Gln, the levels of Orn also increased in response to MHW. Orn is a non-protein AA and an intermediate of the urea cycle [87]. The urea cycle is an endogenous source of arginine that also supports the removal of nitrogenous waste following protein metabolism [88]. Both Orn and urea are products of arginine breakdown by arginase in the fish liver [89]. The obtained Orn can be used to generate proline or polyamines (e.g., putrescine). The increase in Orn in *D. sargus* liver indicates that the fish needed to excrete nitrogenous waste resulting from protein and AA catabolism. Supporting our results, Jiang et al. [90] found significantly higher levels of Orn and Gln in the liver tissue of Qingtian paddy field carp (*Cyprinus carpio var qingtianensis*) after acute exposure to heat stress (ΔT = +10 °C for 24 h). Moreover, Pyr was significantly increased under MHW exposure. Pyr is an intermediary in the GSH biosynthesis pathway [91]. In this metabolic pathway, gamma-glutamylcysteine is converted to Pyr by ATP-dependent 5-oxoprolinase, which further converts Pyr into glutamate (Glu) and can restore GSH levels under oxidative stress conditions [91,92]. Thus, the increase in the levels of Pyr and no changes in Glu may indicate the inability to further proceed to GSH synthesis and concomitant oxidative stress and cellular damage. Once again, these results are in accordance with Pereira et al. [52] findings, as oxidative stress and cellular damage (i.e., increased CAT and GST activities, and LPO levels) were observed in non-supplemented fish after exposure to MHW. Alongside, MHW exposure led to an increase in Ser in *D. sargus* juveniles' liver tissue. Ser participates in fat digestion and in the one-carbon metabolism pathway [93]. D-serine is related to fish locomotor activity induced by environmental stressors. Aguilar and colleagues [94] observed an increase in D-serine in zebrafish (*Danio rerio*) after chronic exposure to thermal stress (ΔT = +3.1 °C and +4.6 °C for 270 days). Moreover, it was demonstrated that exogenous Ser promotes the synthesis of GSH, which downregulates ROS to dampen immune responses [95]. Hence, an increase in Ser might be related to the upsurge in metabolism, and consequently, an increase in oxidative stress and locomotor activity.

Besides the increase in AAs, an increase in the levels of the polyol myo-inositol was also observed after exposure to heat stress. This can imply that the TCA cycle was weakened, given that myo-inositol inhibits the activity of succinate dehydrogenase and malate dehydrogenase [95]. This result is supported by Kim et al. [96], where an increase in myo-inositol was also observed in olive flounder (*Paralichthys olivaceus*) juveniles after exposure to heat stress (ΔT = +4 °C for 7 days). Threonate is an oxidized derivative of ascorbate (vitamin C) [97] and, thus, its increase at higher temperatures further reinforces the statement that heat stress resulted in oxidative stress. Both niacin (vitamin B3) and adenine (vitamin B4) levels were observed to increase at higher temperatures. Tian et al. [98] also observed a significant high abundance of vitamins (B6 and C) following rainbow trout (*Oncorhynchus mykiss*) chronic exposure to warmer temperatures (ΔT = +4 °C for 56 days), most likely as a result of the fish homeostasis maintenance process. The observed changes in carbohydrate levels elicited by MHW are likely due to increased cellular energy demands prompted by thermal and/or oxidative stress. Noteworthy, the significant decrease in the levels of the maltose and trehalose might have been compensated by the increase in Orn, Gln, and Ser, as these glucogenic AAs serve as important carbon sources for gluconeogenesis (review in Falco et al. [99]). For instance, Zhao et al. [100] stated that *Gymnocypris chilianensis* specimens can maintain their physiological state under high temperature stress by enhancing glucose metabolism and regulating lipid and AA metabolic pathways. Alongside, in a study on American shad (*Alosa sapidissima*), it was hypothesized that the metabolism of carbohydrates, proteins, and fatty acids was accelerated under heat stress to ensure sufficient energy allocation towards fish immune responses and homeostasis maintenance [101]. In the dynamic energy budget (DEB) theory presented by Kooijman et al. [102], a certain amount of energy assimilated by organisms is reserved for fast providing the extra energy demand during stress exposure, but this consumption for self-maintenance will compete with other energy provisions for reproduction, development, and growth. The depletion of energy-related compounds, such as sugars, might further affect the self-maintenance, growth, and development of *D. sargus* juveniles.

4.3. Influence of A. taxiformis Supplementation on Fish Metabolic Responses to MHW (CTR+HW Versus 1.5-AT-HW, 3.0-AT-HW, and 6.0-AT-HW)

Upon exposure to an MHW, supplementation with 3.0% of *A. taxiformis* did not trigger significant changes in *D. sargus* primary metabolome, compared to non-supplemented fish; i.e., CTR+HW, except for the increase in the levels of a few beneficial AAs (Ala, Gln, and Thr). Diets enriched with Gln have been reported to improve fish feed efficiency, enhance intestinal function, and bolster innate immune responses [103,104]. Moreover, Ala is used as a substrate for glycogen and/or glucose production in the liver but may also be oxidized in the liver and used as a direct energy source [105], while the key roles of Gln and Thr have been discussed above. Indeed, Pereira et al. [52] reported increased GST and superoxide dismutase (SOD) activity, as well as decreased LPO in the liver of fish from 3.0-AT-HW treatment, therefore, confirming that this level of seaweed inclusion renders a better animal physiological condition. In alignment with these findings, results from this study also show that this intermediate level of supplementation corresponds to a nutritionally balanced feed formulation that meets the higher metabolic requirements of fish exposed to acute thermal stress.

In contrast, the significant reduction in various primary metabolite levels in fish supplemented with 1.5% and 6.0% of *A. taxiformis* evidence that these feed formulations do not constitute an effective solution to counteract the negative effects prompted by MHWs. In 1.5-AT-HW treatment, a metabolic reprogramming related to stress occurred; i.e., a substantial decrease was observed in the levels of several AAs that play important roles in fish nutrition and metabolism (e.g., aspartate-Asp, 4-Hyp, lysine-Lys, threonine-Thr; [72]). These key roles include the following: (i) major glucogenic precursor and important energy substrate (Asp; [72]); (ii) growth, developmental regulation and reproduction (e.g., Hyp; [68,72]); and (iii) immune responses and gastrointestinal development (e.g., Lys and Thr; [106,107]).

Supplementation with 6.0% of *A. taxiformis* also resulted in the reduction in several AAs, including Glu, Gln, 4-Hyp, and methionine (Met). Glu, Gln, and Met are functional AAs; i.e., they participate and control vital metabolic pathways in fish associated with health, growth, development, reproduction, antioxidant defenses, and survival [108]. In fish, Gln plays an important role in nutrition and metabolism. It participates in cell signaling, growth and development regulation, immunity, ammonia detoxification, antioxidant defenses, gut development, and stress responses [109]. Regarding Met, it is a glucogenic AA that produces glucose as an energy source, thus having a key role in the one-carbon metabolism pathway [110,111]. Under Met-limiting conditions, an excess of branched chain AAs reduces Met oxidation that will adversely affect gluconeogenesis [110–112], being reported to decrease percent weight gain, feed efficiency, and feed intake [113–115] as well as deposition of muscle protein [116]. Moreover, previous studies have demonstrated the role of Met in enhancing the fish immune system [117–119] and its antioxidant potential [120,121]. In addition to the reduction in several AAs, supplementation with 6.0% of *A. taxiformis* also caused an increase in the levels of several sugars with important roles in fish metabolism (e.g., fructose, maltose, mannose, trehalose, and mannitol). This result may indicate protection against heat stress, as carbohydrates work as available energy reserves and precursors for the biosynthesis of AAs. Previous studies have already pointed out their beneficial effects in fish fitness; these include increased survival rate and serum resistance against bacterial infection at control and/or suboptimal rearing conditions, as well as alleviation of hepatic cholesterol accumulation [122–125]. The drastic changes in both AAs and sugars observed in *D. sargus*'s primary metabolome at 6.0-AT-HW treatment imply that the fish were not physiologically well prepared to deal with thermal stress.

Altogether, data reported in this study coupled to previous studies [43,52] evidence that the optimal, or most-effective, level of seaweed supplementation depends on both: (i) the intended physiological outcome (e.g., improvement of metabolic or nutrient conversion efficiency, immune stimulation, antioxidant capacity enhancement); and (ii) the surrounding environmental conditions (e.g., optimal versus suboptimal). Hence, these findings emphasize the need to address the challenges posed in contemporary aquaculture nutrition from a holistic perspective and to further invest in effective, tailor-made solutions that meet specific animal welfare goals, under specific production contexts.

5. Conclusions

The present study confirmed that (i) *A. taxiformis* supplementation at low (i.e., 1.5%; under optimal temperature conditions) or intermediate (i.e., 3.0%, under suboptimal temperature conditions) inclusion levels have a beneficial effect in juvenile *D. sargus* primary metabolome, leading to an increase in the levels of functional AAs, sugars and other primary metabolites; (ii) the thermal stress induced by MHWs results in a general shutdown of central metabolism leading to a reduction in the levels of key primary metabolites; and (iii) the negative outcomes elicited by a MHW can be partially mitigated with *A. taxiformis* supplementation at an intermediate level of inclusion (3.0%). Yet, it should be noticed that these findings can be limited to the magnitude and duration of the MHW event, as well as to the fish species and life stage.

Although incorporating seaweeds into aquafeed formulations is a promising strategy to enhance the resilience and economic feasibility of the aquaculture sector, the suitability of these eco-innovative diets should first be validated in different contexts (e.g., considering distinct species' nutritional requirements and climate conditions of specific regions of the planet). Additionally, validation should follow holistic nutrigenomics approaches that assess animal responses at multiple levels of biological organization, including molecular, cellular, tissue, and whole-organism scales.

Supplementary Materials: The following supporting information can be downloaded at https://www.mdpi.com/article/10.3390/fishes10070350/s1, Figure S1: Central metabolic pathways representing the changes in the relative levels of primary metabolites in liver tissues of *Diplodus sargus* grown under optimal conditions and upon supplementation with *Asparagopsis taxiformis*; Figure S2: Central metabolic pathways representing the changes in the relative levels of primary metabolites in liver tissues of *D. sargus* grown under suboptimal temperature conditions; Figure S3: Central metabolic pathways representing the changes in the relative levels of primary metabolites in liver tissues of *Diplodus sargus* grown under suboptimal temperature conditions and upon supplementation with *Asparagopsis taxiformis*; Table S1: Normalized raw data (dry weight and internal standard ribitol) obtained in the present study; Table S2: VIP (Variable Importance in Projection) scores for each metabolite across the first and second components of the PLS-DA model, based on the analysis of the primary metabolite profile in fish liver tissues at Phase I; Table S3: VIP (Variable Importance in Projection) scores for each metabolite across the first and second components of the PLS-DA model, based on the analysis of the primary metabolite profile in fish liver tissues at Phase II; Table S4: Fold changes in relative levels of primary metabolites in fish liver tissues, in relation to the respective control (CTR) condition. Fish were kept at the optimal temperature of 24 °C (CTR) and fed a supplemented diet with 1.5, 3.0, and 6.0% *Asparagopsis taxiformis* for 30 days (1.5-AT, 3.0-AT, and 6.0-AT, respectively). Relative values were normalized to the fresh weight (FW) of the samples and IS ribitol. Data are presented as means ± SE of 3 independent measurements. Significant changes of log10-transformed data using one-way ANOVA ($p < 0.05$) are presented in bold. Metabolites grouped in amino acids and derivatives (AA), sugars and derivatives (sugar phosphate; sugar alcohol; sugar acid) (SS), organic acids (OA), and others (O); Table S5A: Fold changes in relative levels of primary metabolites in fish liver tissues, after a marine heatwave simulation, during which fish were exposed to a category II Mediterranean heatwave (CTR+HW), including a ramp temperature increase period of 8 days (+0.5 °C/day) up to 28 °C [50], followed by a 15-day period of exposure to the peak temperature (CTR+HW), in relation to the respective control condition at 24 °C (CTR). Relative values were normalized to the fresh weight (FW) of the samples and IS ribitol. Data are presented as means ± SE of 3 independent measurements. Significant changes of log10-transformed data using one-way ANOVA ($p < 0.05$) are presented in bold. Metabolites grouped in amino acids and derivatives (AA), sugars and derivatives (sugar phosphate; sugar alcohol; sugar acid) (SS), organic acids (OA), and others (O); Table S5B: Fold changes in relative levels of primary metabolites in fish liver tissues. Fish were exposed to a category II Mediterranean marine heatwave after a 30-day supplementation period and fed with a non-supplemented/commercial diet (CTR+HW) or a diet supplemented with 1.5, 3.0, and 6.0% *Asparagopsis taxiformis* (1.5-AT-HW, 3.0-AT-HW, and 6.0-AT-HW, respectively). Relative values were normalized to the fresh weight (FW) of the samples and IS ribitol. Data are presented as means ± SE of 3 independent measurements. Significant changes of log10-transformed data using one-way ANOVA ($p < 0.05$) are presented in bold. Metabolites grouped in amino acids and derivatives (AA), sugars and derivatives (sugar phosphate; sugar alcohol; sugar acid) (SS), organic acids (OA), and others (O).

Author Contributions: Conceptualization, A.L.M.; Data curation, C.A., A.M.R. and A.L.M.; Formal analysis, C.A., A.M.R. and A.L.M.; Funding acquisition, M.S.D. and A.L.M.; Investigation, I.M., M.S.D. and A.L.M.; Methodology, M.D., I.M. and A.L.M.; Resources, A.M. and A.L.M.; Software, C.A. and A.M.R.; Supervision, A.L.M.; Validation, A.L.M.; Visualization, M.D. and A.L.M.; Writing—original draft, M.D.; Writing—review and editing, M.D., I.M., C.A., A.M.R., A.M., M.S.D. and A.L.M. All authors have read and agreed to the published version of the manuscript.

Funding: This research was funded by Fundação Portuguesa para a Ciência e a Tecnologia (FCT I.P.), under the framework of the project Aqua-CLIMADAPT (PTDC/CTA-AMB/0592/2021, https://doi.org/10.54499/PTDC/CTA-AMB/0592/2021) and the "Applied Molecular Biosciences Unit" (UCIBIO), which is financed by national funds from FCT/MCTES (UIDB/04378/2020), the project LA/P/0140/2020 of the Associate Laboratory Institute for Health and Bioeconomy—i4HB. The project Aqua-CLIMADAPT (PTDC/CTA-AMB/0592/2021) also supported the research contract of Marta Dias. Isa Marmelo also acknowledges FCT I.P. for PhD grant (2020.04413.BD, https://doi.org/10.54499/2020.04413.BD).

Institutional Review Board Statement: The animal study protocol was approved by the Institutional Review Board (or Ethics Committee) of IPMA's Animal Welfare and Ethics Body (ORBEA; authorization number 001/2023, 2 November 2023), under the oversight of the National Authority for the use of live animals, also recognized as the Directorate-General for Food and Veterinary (DGAV).

Data Availability Statement: Data is contained within the article or Supplementary Material.

Acknowledgments: The authors acknowledge the team from the Aquaculture Research Station of Olhão (EPPO-IPMA I.P.) for the technical support provided during animal rearing and transplantation, as well as GOPSIS Lda. for supplying the macroalga *Asparagopsis taxifomis* and SPAROS Lda. for supplying the experimental feeds.

Conflicts of Interest: The authors declare no conflicts of interest. The funders had no role in the design of the study; in the collection, analyses, or interpretation of data; in the writing of the manuscript; or in the decision to publish the results.

Appendix A

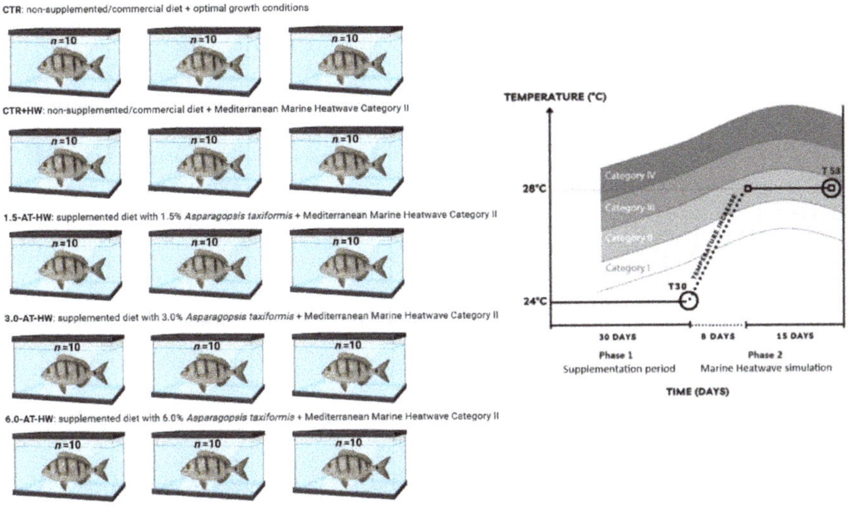

Figure A1. Experimental setup and illustration of the category II Mediterranean heatwave simulation. The fish were randomly and equally distributed in 15 tanks (5 treatments in triplicate, n = 10 fish in each tank). The experimental trial involved two phases: Phase 1—Supplementation period for 30 days, during which fish were kept at the optimal temperature of 24 °C (T30); and Phase 2—Marine heatwave simulation, during which fish were exposed to a category II Mediterranean heatwave, including a ramp temperature increase period of 8 days (+0.5 °C/day) up to 28 °C [53], followed by a 15-day period of exposure to the peak temperature (i.e., 28 °C, T53).

References

1. FAO. *The State of World Fisheries and Aquaculture 2022. Towards Blue Transformation*; Academic Press: London, UK; San Diego, CA, USA, 2022.
2. Gephart, J.A.; Golden, C.D.; Asche, F.; Belton, B.; Brugere, C.; Froehlich, H.E.; Fry, J.P.; Halpern, B.S.; Hicks, C.C.; Jones, R.C.; et al. Scenarios for Global Aquaculture and Its Role in Human Nutrition. *Rev. Fish. Sci. Aquac.* **2021**, *29*, 122–138. [CrossRef]
3. Der Poel, A.F.B.V.; Abdollahi, M.R.; Cheng, H.; Colovic, R.; Den Hartog, L.A.; Miladinovic, D.; Page, G.; Sijssens, K.; Smillie, J.F.; Thomas, M.; et al. Future Directions of Animal Feed Technology Research to Meet the Challenges of a Changing World. *Anim. Feed Sci. Technol.* **2020**, *270*, 114692. [CrossRef]
4. Hardy, R.W.; Kaushik, S.J. (Eds.) *Fish Nutrition*, 4th ed.; Academic Press: London, UK; San Diego, CA, USA; Cambridge, MA, USA; Kidlington, UK, 2022; ISBN 978-0-12-819587-1.
5. Jimenez-Lopez, C.; Pereira, A.G.; Lourenço-Lopes, C.; Garcia-Oliveira, P.; Cassani, L.; Fraga-Corral, M.; Prieto, M.A.; Simal-Gandara, J. Main Bioactive Phenolic Compounds in Marine Algae and Their Mechanisms of Action Supporting Potential Health Benefits. *Food Chem.* **2021**, *341*, 128262. [CrossRef]
6. Naiel, M.A.E.; Alagawany, M.; Patra, A.K.; El-Kholy, A.I.; Amer, M.S.; Abd El-Hack, M.E. Beneficial Impacts and Health Benefits of Macroalgae Phenolic Molecules on Fish Production. *Aquaculture* **2021**, *534*, 736186. [CrossRef]

7. Stadtlander, T.; Khalil, W.K.B.; Focken, U.; Becker, K. Effects of Low and Medium Levels of Red Alga Nori (*Porphyra yezoensis* Ueda) in the Diets on Growth, Feed Utilization and Metabolism in Intensively Fed Nile Tilapia, *Oreochromis niloticus* (L.). *Aquacult. Nutr.* **2013**, *19*, 64–73. [CrossRef]
8. Cian, R.E.; Bacchetta, C.; Rossi, A.; Cazenave, J.; Drago, S.R. Red Seaweed *Pyropia columbina* as Antioxidant Supplement in Feed for Cultured Juvenile Pacú (*Piaractus mesopotamicus*). *J. Appl. Phycol.* **2019**, *31*, 1455–1465. [CrossRef]
9. Xuan, X.; Li, W.; Zhu, W.; Wang, S. Effects of Different Levels of Macroalga *Gracilaria lemaneiformis* on Growth Performance and Feed Utilization on the Red Sea Bream, *Pagrosomus major*. *J. Appl. Phycol.* **2019**, *31*, 3213–3222. [CrossRef]
10. Peixoto, M.J.; Svendsen, J.C.; Malte, H.; Pereira, L.F.; Carvalho, P.; Pereira, R.; Gonçalves, J.F.M.; Ozório, R.O.A. Diets Supplemented with Seaweed Affect Metabolic Rate, Innate Immune, and Antioxidant Responses, but Not Individual Growth Rate in European Seabass (*Dicentrarchus labrax*). *J. Appl. Phycol.* **2016**, *28*, 2061–2071. [CrossRef]
11. Peixoto, M.J.; Salas-Leitón, E.; Brito, F.; Pereira, L.F.; Svendsen, J.C.; Baptista, T.; Pereira, R.; Abreu, H.; Reis, P.A.; Gonçalves, J.F.M.; et al. Effects of Dietary *Gracilaria* sp. and *Alaria* sp. Supplementation on Growth Performance, Metabolic Rates and Health in Meagre (*Argyrosomus regius*) Subjected to Pathogen Infection. *J. Appl. Phycol.* **2017**, *29*, 433–447. [CrossRef]
12. Marmelo, I.; Dias, M.; Grade, A.; Pousão-Ferreira, P.; Diniz, M.S.; Marques, A.; Maulvault, A.L. Immunomodulatory and Antioxidant Effects of Functional Aquafeeds Biofortified with Whole *Laminaria digitata* in Juvenile Gilthead Seabream (*Sparus aurata*). *Front. Mar. Sci.* **2024**, *11*, 1325244. [CrossRef]
13. Pereira, V.; Marques, A.; Gaivão, I.; Rego, A.; Abreu, H.; Pereira, R.; Santos, M.A.; Guilherme, S.; Pacheco, M. Marine Macroalgae as a Dietary Source of Genoprotection in Gilthead Seabream (*Sparus aurata*) against Endogenous and Exogenous Challenges. *Comp. Biochem. Physiol. C Toxicol. Pharmacol.* **2019**, *219*, 12–24. [CrossRef] [PubMed]
14. Bansemir, A.; Blume, M.; Schröder, S.; Lindequist, U. Screening of Cultivated Seaweeds for Antibacterial Activity against Fish Pathogenic Bacteria. *Aquaculture* **2006**, *252*, 79–84. [CrossRef]
15. Vatsos, I.N.; Rebours, C. Seaweed Extracts as Antimicrobial Agents in Aquaculture. *J. Appl. Phycol.* **2015**, *27*, 2017–2035. [CrossRef]
16. Lozano, I.; Wacyk, J.M.; Carrasco, J.; Cortez-San Martín, M.A. Red Macroalgae *Pyropia columbina* and *Gracilaria chilensis*: Sustainable Feed Additive in the *Salmo salar* Diet and the Evaluation of Potential Antiviral Activity against Infectious Salmon Anemia Virus. *J. Appl. Phycol.* **2016**, *28*, 1343–1351. [CrossRef]
17. Gora, A.H.; Sahu, N.P.; Sahoo, S.; Rehman, S.; Dar, S.A.; Ahmad, I.; Agarwal, D. Effect of Dietary *Sargassum wightii* and Its Fucoidan-Rich Extract on Growth, Immunity, Disease Resistance and Antimicrobial Peptide Gene Expression in *Labeo rohita*. *Int. Aquat. Res.* **2018**, *10*, 115–131. [CrossRef]
18. Shi, Q.; Rong, H.; Hao, M.; Zhu, D.; Aweya, J.J.; Li, S.; Wen, X. Effects of Dietary *Sargassum horneri* on Growth Performance, Serum Biochemical Parameters, Hepatic Antioxidant Status, and Immune Responses of Juvenile Black Sea Bream *Acanthopagrus schlegelii*. *J. Appl. Phycol.* **2019**, *31*, 2103–2113. [CrossRef]
19. Xuan, X.; Wen, X.; Li, S.; Zhu, D.; Li, Y. Potential Use of Macro-Algae *Gracilaria lemaneiformis* in Diets for the Black Sea Bream, *Acanthopagrus schlegelii*, Juvenile. *Aquaculture* **2013**, *412–413*, 167–172. [CrossRef]
20. Kamunde, C.; Sappal, R.; Melegy, T.M. Brown Seaweed (AquaArom) Supplementation Increases Food Intake and Improves Growth, Antioxidant Status and Resistance to Temperature Stress in Atlantic Salmon, *Salmo salar*. *PLoS ONE* **2019**, *14*, e0219792. [CrossRef]
21. Cascarano, M.C.; Stavrakidis-Zachou, O.; Mladineo, I.; Thompson, K.D.; Papandroulakis, N.; Katharios, P. Mediterranean Aquaculture in a Changing Climate: Temperature Effects on Pathogens and Diseases of Three Farmed Fish Species. *Pathogens* **2021**, *10*, 1205. [CrossRef]
22. Hobday, A.J.; Alexander, L.V.; Perkins, S.E.; Smale, D.A.; Straub, S.C.; Oliver, E.C.J.; Benthuysen, J.A.; Burrows, M.T.; Donat, M.G.; Feng, M.; et al. A Hierarchical Approach to Defining Marine Heatwaves. *Prog. Oceanogr.* **2016**, *141*, 227–238. [CrossRef]
23. Garrabou, J.; Gómez-Gras, D.; Medrano, A.; Cerrano, C.; Ponti, M.; Schlegel, R.; Bensoussan, N.; Turicchia, E.; Sini, M.; Gerovasileiou, V.; et al. Marine Heatwaves Drive Recurrent Mass Mortalities in the Mediterranean Sea. *Glob. Change Biol.* **2022**, *28*, 5708–5725. [CrossRef] [PubMed]
24. Martínez, J.; Leonelli, F.E.; García-Ladona, E.; Garrabou, J.; Kersting, D.K.; Bensoussan, N.; Pisano, A. Evolution of Marine Heatwaves in Warming Seas: The Mediterranean Sea Case Study. *Front. Mar. Sci.* **2023**, *10*, 1193164. [CrossRef]
25. Islam, M.J.; Kunzmann, A.; Slater, M.J. Responses of Aquaculture Fish to Climate Change-induced Extreme Temperatures: A Review. *J. World Aquac. Soc.* **2022**, *53*, 314–366. [CrossRef]
26. Barbosa, V.; Maulvault, A.L.; Anacleto, P.; Santos, M.; Mai, M.; Oliveira, H.; Delgado, I.; Coelho, I.; Barata, M.; Araújo-Luna, R.; et al. Enriched Feeds with Iodine and Selenium from Natural and Sustainable Sources to Modulate Farmed Gilthead Seabream (*Sparus aurata*) and Common Carp (*Cyprinus carpio*) Fillets Elemental Nutritional Value. *Food Chem. Toxicol.* **2020**, *140*, 111330. [CrossRef]
27. Silva-Brito, F.; Timóteo, F.; Esteves, Â.; Peixoto, M.J.; Ozorio, R.; Magnoni, L. Impact of the Replacement of Dietary Fish Oil by Animal Fats and Environmental Salinity on the Metabolic Response of European Seabass (*Dicentrarchus labrax*). *Comp. Biochem. Physiol. B Biochem. Mol. Biol.* **2019**, *233*, 46–59. [CrossRef] [PubMed]

28. Oliva-Teles, A. Nutrition and Health of Aquaculture Fish. *J. Fish Dis.* **2012**, *35*, 83–108. [CrossRef]
29. Hixson, S.H. Fish Nutrition and Current Issues in Aquaculture: The Balance in Providing Safe and Nutritious Seafood, in an Environmentally Sustainable Manner. *J. Aquac. Res. Development* **2014**, *3*, 234. [CrossRef]
30. Jobling, M. Fish Nutrition Research: Past, Present and Future. *Aquacult. Int.* **2016**, *24*, 767–786. [CrossRef]
31. Mahanty, A.; Yadav, R.P.; Purohit, G.K.; Mohanty, S.; Mohanty, B.P. Metabolomic Response to High Temperature Stress in Murrel *Channa striatus* and Insights for Designer Feeds. In *Outlook of Climate Change and Fish Nutrition*; Sinha, A., Kumar, S., Kumari, K., Eds.; Springer Nature: Singapore, 2022; pp. 197–205. ISBN 978-981-19-5499-3.
32. Viant, M.R.; Rosenblum, E.S.; Tjeerdema, R.S. NMR-Based Metabolomics: A Powerful Approach for Characterizing the Effects of Environmental Stressors on Organism Health. *Environ. Sci. Technol.* **2003**, *37*, 4982–4989. [CrossRef]
33. Alfaro, A.C.; Young, T. Showcasing Metabolomic Applications in Aquaculture: A Review. *Rev. Aquac.* **2018**, *10*, 135–152. [CrossRef]
34. Fiehn, O. Metabolomics—The Link between Genotypes and Phenotypes. In *Functional Genomics*; Town, C., Ed.; Springer: Dordrecht, The Netherlands, 2002; pp. 155–171. ISBN 978-94-010-3903-1.
35. Fernie, A.R.; Trethewey, R.N.; Krotzky, A.J.; Willmitzer, L. Metabolite Profiling: From Diagnostics to Systems Biology. *Nat. Rev. Mol. Cell Biol.* **2004**, *5*, 763–769. [CrossRef]
36. Elolimy, A.; Alharthi, A.; Zeineldin, M.; Parys, C.; Helmbrecht, A.; Loor, J.J. Supply of Methionine During Late-Pregnancy Alters Fecal Microbiota and Metabolome in Neonatal Dairy Calves Without Changes in Daily Feed Intake. *Front. Microbiol.* **2019**, *10*, 2159. [CrossRef] [PubMed]
37. Novais, F.J.; Pires, P.R.L.; Alexandre, P.A.; Dromms, R.A.; Iglesias, A.H.; Ferraz, J.B.S.; Styczynski, M.P.-W.; Fukumasu, H. Identification of a Metabolomic Signature Associated with Feed Efficiency in Beef Cattle. *BMC Genomics* **2019**, *20*, 8. [CrossRef]
38. Carmelo, V.A.O.; Banerjee, P.; Da Silva Diniz, W.J.; Kadarmideen, H.N. Metabolomic Networks and Pathways Associated with Feed Efficiency and Related-Traits in Duroc and Landrace Pigs. *Sci. Rep.* **2020**, *10*, 255. [CrossRef] [PubMed]
39. Clemmons, B.A.; Powers, J.B.; Campagna, S.R.; Seay, T.B.; Embree, M.M.; Myer, P.R. Rumen Fluid Metabolomics of Beef Steers Differing in Feed Efficiency. *Metabolomics* **2020**, *16*, 23. [CrossRef] [PubMed]
40. Wu, R.; Chen, J.; Zhang, L.; Wang, X.; Yang, Y.; Ren, X. LC/MS-Based Metabolomics to Evaluate the Milk Composition of Human, Horse, Goat and Cow from China. *Eur. Food Res. Technol.* **2021**, *247*, 663–675. [CrossRef]
41. Ye, X.; Zhou, L.; Zhang, Y.; Xue, S.; Gan, Q.F.; Fang, S. Effect of Host Breeds on Gut Microbiome and Serum Metabolome in Meat Rabbits. *BMC Vet. Res.* **2021**, *17*, 24. [CrossRef]
42. Fiehn, O. Metabolomics by Gas Chromatography–Mass Spectrometry: Combined Targeted and Untargeted Profiling. *Curr. Protoc. Mol. Biol.* **2016**, *114*, 30.4.1–30.4.32. [CrossRef]
43. Marmelo, I.; Lourenço-Marques, C.; Silva, I.A.L.; Soares, F.; Pousão-Ferreira, P.; Mata, L.; Marques, A.; Diniz, M.S.; Maulvault, A.L. Eco-Innovative Aquafeeds Biofortified with *Asparagopsis taxiformis* to Improve the Resilience of Farmed White Seabream (*Diplodus sargus*) to Marine Heatwave Events. *Heliyon* **2024**, *10*, e35135. [CrossRef]
44. Belghit, I.; Rasinger, J.D.; Heesch, S.; Biancarosa, I.; Liland, N.; Torstensen, B.; Waagbø, R.; Lock, E.-J.; Bruckner, C.G. In-Depth Metabolic Profiling of Marine Macroalgae Confirms Strong Biochemical Differences between Brown, Red and Green Algae. *Algal Res.* **2017**, *26*, 240–249. [CrossRef]
45. Mancuso, F.P.; D'Agostaro, R.; Milazzo, M.; Badalamenti, F.; Musco, L.; Mikac, B.; Lo Brutto, S.; Chemello, R. The Invasive Seaweed *Asparagopsis taxiformis* Erodes the Habitat Structure and Biodiversity of Native Algal Forests in the Mediterranean Sea. *Mar. Environ. Res.* **2022**, *173*, 105515. [CrossRef]
46. Vinagre, C.; Madeira, C.; Dias, M.; Narciso, L.; Mendonça, V. Reliance of Coastal Intertidal Food Webs on River Input—Current and Future Perspectives. *Ecol. Indic.* **2019**, *101*, 632–639. [CrossRef]
47. Gamito, S.; Quental-Ferreira, H.; Parejo, A.; Aubin, J.; Christensen, V.; Cunha, M. Integrated Multi-Trophic Aquaculture Systems: Energy Transfers and Food Web Organization in Coastal Earthen Ponds. *Aquacult. Environ. Interact.* **2020**, *12*, 457–470. [CrossRef]
48. Morales-Nin, B.; Moranta, J.; García, C.; Tugores, M.P.; Grau, A.M.; Riera, F.; Cerdà, M. The Recreational Fishery off Majorca Island (Western Mediterranean): Some Implications for Coastal Resource Management. *ICES J. Mar. Sci.* **2005**, *62*, 727–739. [CrossRef]
49. Veiga, P.; Ribeiro, J.; Gonçalves, J.M.S.; Erzini, K. Quantifying Recreational Shore Angling Catch and Harvest in Southern Portugal (North-east Atlantic Ocean): Implications for Conservation and Integrated Fisheries Management. *J. Fish Biol.* **2010**, *76*, 2216–2237. [CrossRef]
50. FAO. *Fisheries and Aquaculture Department, FAO FishFinder Aquatic Species—Fact Sheets, Diplodus sargus*; FAO: Rome, Italy, 2020.
51. Santos, C.; Soares, F.; Candeias-Mendes, A.; Pousão-Ferreira, P.; Dinis, M.T.; Cortes Valente Oliveira, C. Characterization of Spawning Rhythms of a Sparidae Aquaculture Species, the White Seabream (*Diplodus sargus*), in the South of Portugal. *Aquac. Res.* **2022**, *53*, 1424–1434. [CrossRef]

52. Pereira, A.; Marmelo, I.; Dias, M.; Silva, A.C.; Grade, A.C.; Barata, M.; Pousão-Ferreira, P.; Dias, J.; Anacleto, P.; Marques, A.; et al. *Asparagopsis taxiformis* as a Novel Antioxidant Ingredient for Climate-Smart Aquaculture: Antioxidant, Metabolic and Digestive Modulation in Juvenile White Seabream (*Diplodus sargus*) Exposed to a Marine Heatwave. *Antioxidants* **2024**, *13*, 949. [CrossRef]
53. Hobday, A.J.; Oliver, E.C.; Gupta, A.S.; Benthuysen, J.A.; Burrows, M.T.; Donat, M.G.; Holbrook, N.J.; Moore, P.J.; Thomsen, M.S.; Wernberg, T.; et al. Categorizing and Naming Marine Heatwaves. *Oceanography* **2018**, *31*, 162–173. [CrossRef]
54. Lisec, J.; Schauer, N.; Kopka, J.; Willmitzer, L.; Fernie, A.R. Gas Chromatography Mass Spectrometry–Based Metabolite Profiling in Plants. *Nat. Protoc.* **2006**, *1*, 387–396. [CrossRef]
55. Luedemann, A.; Strassburg, K.; Erban, A.; Kopka, J. TagFinder for the Quantitative Analysis of Gas Chromatography—Mass Spectrometry (GC-MS)-Based Metabolite Profiling Experiments. *Bioinformatics* **2008**, *24*, 732–737. [CrossRef]
56. Kopka, J.; Schauer, N.; Krueger, S.; Birkemeyer, C.; Usadel, B.; Bergmuller, E.; Dormann, P.; Weckwerth, W.; Gibon, Y.; Stitt, M.; et al. GMD@CSB.DB: The Golm Metabolome Database. *Bioinformatics* **2005**, *21*, 1635–1638. [CrossRef]
57. Schauer, N.; Steinhauser, D.; Strelkov, S.; Schomburg, D.; Allison, G.; Moritz, T.; Lundgren, K.; Roessner-Tunali, U.; Forbes, M.G.; Willmitzer, L.; et al. GC–MS Libraries for the Rapid Identification of Metabolites in Complex Biological Samples. *FEBS Letters* **2005**, *579*, 1332–1337. [CrossRef] [PubMed]
58. RStudio Team. RStudio: Integrated Development for R. RStudio, PBC, Boston, MA, USA. 2020. Available online: http://www.rstudio.com/ (accessed on 8 July 2024).
59. de Mendiburu, F. *Agricolae: Statistical Procedures for Agricultural Research, Version 1.3-3*; R Package: Madison, WI, USA, 2020.
60. Warnes, G.R.; Bolker, B.; Bonebakker, L.; Gentleman, R.; Huber, W.; Liaw, A.; Lumley, T.; Maechler, M.; Magnusson, A.; Moeller, S.; et al. Gplots: Various R Programming Tools for Plotting Data. R Package Version 2.12.1. 2020. Available online: http://CRAN.R-project.org/package=gplots (accessed on 13 March 2025).
61. Rohart, F.; Gautier, B.; Singh, A.; Lê Cao, K.-A. mixOmics: An R Package for 'omics Feature Selection and Multiple Data Integration. *PLoS Comput. Biol.* **2017**, *13*, e1005752. [CrossRef]
62. Machado, M.; Machado, S.; Pimentel, F.B.; Freitas, V.; Alves, R.C.; Oliveira, M.B.P.P. Amino Acid Profile and Protein Quality Assessment of Macroalgae Produced in an Integrated Multi-Trophic Aquaculture System. *Foods* **2020**, *9*, 1382. [CrossRef] [PubMed]
63. Félix, R.; Dias, P.; Félix, C.; Cerqueira, T.; Andrade, P.B.; Valentão, P.; Lemos, M.F.L. The Biotechnological Potential of *Asparagopsis armata*: What Is Known of Its Chemical Composition, Bioactivities and Current Market? *Algal Res.* **2021**, *60*, 102534. [CrossRef]
64. De Bhowmick, G.; Hayes, M. In Vitro Protein Digestibility of Selected Seaweeds. *Foods* **2022**, *11*, 289. [CrossRef]
65. Zhou, A.Y.; Robertson, J.; Hamid, N.; Ma, Q.; Lu, J. Changes in Total Nitrogen and Amino Acid Composition of New Zealand *Undaria pinnatifida* with Growth, Location and Plant Parts. *Food Chem.* **2015**, *186*, 319–325. [CrossRef]
66. Aksnes, A.; Mundheim, H.; Toppe, J.; Albrektsen, S. The Effect of Dietary Hydroxyproline Supplementation on Salmon (*Salmo salar* L.) Fed High Plant Protein Diets. *Aquaculture* **2008**, *275*, 242–249. [CrossRef]
67. Liu, X.; Mai, K.; Liufu, Z.; Ai, Q. Effects of Dietary Protein and Lipid Levels on Growth, Nutrient Utilization, and the Whole-body Composition of Turbot, *Scophthalmus maximus*, Linnaeus 1758, at Different Growth Stages. *J. World Aquac. Soc.* **2014**, *45*, 355–366. [CrossRef]
68. Rong, H.; Zhang, Y.; Hao, M.; Lin, F.; Zou, W.; Zhang, H.; Yu, C.; Yu, J.; Shi, Q.; Aweya, J.J.; et al. Effect of Hydroxyproline Supplementation on Growth Performance, Body Composition, Amino Acid Profiles, Blood-biochemistry and Collagen Synthesis of Juvenile Chu's Croaker (*Nibea coibor*). *Aquac. Res.* **2020**, *51*, 1264–1275. [CrossRef]
69. Salamanca, N.; Giráldez, I.; Morales, E.; De La Rosa, I.; Herrera, M. Phenylalanine and Tyrosine as Feed Additives for Reducing Stress and Enhancing Welfare in Gilthead Seabream and Meagre. *Animals* **2020**, *11*, 45. [CrossRef] [PubMed]
70. Wu, Z.; Hou, Y.; Dai, Z.; Hu, C.-A.A.; Wu, G. Metabolism, Nutrition, and Redox Signaling of Hydroxyproline. *Antioxid. Redox Signal.* **2019**, *30*, 674–682. [CrossRef] [PubMed]
71. Ma, F.; Zhao, L.; Ma, R.; Wang, J.; Du, L. FoxO Signaling and Mitochondria-Related Apoptosis Pathways Mediate Tsinling Lenok Trout (*Brachymystax lenok tsinlingensis*) Liver Injury under High Temperature Stress. *Int. J. Biol. Macromol.* **2023**, *251*, 126404. [CrossRef]
72. Li, P.; Mai, K.; Trushenski, J.; Wu, G. New Developments in Fish Amino Acid Nutrition: Towards Functional and Environmentally Oriented Aquafeeds. *Amino Acids* **2009**, *37*, 43–53. [CrossRef]
73. Jasour, M.S.; Wagner, L.; Sundekilde, U.K.; Larsen, B.K.; Greco, I.; Orlien, V.; Olsen, K.; Rasmussen, H.T.; Hjermitslev, N.H.; Hammershøj, M.; et al. A Comprehensive Approach to Assess Feathermeal as an Alternative Protein Source in Aquafeed. *J. Agric. Food Chem.* **2017**, *65*, 10673–10684. [CrossRef] [PubMed]
74. Phang, J.M.; Donald, S.P.; Pandhare, J.; Liu, Y. The Metabolism of Proline, a Stress Substrate, Modulates Carcinogenic Pathways. *Amino Acids* **2008**, *35*, 681–690. [CrossRef]
75. Hu, S.; He, W.; Wu, G. Hydroxyproline in Animal Metabolism, Nutrition, and Cell Signaling. *Amino Acids* **2022**, *54*, 513–528. [CrossRef]

76. Nisar, N.; Mustafa, F.; Tahir, A.; Qadri, R.; Yang, Y.; Khan, M.I.; Wang, F. Proximate Composition, Functional Properties and Quantitative Analysis of Benzoyl Peroxide and Benzoic Acid in Wheat Flour Samples: Effect on Wheat Flour Quality. *PeerJ* **2020**, *8*, e8788. [CrossRef]
77. Libanori, M.C.M.; Santos, G.G.; Pereira, S.A.; Lopes, G.R.; Owatari, M.S.; Soligo, T.A.; Yamashita, E.; Pereira, U.P.; Martins, M.L.; Mouriño, J.L.P. Dietary Supplementation with Benzoic Organic Acid Improves the Growth Performance and Survival of Nile Tilapia (*Oreochromis niloticus*) after Challenge with *Streptococcus agalactiae* (Group B). *Aquaculture* **2021**, *545*, 737204. [CrossRef]
78. Hussein, E.E.; Habiba, M.M.; Ashry, A.M.; Al-Zayat, A.M.; Teiba, I.I.; Shehata, A.I.; Shahin, S.A.; El-Ratel, I.T.; Mzengereza, K.; Tembo, M.; et al. Effects of Dietary Supplementation with Organic Acids Mixture on Growth, Feed Efficiency, Hematobiochemical Parameters, Immunity, and Intestinal Microbiota of Gilthead Seabream (*Sparus aurata*) Juveniles. *Aquac. Rep.* **2023**, *33*, 101846. [CrossRef]
79. Santos, G.G.; Libanori, M.C.M.; Pereira, S.A.; Ferrarezi, J.V.S.; Ferreira, M.B.; Soligo, T.A.; Yamashita, E.; Martins, M.L.; Mouriño, J.L.P. Probiotic Mix of *Bacillus* spp. and Benzoic Organic Acid as Growth Promoter against *Streptococcus agalactiae* in Nile Tilapia. *Aquaculture* **2023**, *566*, 739212. [CrossRef]
80. Hu, Z.; Zhu, Y.; Chen, J.; Chen, J.; Li, C.; Gao, Z.; Li, J.; Liu, L. Discovery of Novel Bactericides from *Aspergillus alabamensis* and Their Antibacterial Activity against Fish Pathogens. *J. Agric. Food Chem.* **2023**, *71*, 4298–4305. [CrossRef]
81. Long, Z.; Qin, H.; Huang, Z.; Xu, A.; Ye, Y.; Li, Z. Effects of Heat Stress on Physiological Parameters, Biochemical Parameters and Expression of Heat Stress Protein Gene in *Lateolabrax maculatus*. *J. Therm. Biol.* **2023**, *115*, 103606. [CrossRef]
82. Hao, R.; Li, H.; Tian, Y.; Ru, X.; Deng, Q.; Zhu, K.; Yang, T.; Huang, Y.; Zhu, C. The Effect of Heat Stress on Energy Metabolism, Immune Function, and Oxidative Stress of Juvenile Greater Amberjack *Seriola dumerili*. *Aquac. Res.* **2024**, *2024*, 1–9. [CrossRef]
83. Yan, H.; Du, J.; Li, S.; Lei, C.; Zhu, T.; Han, L.; Song, H. Chronic Heat Stress Is Capable of Reducing the Growth Performance, Causing Damage to the Liver Structure, and Altering the Liver Glucose Metabolism and Lipid Metabolism in Largemouth Bass (*Micropterus salmoides* L.). *Fish Physiol. Biochem.* **2025**, *51*, 24. [CrossRef]
84. Currie, S.; Bagatto, B.; DeMille, M.; Learner, A.; LeBlanc, D.; Marks, C.; Ong, K.; Parker, J.; Templeman, N.; Tufts, B.L.; et al. Metabolism, Nitrogen Excretion, and Heat Shock Proteins in the Central Mudminnow (*Umbra limi*), a Facultative Air-Breathing Fish Living in a Variable Environment. *Can. J. Zool.* **2010**, *88*, 43–58. [CrossRef]
85. Song, M.; Zhao, J.; Wen, H.-S.; Li, Y.; Li, J.-F.; Li, L.-M.; Tao, Y.-X. The Impact of Acute Thermal Stress on the Metabolome of the Black Rockfish (*Sebastes schlegelii*). *PLoS ONE* **2019**, *14*, e0217133. [CrossRef]
86. Liu, Y.; Liu, J.; Ye, S.; Bureau, D.P.; Liu, H.; Yin, J.; Mou, Z.; Lin, H.; Hao, F. Global Metabolic Responses of the Lenok (*Brachymystax lenok*) to Thermal Stress. *Comp. Biochem. Physiol. Part D Genomics Proteomics* **2019**, *29*, 308–319. [CrossRef]
87. Hillyer, K.E.; Beale, D.J.; Shima, J.S. Artificial Light at Night Interacts with Predatory Threat to Alter Reef Fish Metabolite Profiles. *Sci. Total Environ.* **2021**, *769*, 144482. [CrossRef]
88. Clark, T.C.; Tinsley, J.; Macqueen, D.J.; Martin, S.A.M. Rainbow Trout (*Oncorhynchus mykiss*) Urea Cycle and Polyamine Synthesis Gene Families Show Dynamic Expression Responses to Inflammation. *Fish Shellfish Immunol.* **2019**, *89*, 290–300. [CrossRef] [PubMed]
89. Engelking, L.R. Chapter 10: Urea Cycle (Krebs-Henseleit Ornithine Cycle). In *Textbook of Veterinary Physiological Chemistry*; Academic Press: Cambridge, MA, USA, 2015; pp. 58–64. ISBN 978-0-12-391909-0.
90. Jiang, Y.; Cheng, X.; Lu, J.; Xu, G.; Liu, Q.; Sun, J. Thermal Stress Induces Metabolic Responses in Juvenile Qingtian Paddy Field Carp *Cyprinus carpio var qingtianensis*. *Animals* **2022**, *12*, 3395. [CrossRef] [PubMed]
91. Rehman, A.; Huang, F.; Zhang, Z.; Habumugisha, T.; Yan, C.; Shaheen, U.; Zhang, X. Nanoplastic Contamination: Impact on Zebrafish Liver Metabolism and Implications for Aquatic Environmental Health. *Environ. Int.* **2024**, *187*, 108713. [CrossRef]
92. Gamarra, Y.; Santiago, F.C.; Molina-López, J.; Castaño, J.; Herrera-Quintana, L.; Domínguez, Á.; Planells, E. Pyroglutamic Acidosis by Glutathione Regeneration Blockage in Critical Patients with Septic Shock. *Crit. Care* **2019**, *23*, 162. [CrossRef]
93. Fang, Y.-Z.; Yang, S.; Wu, G. Free Radicals, Antioxidants, and Nutrition. *Nutrition* **2002**, *18*, 872–879. [CrossRef]
94. Aguilar, A.; Mattos, H.; Carnicero, B.; Sanhueza, N.; Muñoz, D.; Teles, M.; Tort, L.; Boltaña, S. Metabolomic Profiling Reveals Changes in Amino Acid and Energy Metabolism Pathways in Liver, Intestine and Brain of Zebrafish Exposed to Different Thermal Conditions. *Front. Mar. Sci.* **2022**, *9*, 835379. [CrossRef]
95. Yang, D.-X.; Yang, M.-J.; Yin, Y.; Kou, T.-S.; Peng, L.-T.; Chen, Z.-G.; Zheng, J.; Peng, B. Serine Metabolism Tunes Immune Responses To Promote *Oreochromis niloticus* Survival upon *Edwardsiella tarda* Infection. *mSystems* **2021**, *6*, e00426-21. [CrossRef] [PubMed]
96. Kim, S.; Kim, A.; Ma, S.; Lee, W.; Lee, S.; Yoon, D.; Kim, D.-H.; Kim, S. Glutathione Injection Alleviates the Fluctuation of Metabolic Response under Thermal Stress in Olive Flounder, *Paralichthys olivaceus*. *Metabolites* **2019**, *10*, 3. [CrossRef]
97. Adam, A.-C.; Lie, K.K.; Moren, M.; Skjærven, K.H. High Dietary Arachidonic Acid Levels Induce Changes in Complex Lipids and Immune-Related Eicosanoids and Increase Levels of Oxidised Metabolites in Zebrafish (*Danio rerio*). *Br. J. Nutr.* **2017**, *117*, 1075–1085. [CrossRef]

98. Tian, Y.; Wang, W.; Jiang, W.; Zhang, G.; He, J.; Dong, S.; Zhou, Y.; Yang, W.; Tang, Q.; Yu, Y.; et al. Non-Targeted Metabolomics Provides Insights into the Distinct Amino Acid and Lipid Metabolism in Liver Tissues of Rainbow Trout (*Oncorhynchus mykiss*) Cultured in Seawater at Different Temperatures. *Aquaculture* **2024**, *579*, 740188. [CrossRef]
99. Falco, F.; Stincone, P.; Cammarata, M.; Brandelli, A. Amino Acids as the Main Energy Source in Fish Tissues. *Aquac. Fish Stud.* **2020**, *3*, 1–11. [CrossRef]
100. Zhao, H.; Ke, H.; Zhang, L.; Zhao, Z.; Lai, J.; Zhou, J.; Huang, Z.; Li, H.; Du, J.; Li, Q. Integrated Analysis about the Effects of Heat Stress on Physiological Responses and Energy Metabolism in *Gymnocypris chilianensis*. *Sci. Total Environ.* **2022**, *806*, 151252. [CrossRef] [PubMed]
101. Luo, M.; Feng, B.; Zhu, W.; Liang, Z.; Xu, W.; Fu, J.; Miao, L.; Dong, Z. Proteomics and Metabolomics Analysis of American Shad (*Alosa sapidissima*) Liver Responses to Heat Stress. *Comp. Biochem. Physiol. A Mol. Integr. Physiol.* **2024**, *296*, 111686. [CrossRef]
102. Kooijman, S.A.L.M.; Baas, J.; Bontje, D.; Broerse, M.; Van Gestel, C.A.M.; Jager, T. Ecotoxicological Applications of Dynamic Energy Budget Theory. In *Ecotoxicology Modeling*; Devillers, J., Ed.; Emerging Topics in Ecotoxicology; Springer: Boston, MA, USA, 2009; Volume 2, pp. 237–259. ISBN 978-1-4419-0196-5.
103. Yan, L.; Qiu-Zhou, X. Dietary Glutamine Supplementation Improves Structure and Function of Intestine of Juvenile Jian Carp (*Cyprinus carpio* var. Jian). *Aquaculture* **2006**, *256*, 389–394. [CrossRef]
104. Cheng, Z.; Buentello, A.; Gatlin, D.M. Effects of Dietary Arginine and Glutamine on Growth Performance, Immune Responses and Intestinal Structure of Red Drum, *Sciaenops ocellatus*. *Aquaculture* **2011**, *319*, 247–252. [CrossRef]
105. Larsen, T.; Wang, Y.V.; Wan, A.H.L. Tracing the Trophic Fate of Aquafeed Macronutrients With Carbon Isotope Ratios of Amino Acids. *Front. Mar. Sci.* **2022**, *9*, 813961. [CrossRef]
106. Zhou, X. Use of Synthetic Lysine in Fish Feeds: A Review on Research and Application. *Feed Ind.* **2005**, *27*, 1–7.
107. Ramos-Pinto, L.; Machado, M.; Calduch-Giner, J.; Pérez-Sánchez, J.; Dias, J.; Conceição, L.E.C.; Silva, T.S.; Costas, B. Dietary Histidine, Threonine, or Taurine Supplementation Affects Gilthead Seabream (*Sparus aurata*) Immune Status. *Animals* **2021**, *11*, 1193. [CrossRef]
108. Wu, G. Functional Amino Acids in Nutrition and Health. *Amino Acids* **2013**, *45*, 407–411. [CrossRef]
109. Li, X.; Zheng, S.; Wu, G. Nutrition and Metabolism of Glutamate and Glutamine in Fish. *Amino Acids* **2020**, *52*, 671–691. [CrossRef]
110. Gillis, T.E.; Ballantyne, J.S. The Effects of Starvation on Plasma Free Amino Acid and Glucose Concentrations in Lake Sturgeon. *J. Fish Biol.* **1996**, *49*, 1306–1316. [CrossRef]
111. Hassel, B. Pyruvate Carboxylation in Neurons. *J. Neurosci. Res.* **2001**, *66*, 755–762. [CrossRef] [PubMed]
112. Wang, L.; Gao, C.; Wang, B.; Wang, C.; Sagada, G.; Yan, Y. Methionine in Fish Health and Nutrition: Potential Mechanisms, Affecting Factors, and Future Perspectives. *Aquaculture* **2023**, *568*, 739310. [CrossRef]
113. Tulli, F.; Messina, M.; Calligaris, M.; Tibaldi, E. Response of European Sea Bass (*Dicentrarchus labrax*) to Graded Levels of Methionine (Total Sulfur Amino Acids) in Soya Protein-Based Semi-Purified Diets. *Br. J. Nutr.* **2010**, *104*, 664–673. [CrossRef]
114. He, J.-Y.; Tian, L.-X.; Lemme, A.; Gao, W.; Yang, H.-J.; Niu, J.; Liang, G.-Y.; Chen, P.-F.; Liu, Y.-J. Methionine and Lysine Requirements for Maintenance and Efficiency of Utilization for Growth of Two Sizes of Tilapia (*Oreochromis niloticus*). *Aquacult. Nutr.* **2013**, *19*, 629–640. [CrossRef]
115. Rolland, M.; Dalsgaard, J.; Holm, J.; Gómez-Requeni, P.; Skov, P.V. Dietary Methionine Level Affects Growth Performance and Hepatic Gene Expression of GH–IGF System and Protein Turnover Regulators in Rainbow Trout (*Oncorhynchus mykiss*) Fed Plant Protein-Based Diets. *Comp. Biochem. Physiol. B Biochem. Mol. Biol.* **2015**, *181*, 33–41. [CrossRef] [PubMed]
116. Belghit, I.; Skiba-Cassy, S.; Geurden, I.; Dias, K.; Surget, A.; Kaushik, S.; Panserat, S.; Seiliez, I. Dietary Methionine Availability Affects the Main Factors Involved in Muscle Protein Turnover in Rainbow Trout (*Oncorhynchus mykiss*). *Br. J. Nutr.* **2014**, *112*, 493–503. [CrossRef]
117. Kuang, S.-Y.; Xiao, W.-W.; Feng, L.; Liu, Y.; Jiang, J.; Jiang, W.-D.; Hu, K.; Li, S.-H.; Tang, L.; Zhou, X.-Q. Effects of Graded Levels of Dietary Methionine Hydroxy Analogue on Immune Response and Antioxidant Status of Immune Organs in Juvenile Jian Carp (*Cyprinus carpio* var. Jian). *Fish Shellfish Immunol.* **2012**, *32*, 629–636. [CrossRef]
118. Machado, M.; Azeredo, R.; Díaz-Rosales, P.; Afonso, A.; Peres, H.; Oliva-Teles, A.; Costas, B. Dietary Tryptophan and Methionine as Modulators of European Seabass (*Dicentrarchus labrax*) Immune Status and Inflammatory Response. *Fish Shellfish Immunol.* **2015**, *42*, 353–362. [CrossRef]
119. Pan, F.-Y.; Feng, L.; Jiang, W.-D.; Jiang, J.; Wu, P.; Kuang, S.-Y.; Tang, L.; Tang, W.-N.; Zhang, Y.-A.; Zhou, X.-Q.; et al. Methionine Hydroxy Analogue Enhanced Fish Immunity via Modulation of NF-κB, TOR, MLCK, MAPKs and Nrf2 Signaling in Young Grass Carp (*Ctenopharyngodon idella*). *Fish Shellfish Immunol.* **2016**, *56*, 208–228. [CrossRef]
120. Elmada, C.Z.; Huang, W.; Jin, M.; Liang, X.; Mai, K.; Zhou, Q. The Effect of Dietary Methionine on Growth, Antioxidant Capacity, Innate Immune Response and Disease Resistance of Juvenile Yellow Catfish (*Pelteobagrus fulvidraco*). *Aquacult. Nutr.* **2016**, *22*, 1163–1173. [CrossRef]

121. Noor, Z.; Noor, M.; Khan, S.A.; Younas, W.; Ualiyeva, D.; Hassan, Z.; Yousafzai, A.M. Dietary Supplementations of Methionine Improve Growth Performances, Innate Immunity, Digestive Enzymes, and Antioxidant Activities of Rohu (*Labeo rohita*). *Fish Physiol. Biochem.* **2021**, *47*, 451–464. [CrossRef] [PubMed]
122. Jiang, M.; Yang, L.-F.; Zheng, J.; Chen, Z.-G.; Peng, B. Maltose Promotes Crucian Carp Survival against *Aeromonas sobrial* Infection at High Temperature. *Virulence* **2020**, *11*, 877–888. [CrossRef] [PubMed]
123. Cao, Y.; Kou, T.; Peng, L.; Munang'andu, H.M.; Peng, B. Fructose Promotes Crucian Carp Survival Against *Aeromonas hydrophila* Infection. *Front. Immunol.* **2022**, *13*, 865560. [CrossRef] [PubMed]
124. Kou, T.; Wu, J.; Chen, X.; Peng, B. Functional Proteomics Identify Mannitol Metabolism in Serum Resistance and Therapeutic Implications in *Vibrio alginolyticus*. *Front. Immunol.* **2022**, *13*, 1010526. [CrossRef]
125. Li, R.-X.; Chen, L.-Y.; Yao, B.; Rahimnejad, S.; Ren, J.; Luo, Y.; Qiao, F.; Zhang, M.-L.; Du, Z.-Y. Trehalose Alleviated Hepatic Cholesterol Accumulation via Inhibiting Transformation from Glucose-Derived Acyl-CoA to Cholesterol Synthesis in Nile Tilapia. *Aquaculture* **2022**, *560*, 738600. [CrossRef]

fishes

Article

Effects of Dietary β-Carotene on the Gonadal Color, Pigmentation, and Regulation Mechanisms in Sea Urchin Strongylocentrotus Intermedius

Weixiao Di [1,2,†], Yinuo Zhang [1,2,†], Huinan Zuo [1,2], Haijing Liu [1,2], Lina Wang [1,2], Jun Ding [1], Yaqing Chang [1] and Rantao Zuo [1,2,*]

1 Key Laboratory of Mariculture and Stock Enhancement in North China's Sea (Ministry of Agriculture and Rural Affairs), Dalian Ocean University, Dalian 116023, China; dwx980914@gmail.com (W.D.); 15531823963@163.com (Y.Z.); zhn13614042329@163.com (H.Z.); lhaijing@hotmail.com (H.L.); lnwang2024@163.com (L.W.); dingjun1119@dlou.edu.cn (J.D.); yqchang@dlou.edu.cn (Y.C.)

2 Dalian Jinshiwan Laboratory, Dalian 116023, China

* Correspondence: rtzuo@dlou.edu.cn; Tel.: +86-411-84763513

† These authors contributed equally to this work.

Abstract: This study aims to clarify the dose–response relationship between dietary β-carotene levels and gonadal pigment deposition and regulation mechanisms related to the carotenoid synthesis of *Strongylocentrotus intermedius* based on a 60-day feeding trial and subsequent transcriptome analysis. Adult sea urchins (initial weight: 9.33 ± 0.21 g) of three cages were given one of the dry feeds with different doses of β-carotene (0 mg/kg, 150 mg/kg, 300 mg/kg) or fresh kelp (*Saccharina japonica*). The results indicated that the weight gain rate (WGR) of sea urchins increased with the addition of β-carotene, with that of the C300 group being markedly higher than that of the C0 group. The addition of β-carotene significantly improved the redness (a*) and yellowness (b*) values of the gonads, with sea urchins in the C300 group exhibiting closest gonad coloration to those in the kelp-fed group. Meanwhile, β-carotene and echinenone in the gonads of the C300 group showed the highest contents, reaching 1.96 μg/kg and 11.97 μg/kg, respectively. Several differential genes, enriched in the pathways of steroid biosynthesis, oxidative phosphorylation, and ubiquitination, were screened based on transcriptome analysis. Real-time PCR further demonstrated that β-carotene significantly upregulated the expression of *cholesterol 25-hydroxylase* (*CH25H*), *NADH dehydrogenase subunit 1* (*ND1*), *NADH dehydrogenase subunit 2* (*ND2*), and *NADH dehydrogenase subunit 4* (*ND4*) while it downregulated the expression of *24-dehydrocholesterol reductase* (*DHCR24*). These results showed that 300 mg/kg β-carotene significantly increased the WGR, redness, and yellowness values, as well as the contents of β-carotene and echinenone in the gonads of *S. intermedius*. On the one hand, dietary β-carotene increased NADH enzyme activity, which participates in echinenone synthesis by donating electrons for the transformation of β-carotene to echinenone synthesis. On the other hand, the addition of β-carotene inhibited cholesterol synthesis by increasing the expression of *CH25H* and decreasing the expression of *DHCR24*, which could in turn increase the fluidity and permeability of the cell membranes and the transport efficiency of β-carotene and echinenone from the digestive tract to the gonads. These results provided fundamental insights into the production of sea urchin gonads with market-favored colors.

Keywords: sea urchin; carotenoids; pigmentation; gonad color

Key Contribution: This study reveals that dietary β-carotene supplementation promotes the growth of *Strongylocentrotus intermedius*, optimizes gonadal color, and increases the content of β-carotene and echinenone in the gonads. Through transcriptome analysis

https://doi.org/10.3390/fishes10070304

and real-time PCR, it identifies related differential genes and their expression patterns, elucidating the underlying regulatory mechanisms related to echinenone synthesis and offering insights into the production of sea urchin gonads with market-favored colors.

1. Introduction

Sea urchin roe (gonads), highly regarded as gourmet, is particularly popular in some countries and regions, including Asia, the Mediterranean, and the Americas [1] Nevertheless, the rapidly increasing international demand for gonads has led to the overexploitation of natural populations of edible sea urchins [2]. In this context, echinoculture is regarded as an advantageous countermeasure to meet the scarcity of wild resources [3]. In 1989, *Strongylocentrotus intermedius* was first imported to China from Japan by our research group [4]. *S. intermedius* has become lucrative in northern China [5]. Sea urchins naturally live mainly on fresh macroalgae. However, the availability and nutrition of macroalgae vary at different seasons all year round, which has limited the massive use for large-scale aquaculture [6]. The gonadal color of sea urchins is an important determinant of their commercial value [7], with bright yellow and yellow-orange widely accepted as premium colors by consumers. Compared to kelp, formulated feeds usually produce sea urchins with pale gonads [1,8]. Thus, it becomes critically imperative to clarify the underlying mechanisms and develop efficient feed optimization strategies to enhance the gonadal color of sea urchins.

Sea urchin gonadal color is attributed to the accumulation of carotenoids [9]. Echinenone has proven to correlate positively with the occurrence of gonads with bright yellow and yellow-orange color [10–12]. Sea urchins have been reported to have the ability to synthesize echinenone by oxidizing β-carotene [11,12]. Previous studies have showed that β-carotene at an addition level of 100–250 mg/kg can effectively provide a more commercially desirable color to the gonads of sea urchin *S. droebachiensis* [13]. Additionally, with the advancement of high-throughput sequencing technology, a growing number of researchers are using transcriptome analysis to investigate molecular mechanisms [14]. New color-related genes were discovered by sequencing color-different aquatic animal samples and combining them with biochemical analyses [15–17]. Two apolipoprotein genes involved in carotenoid transport and storage were identified by the transcriptome analysis of Pacific oyster *Crassostrea gigas* fed different levels of β-carotene [16]. These results have significantly enhanced our understanding of pigmentation processes and their regulatory mechanisms in aquatic animals. However, from what we know, no studies have been published on the impact of dietary β-carotene on the distribution of major carotenoids and the associated mechanisms about echinenone biosynthesis in *S. intermedius*.

Therefore, the aims of this research are to determine (i) how β-carotene affects the distribution of major carotenoids in *S. intermedius* and (ii) the specific genes and pathways that participate in the echinenone biosynthesis in the gonads of *S. intermedius*. These results could help provide some basic clues for producing gonads with market-favored colors of *S. intermedius*.

2. Materials and Methods

2.1. Diet Preparation and Feeding Experiment

Kelp (*Saccharina japonica*) and three iso-proteinic (20.13%) and iso-lipidic (5.92%) formulated feeds were used for the feeding experiment. The formulated feeds were produced with the addition of β-carotene at different concentrations (0 mg/kg, 150 mg/kg, 300 mg/kg) (Table 1), which were named C0, C150, and C300, respectively.

Table 1. Formulation and proximate composition of the experimental diets (% dry weight).

Ingredients (%)	Formulated Feed		
	C0	C150	C300
Casein	3	3	3
Soybean meal [1]	10	10	10
Wheat meal [2]	15	15	15
Wheat gluten [3]	3	3	3
wheat bran [4]	40	40	40
Corn Starch	18.71	18.69	18.68
Palm oil	2	2	2
Soybean lecithin	2	2	2
Choline chloride	0.1	0.1	0.1
Vitamin premix [5]	2	2	2
Mineral premix [6]	2	2	2
Calcium propionate	0.18	0.18	0.18
Ethoxyquin	0.01	0.01	0.01
Monocalcium phosphate	2	2	2
β-carotene [7]	0	0.0161	0.0323
Proximate composition			
Crude protein	20.1	20.13	20.08
Crude lipid	5.95	5.93	5.92
β-carotene (mg/kg DM)			
β-carotene targets	0	150	300
β-carotene measure contention	0.9	58.3	121.2

Note: [1] Soybean meal: crude protein, 51.56%, crude lipid, 0.9% dry matter; [2] wheat meal: crude protein 13.88% dry matter, crude lipid 1.0% dry matter; [3] gluten: crude protein, 68.99%, crude lipid, 2.8% dry matter; [4] wheat bran: crude protein, 19.85%; crude lipid, 4% dry matter; [5] vitamin premix (mg or g kg^{-1} diet): vitamin D, 5 mg; vitamin K, 10 mg; vitamin B_{12}, 10 mg; vitamin B_6, 20 mg; folic acid, 20 mg; vitamin B_1, 25 mg; vitamin A, 32 mg; vitamin B_2, 45 mg; pantothenic acid, 60 mg; biotin, 60 mg; niacin acid, 200 mg; α-tocopherol, 240 mg; inositol, 800 mg; ascorbic acid, 2000 mg; microcrystalline cellulose, 16.47 g; [6] mineral premix (mg or g kg^{-1} diet): $CuSO_4 \cdot 5H_2O$, 10 mg; Na_2SeO_3 (1%), 25 mg; $ZnSO_4 \cdot H_2O$, 50 mg; $CoCl_2 \cdot 6H_2O$ (1%), 50 mg; $MnSO_4 \cdot H_2O$,60 mg; $FeSO_4 \cdot H_2O$, 80 mg; Ca $(IO_3)_2$, 180 mg; $MgSO_4 \cdot 7H_2O$, 1200 mg; zeolite, 18.35 g; [7] purchased from sigma; CAS No. 7235-40-7; the purity of β-carotene is 93%.

All powder ingredients (<150 μm) of each feed were accurately weighed and mixed. Subsequently, the oils were blended with the mixture. After that, 30% water was added before the diets were processed into pellets using a pelletizer (DES-TS1280, Jinan, China). Feed pellets were dried, sealed, and placed in a refrigerator (−20 °C).

After acclimation to the experimental conditions, 120 healthy sea urchins (9.33 ± 0.12 g) were randomly divided among 12 floating cages (30 cm × 15 cm × 45 cm). Then, three cages of sea urchins were randomly fed one of the test diets until they showed obvious satiation thrice daily (8:00, 12:00, and 18:00) for 60 days. On average, 30% of water was replaced daily by siphoning the remaining dietary residues and excrement from the bottom of the tank. During the feeding period, the water temperature, pH, salinity, and dissolved oxygen were kept at 13 ± 1 °C, 8.0 ± 0.1, 33 ± 1‰, and above 8.0 mg/L.

2.2. Sampling

After the feeding test ended, the experimental animals were starved overnight before being counted and weighed separately to obtain the survival rate (SR) and weight gain rate (WGR). Then, sea urchins were individually dissected and sampled to obtain digestive tracts, gonads, and skeletons. The weights of the wet body mass, digestive tract, and gonads for every individual sea urchin were taken to compute the digestive tract index (DTI) and the gonadosomatic index (GSI). Subsequently, the digestive tracts of the sea urchins were preserved at −80 °C for subsequent analysis of carotenoid levels and gene expression levels. The gonads of each sea urchin were separated into four sections: one section for the histological observation; two sections for color measurement and carotenoid determination; one section for transcriptome analysis; and one section for gene expression. Finally, the skeletons of each sea urchin were comminuted into fine powder and kept at −80 °C for carotenoid determination.

2.3. Gonad Color Measurement

The gonad color was measured using a Color Cue 2 Colorimeter (PANTONE, Carlstadt, NJ, USA). ΔE denoted the distinction between the actual color value and normal color levels. The standards for ΔE1 (L* = 68.9, a* = 28.7, b* = 60.4) and ΔE2 (L* = 74.6, a* = 28.7, b* = 66.1) were reported by McBride [18].

2.4. Determination of Carotenoids

The procedures for extraction and carotenoid analysis were in line with Baião et al. [19]. The contents of β-carotene in the diets and those of echinenone in the sea urchin tissues were determined by a high-performance liquid chromatograph (HPLC) (model Agilent-1290-6470 brand Agilent) (Agilent Technologies, Santa Clara, CA, USA) and a mass spectrometer (model QQQ brand Agilent). The carotenoid extracts of 0.05 mg lyophilized samples were suspended in 0.2 mL of methanol/methyl tert-butyl ether (1:1), and 5 μL was injected into the Agilent C18 column (2.1 mm × 100 mm, 1.8 μm). Elution agents include solvent A (methanol/acetonitrile = 3:1) and solvent B (methyl tertiary-butyl ether). The 0.3 mL/min flow rate was prescribed, the column temperature was stabilized at 30 °C, and the acquisition mode was established as APCI. The mobile phase gradient ratio was 95%A + 5%B for the first 0–1 min, and then the gradient ratio was linearly changed to 5%A + 95%B for 1–3 min and maintained for 1 min. After that, the mobile phase was linearly changed to 95%A + 5%B for 4–4.5 min and held for 1.5 min. Finally, the gradient ratio was linearly changed to 70%A + 30%B for 6–9 min. The atmospheric pressure chemical ion source for mass spectrometry was set as follows: the atomization temperature of vaporization was 325 °C; the gas flow speed was 4 L/min; the pressure of the nebulizer was 20 psi; and the voltage of the capillary voltage was 4500 V. The retention time of each carotenoid was recorded by comparison with pure products of β-carotene (Tan mo, Changzhou City, Jiangsu Province, China) and echinenone (CaroteNature, Münsingen, Switzerland). The specific contents of β-carotene and echinenone were calculated by the calibration curves. The data were expressed as μg/kg dry weight.

2.5. Histological Analysis

Gonadal sections were prepared based on the methodology defined by Santos et al. [20]. To sum up briefly, the fixed tissues were subjected to ethanol treatment for the purpose of dehydration and then embedded in paraffin before being sectioned into thin slices. These sections were stained by immersion in hematoxylin and eosin. Eventually, the slices were analyzed microscopically. For the stages of gonadal maturation (stages I–VI), refer to Santos et al. [20].

2.6. RNA-Seq and Transcriptome Analyses

The Illumina Nova seq 6000 was used to sequence the gonadal transcriptomes of sea urchins of different dietary groups. Trinity was used to complete the transcriptome assembly [21]. DIAMOND was used to annotate the assembled genes based on multiple databases, such as the NCBI non-redundant (Nr) database, Swiss-Prot, KOG/COG (protein homology groups), GO (gene ontology), and Pfam (large collection of protein families) [22]. To annotate potential pathways of metabolism, each gene was aligned to the Kyoto Encyclopedia of Genes and Genomes (KEGG) database. The reads generated by sequencing were aligned with the Unigene library by using Bowtie [23]. On the basis of the findings of the comparison, an expression level estimate was taken in combination with RSEM [22]. Gene sets exhibiting differential expression were identified in distinct samples using the DESeq2 software (version 1.36.0) [24]. To ascertain the differentially expressed genes (DEGs) between samples, the thresholds of multiple changes $\geq$ 1.5 and $p < 0.01$ were set.

2.7. Real-Time Quantitative PCR of Different Expression Genes

The specific description of the operations has been provided by Li et al. [25]. RNA was first extracted to meet the quality of the qPCR. A FastKing gDNA Dispelling RT SuperMix kit (TIANGEN, Beijing, China) was used for the generation of the cDNA. Finally, the cDNA templates were dissolved fivefold before being used for qPCR.

A FastReal qPCR PreMix (SYBR Green) kit (TIANGEN, Beijing, China) was used for the qPCR by following the instructions. The qPCR was accomplished with the application of the LightCycler® 96 real-time PCR system(Roche Diagnostics, Basel, Switzerland). The reaction settings were as follows: 95 °C for 120 s was set first, and then 40 cycles of 5 s at 95 °C and 15 s at 60 °C followed. The relative gene transcriptional level was computed using the formula $2^{-\Delta\Delta CT}$. The sequences of the primers can be found in Table 2.

Table 2. Primer sequences for real-time PCR analysis.

Gene Name	Origin	Primer Sequences (5′→3′)
CH25H	TRINITY_DN1511_c1_g1	F:AGAAGGTCCGCTATTTCGTTTCCC R:CAAGACTGGCATACAGAGGCATCC
ND1	TRINITY_DN8691_c2_g1	F:GCTGCCTACCCACGATTCCG R:AGCCAAAACCGCACACAAAGC
ND2	TRINITY_DN8691_c3_g1	F:GAGTGCCCAATACAGCGGAAAC R:ACACATTATTAGCCTCGCCTCTTG
ND4	TRINITY_DN10351_c0_g1	F:GCAGCACACATTAGAAGTCACCAG R:CGTCTACGAGCGGAGAGGAAC
DHCR24	TRINITY_DN83748_c0_g1	F:CACCTTGCGTGTGGAATCTCTG R:CTGTGGCTGTGTCCCTTTGTTC
18s	[26]	F:GTTCGAAGGCGATCAGATAC R:CTGTCAATCCTCACTGTGTC

2.8. Calculations and Statistical Analyses

$$\text{Survival rate (SR, \%)} = N_f/N_i \times 100$$

$$\text{Weight gain rate (WGR, \%)} = (FW - IW)/IW \times 100$$

$$\text{Gonadosomatic index (GSI, \%)} = GW/FW \times 100$$

$$\text{Digestive tract index (DTI, \%)} = DW/FW \times 100$$

$$\Delta E = SQRT[(L^* - L^*_s)^2 + (a^* - a^*_s)^2 + (b^* - b^*_s)^2].$$

$$\text{Gene expression of transcriptome (log1P)} = \log(1 + FPKM)$$

N_i stands for the initial quantity of sea urchins, while N_f stands for the final quantity; IW and FW denote the initial and final average weight of sea urchin in every cage; GW and DW are the final wet weights of sea urchin gonads and digestive tract of every sea urchin sampled; FPKM (Fragments Per Kilobase of transcript per Million mapped reads) is the number of reads per kilobase of the length of a gene compared to a gene in a million reads [27].

The SPSS software (version 22.0) was employed to conduct the data analysis. The results were compared between the different dietary groups using one-way analysis of variance (ANOVA). Duncan's multiple range test was conducted to assess mean differences between the diets if a significant difference was found. Independent t-tests were used to evaluate data between sexes within the identical dietary group. $p < 0.05$ was regarded as statistically different. The mean ± standard error of the mean (SEM) was adopted for expressing the results.

3. Results

3.1. Growth Performance

As the β-carotene addition level increased, the weight gain rate (WGR) showed an increasing tendency. The C300 group had the highest WGR, which was markedly smaller in comparison with the kelp group ($p < 0.05$) but considerably greater than the C0 and C150 groups ($p < 0.05$) (Table 3).

Table 3. Effects of dietary β-carotene addition on the survival and growth performance of sea urchin (*Strongylocentrotus intermedius*) (mean ± SEM, n = 3) [1].

	Formulated Feed Treatments			Kelp
	C0	C150	C300	
SR(%) [2]	100 ± 0.00	100 ± 0.00	100 ± 0.00	100 ± 0.00
IW(g) [3]	9.43 ± 0.13	9.41 ± 0.20	9.16 ± 0.04	9.33 ± 0.13
FW(g) [3]	19.78 ± 0.75 ^{b}	20.01 ± 0.73 ^{b}	21.43 ± 0.59 ^{b}	26.60 ± 0.76 ^{a}
WGR (%) [4]	109.70 ± 5.63 ^{c}	112.57 ± 4.51 ^{c}	133.87 ± 5.51 ^{b}	185.19 ± 4.18 ^{a}
GW(g) [5]	1.66 ± 0.17	1.64 ± 0.28	1.59 ± 0.06	1.59 ± 0.20
GSI (%) [6]	8.35 ± 0.61	8.14 ± 1.15	7.41 ± 0.22	5.96 ± 0.65
DW(g) [7]	0.76 ± 0.08 ^{a}	0.62 ± 0.08 ^{b}	0.76 ± 0.13 ^{a}	1.06 ± 0.11 ^{a}
DTI (%) [8]	3.85 ± 0.51	3.10 ± 0.27	3.53 ± 0.56	3.99 ± 0.48

[1] Mean values with the different superscript letters within the same row represent significant difference at $p < 0.05$; [2] SR: survival rate; [3] IW: initial body weight; FW: final body weight; [4] WGR: weight gain rate; [5] GW: gonad weight; [6] GSI: gonadosomatic index; [7] DW: digestive tract weight; [8] DTI: digestive tract index.

The digestive tract weight (DW) and digestive index (DTI) showed a "first decreasing and then increasing" tendency with an increasing β-carotene addition level, while the maximum DW was measured in the kelp group. There appeared to be no statistically remarkable differences in the DTI and GSI between experimental groups ($p > 0.05$) (Table 3).

3.2. Gonad Color

With the β-carotene addition level increasing, the gonadal color of sea urchins was changed from white to bright yellow or bright orange, with the most favored color observed in the C300 group. The L, ΔE1, and ΔE2 values of the gonads in both male and female sea urchins showed a decreasing trend with the increase in β-carotene addition, while those in the C300 group were markedly lower than those in the C0 group ($p < 0.05$). The a* and b* values of the gonads in both male and female sea urchins showed an upward trend with the increase in β-carotene addition, with those in the C300 group were markedly higher than those in the C0 group ($p < 0.05$). Compared to those in the dry feed groups, sea urchin gonads in the kelp group displayed more favored colors, bright yellow or bright orange (Figure 1). In the kelp group, the a* and b* values of the gonads were markedly higher than those in all of the dry feed groups ($p < 0.05$), while the gonadal L*, ΔE1, and ΔE2 values were markedly lower than those in the dry feed groups (Table 4).

Figure 1. Effects of dietary β-carotene addition on the gonad color appearance in the gonads of sea urchin (*Strongylocentrotus intermedius*). (**A**) A1~A3: gonads of male sea urchins fed C0; A4~A6: gonads of female sea urchins fed C0; (**B**) B1~B3: gonads of male sea urchins fed C150; B4~B6: gonads of female sea urchins fed C150; (**C**) C1~C3: gonads of male sea urchins fed C300; C4~C6: gonads of female sea urchins fed C300; (**D**) D1~D3: gonads of male sea urchins fed kelp; D4~D6: gonads of female sea urchins fed kelp.

Table 4. Effects of dietary β-carotene addition on the gonad color of sea urchin (*Strongylocentrotus intermedius*) (mean ± SEM, n = 3) [1].

	Male				**Female**			
	C0	**C150**	**C300**	**Kelp**	**C0**	**C150**	**C300**	**Kelp**
L	82.21 ± 0.41 [a]	79.10 ± 1.45 [ab]	77.04 ± 1.40 [bc]	73.89 ± 0.88 [c]	79.58 ± 0.29 [a*]	74.73 ± 2.16 [ab]	76.56 ± 1.82 [ab]	71.33 ± 1.27 [b]
a	9.66 ± 0.60 [c]	14.27 ± 1.03 [bc]	17.10 ± 2.15 [b]	24.76 ± 2.21 [a]	11.75 ± 0.91 [c]	17.59 ± 2.00 [b]	18.47 ± 1.82 [b]	31.29 ± 1.66 [a]
b	32.49 ± 1.44 [c]	39.88 ± 0.87 [b]	43.34 ± 2.97 [b]	52.33 ± 1.41 [a]	33.34 ± 3.16 [c]	40.37 ± 1.95 [bc]	45.41 ± 1.80 [b]	63.72 ± 1.26 [a*]
ΔE1	36.33 ± 1.40 [a]	27.15 ± 1.47 [b]	22.21 ± 3.83 [b]	10.59 ± 2.08 [c]	33.74 ± 2.89 [a]	23.78 ± 3.03 [b]	19.95 ± 2.21 [b]	5.80 ± 1.00 [c]
ΔE2	39.38 ± 1.50 [a]	30.34 ± 1.26 [b]	25.73 ± 3.67 [b]	14.63 ± 1.87 [c]	37.25 ± 3.16 [a]	28.24 ± 2.54 [b]	23.43 ± 1.95 [b]	5.09 ± 2.13 [c*]

[1] Mean values with different superscript letters represent significant differences between different experimental groups of the same gender $p < 0.05$, and mean values with superscript "*" represent significant differences between different genders in the same experimental group $p < 0.05$.

Compared with the gonads of male sea urchins, those of most female sea urchins showed more favored colors. The gonads of female sea urchins exhibited higher a* and b* values compared to those of their male counterparts. On the contrary, the female sea urchin gonad L, ΔE1, and ΔE2 values were lower than those of the male counterparties. The different coloration of gonads between males and females was notable in the kelp group, with the gonads of females showing a markedly higher b* value but lower ΔE2 value than those of males ($p < 0.05$) (Table 4).

3.3. Carotenoids' Concentration

The introduction of β-carotene significantly affected the distribution of β-carotene and echinenone in the digestive tract and gonads of sea urchins. As the amount of β-carotene added improved, the β-carotene and echinenone content in the digestive tract and gonads indicated an increasing trend. The concentrations of β-carotene and echinenone in the

C300 group were substantially elevated compared to those in the C0 and C150 groups ($p < 0.05$), yet they were still substantially depressed in comparison to those in the kelp group ($p < 0.05$). Although the β-carotene and echinenone contents showed no obvious difference in the skeletons of the feed groups ($p > 0.05$), they were all substantially lower than those in the kelp group ($p < 0.05$) (Tables 5–7).

Table 5. Effects of β-carotene addition on the distribution of major carotenoids in digestive tract of sea urchin (*Strongylocentrotus intermedius*) (mean ± SEM, n = 3) [1].

	Digestive Tract			
	C0	**C150**	**C300**	**Kelp**
Echinenone (μg/kg)	0.38 ± 0.05 ^{c}	0.48 ± 0.02 ^{c}	0.84 ± 0.04 ^{b}	6.93 ± 0.11 ^{a}
β-carotene (μg/kg)	0.34 ± 0.01 ^{b}	0.40 ± 0.03 ^{b}	0.51 ± 0.05 ^{c}	9.25 ± 0.46 ^{a}

[1] Mean values of the same tissue with different superscript letters indicate significant differences at the $p < 0.05$ level.

Table 6. Effects of β-carotene addition on the distribution of major carotenoids in the gonads of sea urchin (*Strongylocentrotus intermedius*) (mean ± SEM, n = 3) [1].

	Gonad			
	C0	**C150**	**C300**	**Kelp**
Echinenone (μg/kg)	3.25 ± 0.56 ^{c}	3.27 ± 0.11 ^{c}	11.97 ± 2.27 ^{b}	31.72 ± 1.33 ^{a}
β-carotene (μg/kg)	0.38 ± 0.01 ^{c}	0.52 ± 0.01 ^{c}	1.96 ± 0.04 ^{b}	11.12 ± 0.46 ^{a}

[1] Mean values of the same tissue with different superscript letters indicate significant differences at the $p < 0.05$ level.

Table 7. Effects of β-carotene addition on the distribution of major carotenoids in skeleton of sea urchin (*Strongylocentrotus intermedius*) (mean ± SEM, n = 3) [1].

	Skeleton			
	C0	**C150**	**C300**	**Kelp**
Echinenone (μg/kg)	0.32 ± 0.05 ^{b}	0.27 ± 0.02 ^{b}	0.33 ± 0.06 ^{b}	6.71 ± 0.30 ^{a}
β-carotene (μg/kg)	0.31 ± 0.02 ^{b}	0.27 ± 0.01 ^{b}	0.27 ± 0.03 ^{b}	1.55 ± 0.05 ^{a}

[1] Mean values of the same tissue with different superscript letters indicate significant differences at the $p < 0.05$ level.

3.4. Gonad Section Observation

In the male sea urchins of the C150 group (Figure 2C), the spermatogonial layer was thin, presumably at the early stage of stage II. In contrast, the spermatogonial layers of sea urchins in the C0 group, C300 group, and kelp group (Figure 2A,E,G) were thicker and showed a trend of central migration, suggesting that they were in the mid-stage of stage II.

For the female sea urchins in the C150 group (Figure 2D), the number of oocytes was small and loosely arranged, presumably at the early stage of stage II. Conversely, sea urchins in the C0, C300, and kelp groups (Figure 2B,F,H) had a larger number of tightly packed oocytes, indicating that they were in the mid-stage of stage II (Figure 2). β-carotene addition did not notably influence the gonadal development of sea urchins. Male and female sea urchins in all diet groups were synchronized to stage II (Figure 2).

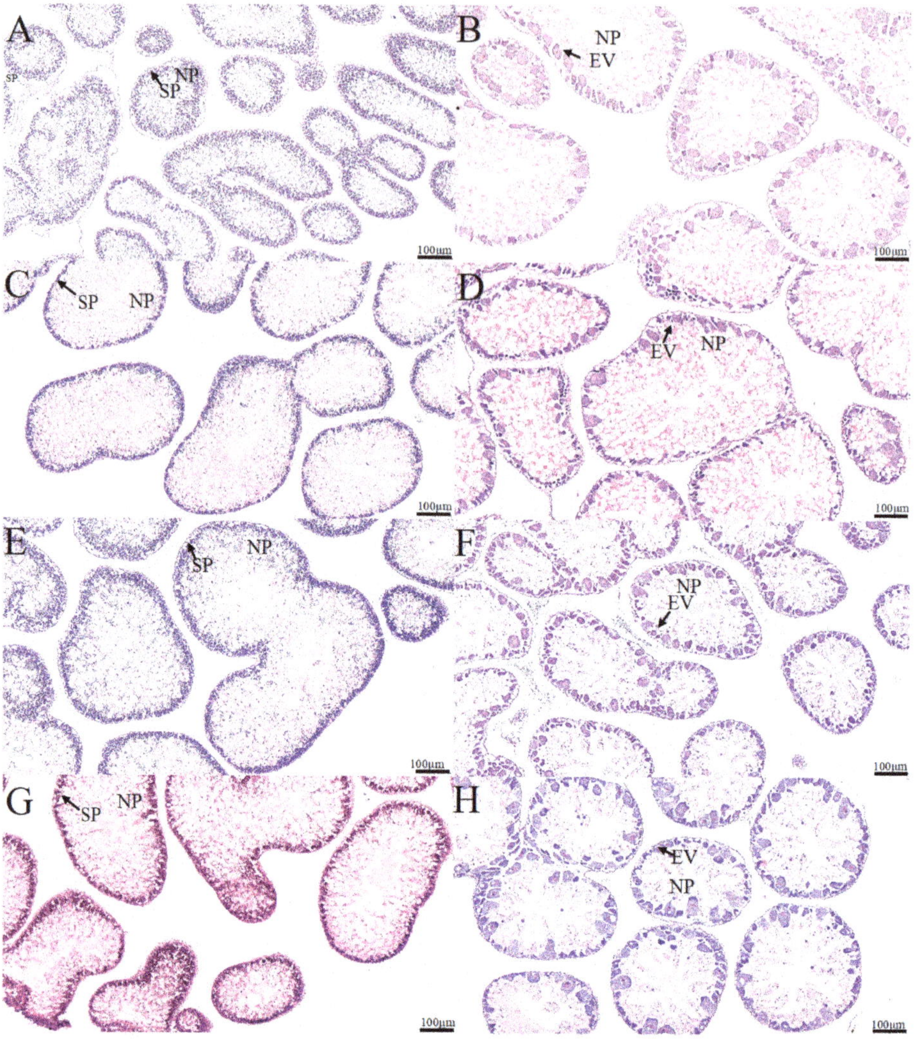

Figure 2. Effects of dietary β-carotene addition on the gonad histology of sea urchin (*Strongylocentrotus intermedius*). (**A**) C0, male, stage II; (**B**) C0, female, stage II; (**C**) C150, male, stage II; (**D**) C150, female, stage II; (**E**) C300, male, stage II; (**F**) C300, female, stage II; (**G**) Kelp, male, stage II; (**H**) kelp, female, stage II. NP: nutritive phagocyte. SP: spermatocyte. EV: early vitellogenic oocyte.

3.5. Transcriptome and Related Gene Expression

Among the three feed groups, 1949 differential expression genes (DEGs) were analyzed. A total of 214 DEGs were shared in all groups, while there were 492 and 1243 DEGs significantly expressed in the C0 vs. C150 and C0 vs. C300 groups, respectively (Figure 3). There were 706 DEGs in the C0 vs. C150 groups, which included 435 upregulated and 271 downregulated genes. There were 1457 DEGs in C0 vs. C300, which included 903 upregulated and 554 downregulated genes (Figure 4). KEGG pathway was used to annotate DEGs. The most abundant pathway was the N metabolism pathway, followed by the apelin and calcium signaling pathways. In addition, the ubiquitin-mediated protein hydrolysis-related pathway was crucially downregulated, and the ribosome-related pathway was crucially upregulated (Figure 5).

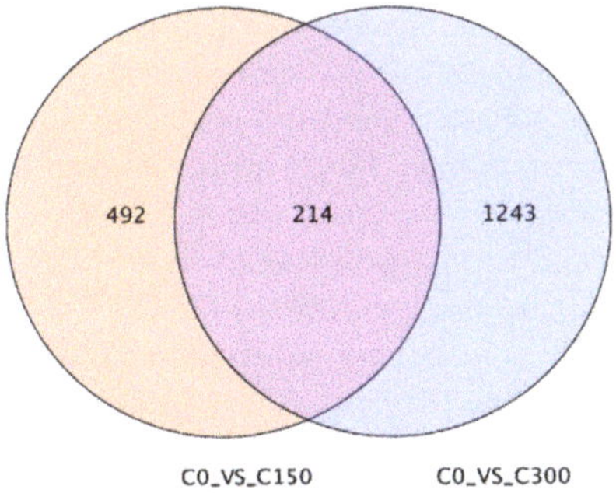

Figure 3. Effects of dietary β-carotene addition on differentially expressed genes in transcriptome levels in sea urchin (*Strongylocentrotus intermedius*).

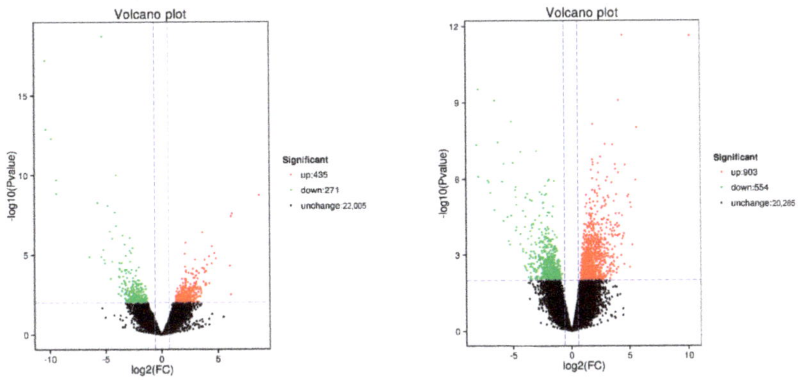

Figure 4. Volcano map showing the number of upregulated and downregulated genes in C0 vs. C150 (**left**) and C0 vs. C300 (**right**). Note: The *X*-axis is represented by log2 (fold change), and the larger the variation, the wider the distribution; the *Y*-axis is represented by −log10 (*p* value); the green dots are used to represent downregulated genes, the red dots are used to represent upregulated genes, and the black dots are used to represent genes that do not differ significantly.

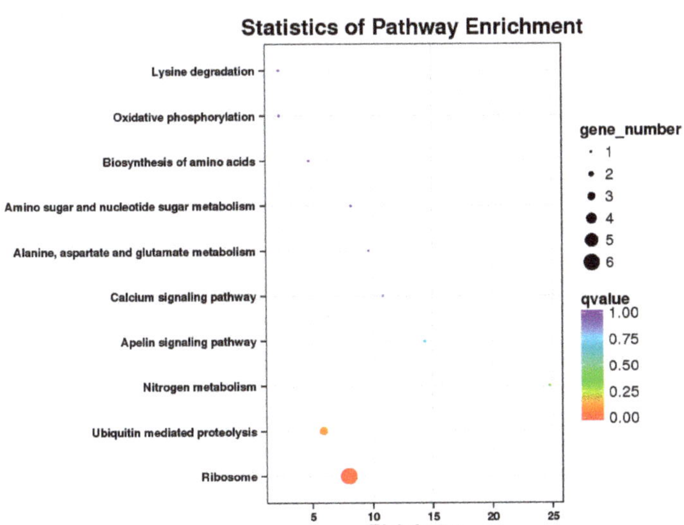

Figure 5. KEGG analyses of differentially expressed genes (DEGs) in C0 vs. C150 ∩ C0 vs. C300.

The results of the qPCR verification data on DEGs were consistent with gene expression data from the transcriptome. With increasing β-carotene supplementation, the transcription of *CH25H*, *ND1*, *ND2*, and *ND4* genes in sea urchin gonads demonstrated a rising trend. The transcription levels of *CH25H*, *ND2*, and *ND4* genes of the C300 group were analogous to the C150 group ($p > 0.05$) but were obviously elevated compared with the C0 group ($p < 0.05$). In the C300 group, the transcription level of the *ND1* gene was obviously higher than that in the C0 and C150 groups ($p < 0.05$). The transcription of the *DHCR24* gene presented a declining tendency. In the C0 group, the transcription level of the *DHCR24* gene resembled that in the C150 group ($p > 0.05$), while it was obviously elevated compared to that in the C300 group ($p < 0.05$) (Figure 6).

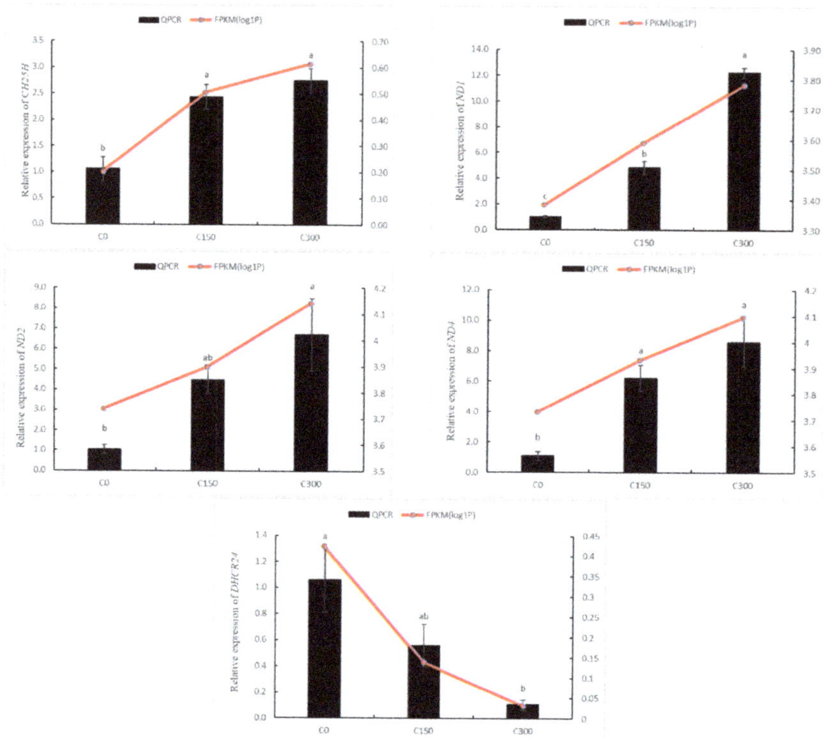

Figure 6. Effect of dietary β-carotene addition on the related gene expression in the gonad of sea urchin (*Strongylocentrotus intermedius*). Mean value bars bearing with different lowercase letters are significantly different at $p < 0.05$ (mean ± SEM, n = 3).

4. Discussion

Jintasataporn and Yuangsoi [28] have highlighted that carotenoids, as sensitive organic compounds, are easily damaged by water, oxygen, and high temperatures. Baião et al. [19] further confirmed the vulnerability of carotenoids to degradation during feed processing. In this study, to minimize the loss of carotene activity, a series of measures have been adopted such as rapid mixing, low-temperature drying, sealing, and freezing storage during feed production. However, the degradation of carotenoids was not avoided based on the fact that the actual values were markedly lower than the targeted values. Therefore, the incorporation method of carotenoids should be improved in the following studies, such as microencapsulating carotenoids, adding higher contents or more potent antioxidants, and using vacuum freeze-drying methods.

In this experiment, the WGR of *S. intermedius* demonstrated a rising trend as the β-carotene addition level increased. Similar results have been obtained in other aquatic animals. Compared to the control diet, specific growth rate and weight gain were obviously higher in Nile tilapia *Oreochromis niloticus* fed a diet containing 50 mg/kg β-carotene [29]. White shrimp *Litopenaeus vannamei* fed β-carotene at a concentration of 300–400 mg/kg had obviously better growth performance [30]. A higher specific growth rate and weight gain were found in juvenile pacu fish *Piaractus mesopotamicus* fed diets supplemented with β-carotene in comparison with those fed extruded feeds without β-carotene [31]. Dietary β-carotene can improve the developmental performance of juvenile green sea urchins *Strongylocentrotus droebachiensis* [32] and the larvae of the sea urchin *Paracentrotus lividus* [33]. However, the elevation in growth performance was not observed in adult *P. lividus* [19]. This difference could result from the larger initial body weight (44.0 g) of the sea urchins used by Baião et al. [19]. In this study, the gonadosomatic index (GSI) and gonadal development were not affected by β-carotene addition. These results were in line with the discoveries of some earlier investigations, which found that dietary β-carotene did not impact the gonad growth and maturation of the sea urchin *Lytechinus variegatus* [34] and *P. lividus* [19,35].

Gonads with bright yellow and bright orange colors are more acceptable by consumers and thus have higher market values [5,18]. The redness (a*) and yellowness (b*) of gonads are caused by the accumulation of carotenoids and metabolism [2,11]. Since most animals are unable to produce carotenoids de novo [36], supplementing with β-carotene in the diet is an effective way to obtain carotenoids and thereby enhance the deposition of pigments in sea urchin gonads [10]. The addition of β-carotene, including both natural and artificial sources, to feed successfully has turned out to be an effective way to improve sea urchin gonadal color [10,13,37,38]. In this study, the favorable color frequency coupled with a* and b* values of the sea urchin gonads enhanced with the increasing addition of β-carotene; female sea urchins exhibited higher a* and b* values than males. Studies have demonstrated that when fed identical diets, female *Paracentrotus lividus* consistently achieve better coloration than males [19]. While the compositional profiles of carotenoids in the gonads of female and male Australian sea urchins and red sea urchins are similar, the carotenoid levels in ovaries are markedly higher than those in testes. During gametocyte development, carotenoids are transferred to developing eggs along with other nutrients, leading to a higher carotenoid concentration in mature oocytes compared to testes [9], a phenomenon not reported in male sea urchins. In addition, according to the unpublished research of our team, this may also be related to the decreased ability of carotenoid absorption, transportation, and metabolism caused by excessive inflammation in the digestive tract and gonads of male sea urchins. Echinenone, as the predominant carotenoid in most sea urchin gonads, is proved to be closely associated with the formation of characteristic color, especially redness [10]. Sea urchins could possess a certain capacity to synthesize echinenone from the substrate β-carotene [11,12]. In the current investigation, the concentrations of β-carotene and echinenone in the digestive tract and gonads of sea urchins were elevated with the increase in β-carotene in diets. Since the experimental feeds did not contain echinenone, the echinenone in the digestive tract and gonads could be synthesized through a series of steps, including the oxidation of β-carotene by *S. intermedius*.

In this study, the variation between the levels of β-carotene and β-carotene was more pronounced in the gonads, which was nearly ten times higher than that in the digestive tract. This indicated that gonad could be the main synthetic site of echinenone in sea urchins. It could also be inferred that the echinenone synthesis could be regulated by steroids because the coloration and echinenone contents of gonads are different between male and female individuals. The synthesis of β-carotene to echinenone requires hydroxylation and then ketonization [9]. However, to our knowledge, there are no reports of β-carotene hydroxylases and ketonases in sea urchins. A large proportion of β-carotene oxidases in other animals belong to the CYP450 family [39], where an electron donor is needed for catalyzing the reaction, usually NADH or NADPH. These electron donors pass electrons through specific reductases (e.g., CYP450 reductases) to the CYP450 enzymes [40]. In this process, NADH or NADPH is oxidized and the CYP450 enzyme is reduced so that it can accept the substrate and carry out the oxidation reaction. The findings of this study indicated that the expression of *ND1*, *ND2*, and *ND4* demonstrated a rising trend with the increase in β-carotene. The *ND1*, *ND2*, and *ND4* genes encode the core subunits of the mitochondrial respiratory chain complex I. The proteins *ND1*, *ND2*, and *ND4* are subunits of the enzyme NADH dehydrogenase, which is situated in the inner membrane of the mitochondrion and is one of the five largest complexes in the electron transport chain. Their functions mainly include the initiation of NADH dehydrogenase (ubiquinone) activity, participation in mitochondrial electron transport, NADH to ubiquinone, and mitochondrial respiratory chain complex I assembly [41]. Thus, increased NADH enzyme activity provides electrons for the transformation of β-carotene to echinenone in the gonad, thereby increasing the amount of echinenone in the gonad.

Furthermore, some echinenone in the gonads was probably transported from the digestive tract, which functions as an important organ for storing lipids and lipid-soluble substances. Carotenoids are mostly lipid-soluble substances and have been reported to cross membranes mainly through diffusion with the help of specific carrier proteins and membrane fluidity [42]. This is a way for substances to pass through cell membranes that depends on specific carrier proteins in the membrane. The fluidity of the membrane allows the carrier proteins to move freely across the membrane. Cholesterol content is an important factor that is negatively correlated with the fluidity and permeability of cell membranes [43]. In this experiment, the supplementation of β-carotene increased the expression of *cholesterol 25-hydroxylase* (*CH25H*) and decreased the expression of the *24-dehydrocholesterol reductase* (*DHCR24*). The *CH25H* could limit the activation of *sterol-regulatory element binding protein-2* (*SREBP2*) and thereby reduce cellular cholesterol synthesis [44]. The *DHCR24* protein is the final step of intracellular cholesterol synthesis by converting desmosterol to cholesterol [45]. Therefore, the addition of β-carotene could inhibit cholesterol synthesis by increasing the expression of *CH25H* and decreasing the expression of *DHCR24*, which in turn increases the fluidity and permeability of the cell membranes. This could increase the transport efficiency of β-carotene and echinenone from the digestive tract to the gonads. However, the cholesterol contents of cell membrane were not assayed in the present study. Future studies are needed to find more direct evidence to support our hypothesis.

5. Conclusions

In conclusion, 300 mg/kg β-carotene significantly increased the WGR, redness and yellowness values, as well as the contents of β-carotene and echinenone in the gonads of *S. intermedius* without affecting gonadal growth and development. Gene expression changes suggested that β-carotene enhanced gonad coloration by inhibiting cholesterol synthesis for carotenoid transport and boosting NADH dehydrogenase activity essential for echinenone production. Further exploration is required to find more evidence to elucidate the regulation mechanisms related to echinenone synthesis in sea urchins. Additionally, it is worth investigating the synergistic effects of phospholipids and antioxidants in adult sea urchins. These results provide fundamental insights for producing sea urchin gonads with market-favored colors.

Author Contributions: W.D. and R.Z. designed and performed the whole experiment. W.D., H.Z., H.L., and L.W. collected and analyzed the data. W.D., Y.Z., and R.Z. drafted the manuscript. J.D., R.Z., and Y.C. revised the manuscript. All authors have read and agreed to the published version of the manuscript.

Funding: This research was funded by Xingliao Talent Plan for Top Young Talents (XLYC2203036) and the Open Foundation of Dalian Jinshiwan Laboratory (Dljswkf202413).

Institutional Review Board Statement: The Ethics Committee of Dalian Ocean University did not require the study to be reviewed or approved because sea urchins are invertebrates.

Informed Consent Statement: Not applicable.

Data Availability Statement: The data of this study can be provided by the corresponding author if they are requested.

Conflicts of Interest: The authors declare no conflicts of interest.

Abbreviations

The following abbreviations are used in this manuscript:

CH25H	*cholesterol 25-hydroxylase*
ND1	*NADH dehydrogenase subunit 1*
ND2	*NADH dehydrogenase subunit 2*
ND4	*NADH dehydrogenase subunit 4*
DHCR24	*24-dehydrocholesterol reductase*
SR	survival rate
WGR	weight gain rate
DTI	digestive tract index
GSI	gonadosomatic index
HPLC	high-performance liquid chromatograph
DEGs	differentially expressed genes
FPKM	Fragments Per Kilobase of transcript per Million mapped reads
ANOVA	analysis of variance
DW	digestive tract weight
DTI	digestive index
SREBP2	*sterol-regulatory element binding protein-2*

References

1. Di, W.; Heqiu, Y.; Gou, D.; Gong, P.; Ding, J.; Chang, Y.; Zuo, R. Effects of Supplementary Kelp Feeding on the Growth, Gonad Yield, and Nutritional and Organoleptic Quality of Subadult Sea Urchin (*Strongylocentrotus intermedius*) with Soya Lecithin Intake History. *Aquac. Nutr.* **2023**, *8894923*, 15. [CrossRef] [PubMed]
2. Lourenço, S.; Raposo, A.; Cunha, B.; Pinheiro, J.; Santos, P.M.; Gomes, A.S.; Ferreira, S.; Gil, M.M.; Costa, J.L.; Pombo, A. Temporal changes in sex-specific color attributes a nd carotenoid concentration in the gonads (roe) of the purple sea urchin (*Paracentrotus lividus*) provided dry feeds supplemented with β-carotene. *Aquaculture* **2022**, *560*, 738608. [CrossRef]
3. Lourenço, S.; Valente, L.M.P.; Andrade, C. Meta-analysis on nutrition studies modulating sea urchin roe growth, colour and taste. *Rev. Aquacult.* **2019**, *11*, 766–781. [CrossRef]
4. Chang, Y.; Zhang, W.; Zhao, C.; Song, J. Estimates of heritabilities and genetic correlations for growth and gonad traits in the sea urchin *Strongylocentrotus intermedius*. *Aquac. Res.* **2012**, *43*, 271–280. [CrossRef]
5. Ning, Y.; Zhang, F.; Tang, L.; Song, J.; Ding, J.; Chang, Y.; Zuo, R. Effects of dietary lipid sources on the growth, gonad development, nutritional and organoleptic quality, transcription of fatty acid synthesis related genes and antioxidant capacity during cold storage in adult sea urchin (*Strongylocentrotus intermedius*). *Aquaculture* **2022**, *548*, 737688. [CrossRef]
6. Raposo, A.; Ferreira, S.; Ramos, R.; Anjos, C.; Pombo, A. Effect of three diets on the gametogenic development and fatty acid profile of *Paracentrotus lividus* (Lamarck, 1816) gonads. *Aquac. Res.* **2019**, *50*, 2023–2038. [CrossRef]
7. Zhang, W.; Zhao, C.; Liu, P.; Chang, Y. First report on tube feet differential pigmentation in the cultivated sea urchin *Strongylocentrotus intermedius* (Agassiz, 1863) and its relationship with growth performance: Tube feet differential pigmentation in sea urchins. *Aquac. Res.* **2010**, *41*, e706–e708. [CrossRef]

8. Onomu, A.J.; Vine, N.G.; Cyrus, M.D.; Macey, B.M.; Bolton, J.J. The effect of fresh seaweed and a formulated diet supplemented with seaweed on the growth and gonad quality of the collector sea urchin, *Tripneustes gratilla*, under farm conditions. *Aquac. Res.* **2020**, *51*, 4087–4102. [CrossRef]
9. Suckling, C.C.; Kelly, M.S.; Symonds, R.C. Carotenoids in sea urchins. In *Developments in Aquaculture and Fisheries Science*; Elsevier: Amsterdam, The Netherlands, 2020; pp. 209–217. [CrossRef]
10. Suckling, C.C.; Symonds, R.C.; Kelly, M.S.; Young, A.J. The effect of artificial diets on gonad colour and biomass in the edible sea urchin *Psammechinus miliaris*. *Aquaculture* **2011**, *318*, 335–342. [CrossRef]
11. Symonds, R.C.; Kelly, M.S.; Caris-Veyrat, C.; Young, A.J. Carotenoids in the sea urchin *Paracentrotus lividus*: Occurrence of 9′-cis-echinenone as the dominant carotenoid in gonad colour determination. *Comp. Biochem. Physiol. Part B Biochem. Mol. Biol.* **2007**, *148*, 432–444. [CrossRef]
12. Symonds, R.C.; Kelly, M.S.; Suckling, C.C.; Young, A.J. Carotenoids in the gonad and gut of the edible sea urchin *Psammechinus miliaris*. *Aquaculture* **2009**, *288*, 120–125. [CrossRef]
13. Robinson, S.M.C.; Castell, J.D.; Kennedy, E.J. Developing suitable colour in the gonads of cultured green sea urchins (*Strongylocentrotus droebachiensis*). *Aquaculture* **2002**, *206*, 289–303. [CrossRef]
14. Qiao, Y.; Zhou, L.; Qu, Y.; Lu, K.; Han, F.; Li, E. Effects of Different Dietary β-Glucan Levels on Antioxidant Capacity and Immunity, Gut Microbiota and Transcriptome Responses of White Shrimp (*Litopenaeus vannamei*) under Low Salinity. *Antioxidants* **2022**, *11*, 2282. [CrossRef]
15. Ahmed, R.O.; Ali, A.; Leeds, T.; Salem, M. Fecal Microbiome Analysis Distinguishes Bacterial Taxa Biomarkers Associated with Red Fillet Color in *RainbowTrout*. *Microorganisms* **2023**, *11*, 2704. [CrossRef]
16. Wan, S.; Li, Q.; Yu, H.; Liu, S.; Kong, L. Transcriptome analysis based on dietary beta-carotene supplement reveals genes potentially involved in carotenoid metabolism in *Crassostrea gigas*. *Gene* **2022**, *818*, 146226. [CrossRef]
17. Wu, B.; Jiao, R.; Cui, D.; Zhao, T.; Song, J.; Zhan, Y.; Chang, Y. Integrated microRNA-mRNA analysis provides new insights into gonad coloration in the sea urchin *Strongylocentrotus intermedius*. *Front. Mar. Sci.* **2023**, *10*, 1247470. [CrossRef]
18. McBride, S. Comparison of gonad quality factors: Color, hardness and resilience, of *Strongylocentrotus franciscanus* between sea urchins fed prepared feed or algal diets and sea urchins harvested from the Northern California fishery. *Aquaculture* **2004**, *233*, 405–422. [CrossRef]
19. Baião, L.F.; Rocha, F.; Sá, T.; Oliveira, A.; Pintado, M.; Lima, R.C.; Cunha, L.M.; Valente, L.M.P. Sensory profiling, liking and gonad composition of sea urchin gonads fed synthetic or natural sources of β-carotene enriched diets. *Aquaculture* **2022**, *549*, 737778. [CrossRef]
20. Santos, P.M.; Albano, P.; Raposo, A.; Ferreira, S.M.F.; Costa, J.L.; Pombo, A. The effect of temperature on somatic and gonadal development of the sea urchin *Paracentrotus lividus* (Lamarck, 1816). *Aquaculture* **2020**, *528*, 735487. [CrossRef]
21. Grabherr, M.G.; Haas, B.J.; Yassour, M.; Levin, J.Z.; Thompson, D.A.; Amit, I.; Adiconis, X.; Fan, L.; Raychowdhury, R.; Zeng, Q.; et al. Full-length transcriptome assembly from RNA-Seq data without a reference genome. *Nat. Biotechnol.* **2011**, *29*, 644–652. [CrossRef]
22. Buchfink, B.; Xie, C.; Huson, D.H. Fast and sensitive protein alignment using DIAMOND. *Nat. Methods* **2015**, *12*, 59–60. [CrossRef]
23. Langmead, B.; Trapnell, C.; Pop, M.; Salzberg, S.L. Ultrafast and memory-efficient alignment of short DNA sequences to the human genome. *Genome Biol.* **2009**, *10*, R25. [CrossRef]
24. Love, M.I.; Huber, W.; Anders, S. Moderated estimation of fold change and dispersion for RNA-seq data with DESeq2. *Genome Biol.* **2014**, *15*, 550. [CrossRef]
25. Li, M.; Gou, D.; Gong, P.; Di, W.; Wang, L.; Ding, J.; Chang, Y.; Zuo, R. An Investigation on the Effects of Dietary Vitamin E on Juvenile Sea Urchin (*Strongylocentrotus intermedius*): Growth, Intestinal Microbiota, Immune Response, and Related Gene Expression. *Biology* **2023**, *12*, 1523. [CrossRef]
26. Zhang, F.; Di, W.X.; Tang, L.; Ning, Y.C.; Ding, J.; Chang, Y.Q.; Zuo, R.T. Effects of soya lecithin on the growth, gonad development, transcription of genes related with fatty acid synthesis and inflammation between sexes of sea urchin *Strongylocentrotus intermedius* (A. Agassiz, 1864). *Aquaculture* **2023**, *575*, 739774. [CrossRef]
27. Trapnell, C.; Williams, B.A.; Pertea, G.; Mortazavi, A.; Kwan, G.; van Baren, M.J.; Salzberg, S.L.; Wold, B.J.; Pachter, L. Transcript assembly and quantification by RNA-Seq reveals unannotated transcripts and isoform switching during cell differentiation. *Nat. Biotechnol.* **2010**, *28*, 511–515. [CrossRef]
28. Jintasataporn, O.; Yuangsoi, B. Stability of carotenoid diets during feed processing and under different storage conditions. *Molecules* **2012**, *17*, 5651–5660. [CrossRef]
29. Hassaan, M.S.; Mohammady, E.Y.; Soaudy, M.R.; Sabae, S.A.; Mahmoud, A.M.A.; El-Haroun, E.R. Comparative study on the effect of dietary β-carotene and phycocyanin extracted from Spirulina platensis on immune-oxidative stress biomarkers, genes expression and intestinal enzymes, serum biochemical in Nile tilapia, Oreochromis niloticus. *Fish Shellfish Immunol.* **2021**, *108*, 63–72. [CrossRef]

30. Fawzy, S.; Wang, W.; Zhou, Y.; Xue, Y.; Yi, G.; Wu, M.; Huang, X. Can dietary β-carotene supplementation provide an alternative to astaxanthin on the performance of growth, pigmentation, biochemical, and immuno-physiological parameters of *Litopenaeus vannamei*. *Aquac. Rep.* **2022**, *23*, 101054. [CrossRef]
31. Bacchetta, C.; Rossi, A.; Cian, R.; Drago, S.; Cazenave, J. Dietary β-carotene improves growth performance and antioxidant status of juvenile *Piaractus mesopotamicus*. *Aquac. Nutr.* **2019**, *25*, 761–769. [CrossRef]
32. Kennedy, E.J.; Robinson, S.M.C.; Parsons, G.J.; Castell, J.D. Effect of Dietary Minerals and Pigment on Somatic Growth of Juvenile Green Sea Urchins, *Strongylocentrotus droebachiensis*. *J. World Aquac. Soc.* **2007**, *38*, 36–48. [CrossRef]
33. Gomes, A.S.; Lourenço, S.; Santos, P.M.; Neves, M.; Adão, P.; Tecelão, C.; Pombo, A. High dietary protein, n − 3/n − 6 ratio and β-carotene enhances *Paracentrotus lividus* (Lamarck, 1816) larval development. *Aquac. Res.* **2022**, *53*, 5398–5412. [CrossRef]
34. Christopher Taylor, J.; Lawrence, A.L.; Watts, S.A. Select light spectra affect gonad color in the sea urchin *Lytechinus variegatus*. *Mar. Freshw. Behav. Physiol.* **2014**, *47*, 415–428. [CrossRef]
35. Shpigel, M.; Schlosser, S.C.; Ben-Amotz, A.; Lawrence, A.L.; Lawrence, J.M. Effects of dietary carotenoid on the gut and the gonad of the sea urchin *Paracentrotus lividus*. *Aquaculture* **2006**, *261*, 1269–1280. [CrossRef]
36. Concepcion, M.R.; Avalos, J.; Bonet, M.L.; Boronat, A.; Gomez-Gomez, L.; Hornero-Mendez, D.; Limon, M.C.; Meléndez-Martínez, A.J.; Olmedilla-Alonso, B.; Palou, A. A global perspective on carotenoids: Metabolism, biotechnology, and benefits for nutrition and health. *Prog. Lipid Res.* **2018**, *70*, 62–93. [CrossRef]
37. Mclaughlin, G. Effect of artificial diets containing cartenoid-rich microalgae on gonad growth and color in the sea urchin *Psammechinus miliaris* (Gmelin). *J. Shellfish Res.* **2001**, *20*, 377–382.
38. Pearce, C.M.; Daggett, T.L.; Robinson, S. Effects of starch type, macroalgal meal source, and β-carotene on gonad yield and quality of the green sea urchin, *Strongylocentrotus droebachiensis* (Müller), fed prepared diets. *J. Shellfish Res.* **2003**, *22*, 505–519. [CrossRef]
39. Mojib, N.; Amad, M.; Thimma, M.; Aldanondo, N.; Kumaran, M.; Irigoien, X. Carotenoid metabolic profiling and transcriptome-genome mining reveal functional equivalence among blue-pigmented copepods and appendicularia. *Mol. Ecol.* **2014**, *23*, 2740–2756. [CrossRef] [PubMed]
40. Alagoz, Y.; Mi, J.; Balakrishna, A.; Almarwaey, L.; Al-Babili, S. Chapter Twenty-Two—Characterizing cytochrome P450 enzymes involved in plant apocarotenoid metabolism by using an engineered yeast system. In *Methods in Enzymology, Carotenoids: Carotenoid and Apocarotenoid Biosynthesis Metabolic Engineering and Synthetic Biology*; Wurtzel, E.T., Ed.; Academic Press: Cambridge, MA, USA, 2022; pp. 527–552. [CrossRef]
41. Lin, X.; Zhou, Y.; Xue, L. Mitochondrial complex I subunit MT-ND1 mutations affect disease progression. *Heliyon* **2024**, *10*, e28808. [CrossRef] [PubMed] [PubMed Central]
42. von Lintig, J.; Moon, J.; Lee, J.; Ramkumar, S. Carotenoid metabolism at the intestinal barrier. Biochimica et Biophysica Acta (BBA)—Molecular and Cell Biology of Lipids. *Carotenoids Recent Adv. Cell Mol. Biol.* **2020**, *1865*, 158580. [CrossRef]
43. Chakraborty, S.; Doktorova, M.; Molugu, T.R.; Heberle, F.A.; Scott, H.L.; Dzikovski, B.; Nagao, M.; Stingaciu, L.-R.; Standaert, R.F.; Barrera, F.N.; et al. How cholesterol stiffens unsaturated lipid membranes. *Proc. Natl. Acad. Sci. USA* **2020**, *117*, 21896–21905. [CrossRef] [PubMed]
44. Mao, S.; Ren, J.; Xu, Y.; Lin, J.; Pan, C.; Meng, Y.; Xu, N. Studies in the antiviral molecular mechanisms of 25-hydroxycholesterol: Disturbing cholesterol homeostasis and post-translational modification of proteins. *Eur. J. Pharmacol.* **2022**, *926*, 175033. [CrossRef] [PubMed]
45. Qi, Z.; Zhang, Y.; Yao, K.; Zhang, M.; Xu, Y.; Zhang, J.; Bai, X.; Zu, H. DHCR24 Knockdown Lead to Hyperphosphorylation of Tau at Thr181, Thr231, Ser262, Ser396, and Ser422 Sites by Membrane Lipid-Raft Dependent PP2A Signaling in SH-SY5Y Cells. *Neurochem. Res.* **2021**, *46*, 1627–1640. [CrossRef]

fishes

MDPI

Article

Evaluation of Apparent Nutrient Digestibility of Novel and Conventional Feed Ingredients in Sobaity Seabream (*Sparidentex hasta*) for Sustainable Aquaculture

Seemab Zehra [1,*], Aboobucker S. Abul Kasim [1], Reda Saleh [1], Paulo De Mello [1], Ali Alshaikhi [2], Joseph Laranja [1], Yousef Alhafedh [2], Brett D. Glencross [3], Majed A. Alghamdi [1] and Asaad Widaa Mohamed [1,*]

[1] Kaust Beacon Development, King Abdullah University of Science and Technology, Thuwal, Jeddah 23955, Saudi Arabia

[2] Ministry of Environment, Water and Agriculture, King Abdul Aziz Rd., Riyadh 11195, Saudi Arabia

[3] Institute of Aquaculture, University of Stirling, Stirling FK9 4LA, UK

* Correspondence: seemab.zehra@kaust.edu.sa (S.Z.); asaad.mohamed@kaust.edu.sa (A.W.M.)

Abstract: This study aimed to evaluate the apparent digestibility coefficients (ADCs) for nutrients and energy of seven conventional and alternative feed ingredients (poultry feather meal, fermented feather meal, mealworm meal, defatted black soldier fly, Chlorella, poultry by-product meal, and corn meal) when fed to Sobaity seabream (*Sparidentex hasta*), with the goal of identifying sustainable, digestible, and nutritionally viable ingredients for aquaculture feed formulations. A reference diet (RF) was formulated to meet the nutrient requirements of Sobaity seabream while test diets were prepared to contain 70% RF and 30% of the test ingredients. Sobaity seabream (200 ± 8.0 g) were fed the diets for seven days before fecal matter was collected by stripping. The whole length of the digestibility trial was 21 days. The ingredient apparent digestibility of dry matter (34.8–70.4%), crude protein (52.8–107.8%), crude lipid (67.7–112.9%), and energy (52.2–86.1%) were affected by test ingredients ($p < 0.01$). The dry matter digestibility of mealworm meal was the highest (70.4%) compared to other ingredients. Feather meal, Chlorella, and black soldier fly meal had significantly lower values of dry matter digestibility. Dry matter and crude protein were significantly more digestible in fermented feather meal than the feather meal without fermentation. The crude protein digestibility was significantly higher (107.8%) for mealworm meal. However, feather meal has shown a significantly lower value (52.8%) for crude protein digestibility compared to other ingredients. Energy digestibility showed a significant positive correlation with dry matter digestibility (r = 0.870). The energy digestibility of mealworm meal was significantly higher (86.1%, $p < 0.05$) than other ingredients. Feather meal had the lowest energy digestibility (52.2%) with no statistically significant difference from Chlorella, corn meal, and black soldier fly meal. This study indicates that mealworm meal is the most easily digestible protein source for Sobaity seabream and should be prioritized in their diets. Fermentation enhances the digestibility of feather meal and is recommended when using it. Ingredients with a lower digestibility, like feather meal, chlorella, and black soldier fly meal, should be used in moderation or undergo further processing to improve nutrient availability.

Keywords: sobaity seabream; protein and energy digestibility; feed ingredients; sustainability

Key Contribution: This study identifies mealworm meal as the most digestible protein source for Sobaity seabream and demonstrates that fermentation enhances the digestibility of feather meal, offering valuable insights for optimizing feed formulations.

1. Introduction

Sparidentex hasta, commonly known as the Sobaity seabream, is a high-value marine species in aquaculture due to its fast growth rate, market demand, and adaptability to intensive farming systems [1]. The Sobaity seabream *S. hasta*, also known as silvery-black porgy or blue-finned seabream, is a native species in the Arabian gulf, the western Indian Ocean, and the coasts of India [2]. Its cultivation holds the potential to contribute significantly to sustainable aquaculture practices, a growing concern in the industry. Traditionally, fish meal has been a primary source of protein in aquafeeds, due to its high digestibility, balanced amino acid profile, and nutrient density [3]. However, the use of fish meal in aquaculture is becoming more challenging due to its increasing price and limited availability. To ensure the long-term viability of aquaculture, it is essential that we explore alternative protein sources that can significantly reduce fish meal [4] in the diets of species like *S. hasta* without compromising their growth performance, health, or product quality.

Several alternative protein sources show promise in aquafeeds. For example, mealworm meal is a promising protein source due to its balanced amino acid profile and digestibility, making it a valuable ingredient in aquafeeds. Chlorella, a microalgae, is high in protein and essential fatty acids, which support fish health and nutrient absorption [5]. Corn meal is widely used as a carbohydrate source, although its digestibility varies among fish species [6]. Fermented feather meal and feather meal provide a sustainable protein option due to their high keratin content, though their use requires processing to enhance digestibility [7]. Black soldier fly larvae, rich in protein and fats, are becoming popular in fish diets due to their favorable nutrient profile and digestibility [8]. Poultry by-product meal, another animal-derived protein source, has been reported to have high digestibility values across species [9].

Selecting alternative ingredients with a high protein digestibility is essential for ensuring optimal nutrient absorption and utilization in aquaculture species while minimizing waste. These alternative protein sources must not only meet the nutritional needs of cultured species but also exhibit favorable palatability and digestibility characteristics, ensuring they are well-accepted and efficiently utilized in aquafeeds. The high digestibility of alternative ingredients enables optimal nutrient absorption, which directly supports growth performance and helps maintain good health in aquaculture species.

The palatability of feed ingredients in aquaculture is a critical aspect that influences the feed intake, growth performance, and overall health of cultured species. The palatability of insect meals for fish remains a concern at higher dietary inclusion levels, with several studies noting a reduction in feed intake when black soldier fly meal is used to replace fish meal protein in species like gilthead seabream *Sparus aurata* [10], Nile tilapia *Oreochromis niloticus* [11], silver perch *Perca maxima* [12], Atlantic salmon *Salmo salar* [13], and rainbow trout *Oncorhynchus mykiss* [14]. Several factors influence the palatability of feed ingredients in aquaculture, including the type of species being cultured, the composition of the feed, feed ingredients, and the presence of attractants or feeding stimulants [15,16]. It has been shown that, in addition to palatability, the nutrient digestibility of an ingredient is among the most influential factors affecting the utility of an ingredient [17]. The digestibility coefficients of feed ingredients provide insight concerning the nutrient qualities of those ingredients and should enable more accurate formulation, allowing better ingredient substitutions in diets designed for target fish species. The nutrient digestibility can vary significantly based on the composition of the ingredients used in the diet. The digestibility of an ingredient reflects the capability of a certain species to utilize the nutrients of that ingredient, predicting its potential as feedstuff [18].

Several studies have reported the potential to advance the field of aquaculture nutrition by providing empirical data on nutrient digestibility, which is critical for optimizing feed formulations, improving feed conversion ratios, and enhancing the overall growth and health of cultured species [19]. Enhanced feed efficiency not only promotes better growth performance and health in farmed species but also reduces nutrient waste, thereby minimizing the environmental impact of aquaculture operations. Furthermore, the development of more sustainable and cost-effective feed formulations will contribute to the economic viability of aquaculture systems. A wide range of ingredients are now used in aquaculture feeds, which are being formulated based on a digestible nutrient and digestible energy basis [5,20–22], and the determination of nutrient digestibility is considered one of the initial steps for evaluating potential ingredients in diets for target fish species [23].

To advance the development of low fish meal diets with high levels of alternative proteins, it is essential that we evaluate their ability to supply essential nutrients. Hence, this study was conducted to evaluate the digestibility of various feed ingredients utilized in aquaculture, with the goal of improving feed efficiency and enhancing the nutritional quality of diets for this fish species. By assessing the digestibility of key nutrients, this study aims to provide a comprehensive understanding of nutrient utilization across different ingredients, leading to the sustainable culture of this fish.

2. Materials and Methods

2.1. Ingredient Sourcing and Diet Development

Fish meal, fish oil, wheat meal, wheat gluten meal, soybean meal, corn meal, vitamin and mineral premixes were obtained from commercial feed mills (ARASCO, Riyadh, Kingdom of Saudi Arabia, KSA) and retained for analysis and feed production for the reference diet. Poultry meal and steam-hydrolyzed feather meal were sourced locally from the Almarai Group in the Riyadh, Kingdom of Saudi Arabia, KSA. Defatted black soldier fly larvae meal and mealworm meal were obtained from Qingdao, Shandong, China. Broken-cell wall Chlorella was procured from Xi'an, China.

A diet-substitution formulation strategy was used as the basis for the design of the experimental diets. For this, a basal mash was formulated based on industrial diet specifications for Sobaity seabream and prepared to include approximately 55% protein, 7% fat, and 0.1% yttrium oxide. The dietary protein was set at 55% according to the previously published requirements for the different stages of this fish species [24]. The basal mash was prepared and thoroughly mixed, forming the basis for all experimental diets in this study. The test ingredients (feather meal, fermented feather meal, mealworm meal, broken cell wall Chlorella, poultry meal, corn meal, and black soldier fly meal) of study for each test diet were added at 30% inclusion, with a concomitant (70%) subsample of the basal mash used to make up 100% of the diet mix, respectively (Table 1). A reference diet was prepared without any further addition of any ingredient. Each of the diets was prepared by using Twin-screw Extruder Feed Mill (Coperion ZSK 27 MvPLUS, Stuttgart, Germany) at King Abdullah University of Science and Technology (KAUST). All twelve diets were processed through a 6 mm diameter die. The diets were stored at −20 °C until used to ensure stability and prevent spoilage. The proximate compositions of the different ingredients used in this study is given in Table 1. The reference diet and test diet formulation and their compositions are presented in Tables 2 and 3, respectively.

Table 1. Nutrient composition of different feed ingredients used to formulate the basal and test diets (all values are % as is basis as used unless otherwise indicated).

	Fish Meal	Wheat Meal	Chlorella	Wheat Gluten Meal	Soybean Meal	Feather Meal	Poultry By-Products Meal	Corn Meal	Fermented Feather Meal	Black Soldier Fly Meal	Mealworm Meal
Dry matter	93.6	90.1	93.9	93.9	91.1	96.5	96.9	87.3	92.1	92.6	98.0
Protein	64.0	14.4	58.4	77.5	45.6	87.0	68.4	8.6	84.5	51.7	58.8
Lipid	9.2	2.6	9.2	7.5	3.3	7.3	16.0	3.5	5.8	14.1	18.9
CHO *	2.0	71.4	25.6	8.1	37.0	0.7	0.1	73.9	0.9	15.3	16.1
Ash	18.4	1.7	0.7	0.8	5.2	1.5	12.4	1.3	0.8	11.4	4.2
Energy	19.0	16.7	21.8	22.6	18.4	23.5	22.3	16.2	22.4	20.3	23.9

* Carbohydrate (CHO) content determined based on dry matter minus protein, ash, and lipids, (CHO% = Dry matter − (crude protein + crude ash + crude lipid)).

Table 2. Diet formulations of basal and test diets (all values are % dry matter as used unless otherwise indicated) to estimate apparent digestibility of different feed ingredients for Sobaity seabream *Sparidentex hasta* over 21 days trial.

	Reference Diet	Feather Meal	Fermented Feather Meal	Mealworm Meal	Chlorella Meal	Poultry By-Products Meal	Corn Meal	Black Soldier Fly Meal
Ingredients (g)	Diet 1	Diet 2	Diet 3	Diet 4	Diet 5	Diet 6	Diet 7	Diet 8
Fish meal	20	14	14	14	14	14	14	14
Fish oil	7.5	5.3	5.3	5.3	5.3	5.3	5.3	5.3
Wheat meal	16.4	11.4	11.4	11.4	11.4	11.4	11.4	11.4
Wheat gluten meal	30	21	21	21	21	21	21	21
Soybean meal	24.9	17.4	17.4	17.4	17.4	17.4	17.4	17.4
Feather meal	0	30	0	0	0	0	0	0
Fermented feather meal	0	0	30	0	0	0	0	0
Mealworm meal	0	0	0	30	0	0	0	0
Chlorella meal	0	0	0	0	30	0	0	0
Poultry meal	0	0	0	0	0	30	0	0
Corn meal	0	0	0	0	0	0	30	0
Black soldier fly meal	0	0	0	0	0	0	0	30
Vit and Min premix	1	0.7	0.7	0.7	0.7	0.7	0.7	0.7
Choline	0.1	0.07	0.07	0.07	0.07	0.07	0.07	0.07
Yttrium oxide	0.1	0.07	0.07	0.07	0.07	0.07	0.07	0.07
Total (g)	100	100	100	100	100	100	100	100

Table 3. Proximate compositions of basal and test diets (all values are in % as is basis unless otherwise indicated) to estimate apparent digestibility of different feed ingredients for Sobaity seabream *Sparidentex hasta* over 21 days trial.

	Reference Diet	Feather Meal	Fermented Feather Meal	Mealworm Meal	Chlorella Meal	Poultry By-Products Meal	Corn Meal	Black Soldier Fly Meal
%dry basis	Diet 1	Diet 2	Diet 3	Diet 4	Diet 5	Diet 6	Diet7	Diet 8
Dry Matter	97	96	94.3	98	97	95	98	95
Crude Protein	54.6	66	67.7	59.9	58.5	60.9	40.3	56.1
Lipids	7.4	10.2	9.9	9.5	9.2	11.4	6.2	8
Ash	6.1	6.2	4.7	5.3	5.7	8.4	4.9	8.7
Carbohydrates *	28.8	13.6	12	23.3	23.6	14.3	46.6	22.3
Gross Energy (kJ/g)	20.7	21.9	21.9	21.8	21.4	21.2	20	20.2

* Carbohydrate (CHO) content determined based on dry matter minus protein, ash, and lipids, (CHO% = Dry matter − (crude protein + ash + lipids)).

2.2. Fish Handling and Fecal Collection

Fish were obtained from the National Mariculture Center, Kingdom of Bahrain, and raised under commercial settings and held in holding tanks (2000 L), being fed a commercial diet (MarineFish; ARASCO, Al Kharj, Riyadh, Kingdom of Saudi Arabia, KSA), for several weeks. The flow-through tank system supplied filtered marine water (salinity ~42 PSU) with dissolved oxygen typically around 7.4 ± 0.3 mg/L (mean ± S.D.) at a flow rate of about 5 L/min to each of the tanks. To ensure that the fish were disease-free, the health status of the fish was determined by submitting fish samples to the Fish Health and Safety Laboratory of the Jeddah Fisheries Research Center, Jeddah, Kingdom of Saudi Arabia for pathogen and parasite analyses before starting the trial. Fish were tested using qPCR for viral nervous necrosis (VNN), red sea bream iridovirus (RSIV), *Streptococcus iniae* (SI), and *Streptococcus agalactiae* (SA), and no infections were recorded. The fish were maintained in indoor holding tanks until they reached the size needed for the digestibility trial. Each of the tanks (850 L) were stocked with 35 individuals of Sobaity bream of 200 ± 8.0 g and fish were fed the reference diet for 1 week to acclimatize under the same conditions.

Following acclimation (1 week) to the system, the fish were then allocated their dietary treatments. Treatments were randomly assigned amongst the tanks for the first replicate. Fish were re-randomized to start second and third replicates of all the treatments by repeating the experimental setup over time. The digestibility experiment was run over a 21-day period. All experimental work was undertaken at the Centre for Marine Oceanographic Research (CMOR) Laboratory (King Abdullah University of Science and Technology, Thuwal, KSA). Water quality parameters were measured daily during the feeding trials [25] and the following data observed: temperature varied from 26.2–28.0 °C, dissolved oxygen 4.8–6.2 mg L^{-1}. Each of the water quality parameters were analyzed using a ProDSS Multiparameter Digital Water Quality Meter (Yellow Springs, OH, USA).

Each tank of fish was hand-fed to satiation twice daily, at 08:00 and 11:00 h. After feeding, the uneaten feed was collected daily and weighed by sieving outflow water from the tank standpipe. Uneaten feed was air-dried and weighed to calculate the actual feed intake of fish to determine the palatability of the diet. The fish were allowed to acclimatize to their allocated diet for seven days before fecal collection commenced, consistent with earlier studies on similar marine species [22,26]. Feces was collected using a stripping technique based on Austreng [27] as adapted earlier [26]. Fish were netted from their respective tank, and placed in a smaller aerated tank containing AQUI-S (20 ppm) until they lost consciousness. The feces was then removed from the distal intestine by applying gentle abdominal pressure. The hands of the person stripping the fish were rinsed with water between each fish to ensure that the feces was not contaminated by urine or mucus. After removal of the feces from the fish, the fecal sample was inspected to ensure it was free from blood and mucus before being placed in a small plastic vial and stored in a freezer at −20 °C. Stripped feces was collected during 20:00 to 22:30 h over a single day for each block, with each fish only being stripped once before being returned to a pooled stock from which the fish were re-randomized for blocks 2 and 3. Fecal samples were kept frozen at −20 °C before being freeze-dried in preparation for analysis.

2.3. Chemical and Digestibility Analysis

All analytical work was undertaken using Corelab facilities of King Abdullah University of Science and Technology (Thuwal, Jeddah, KSA). Six subsamples of the feed ingredients, experimental diets, and three subsamples of fecal matter of each replicate to determine the proximate composition were analyzed using AOAC methods [28]. Diet, ingredients, and fecal samples were analyzed for dry matter, nitrogen, ash, lipid, and energy content. Experimental diets and fecal matter were also analyzed for yttrium. Dry matter was calculated by gravimetric analysis following oven drying at 105 °C for 24 h. Total yttrium concentrations were determined [29] after acid digestion using inductively coupled plasma atomic emission spectrophotometry (G8015 A, Agilent Technologies, Santa Clara, CA, USA). Protein levels were calculated from the determination of total nitrogen by an elemental-analyzer (Flash 2000, Thermo Fisher Scientific, Waltham, MA, USA), based on N × 6.25. Total lipids were determined gravimetrically following extraction of the lipids using the chloroform: methanol solubilization method [30]. Gross ash content was determined gravimetrically following loss of mass after combustion of a sample in a muffle furnace at 550 °C for 12 h (FNC-BX1200-6, Bioevopaek, Jinan, China). Gross energy was determined by adiabatic bomb calorimetry (ACM-6, Bioevopaek, Jinan, China). Total carbohydrates were calculated based on the dry matter content of a sample minus the protein, lipid, and ash. Diet (DADC) and ingredient (IADC) apparent digestibility coefficients were measured and calculated, respectively, according to the following formulae:

$$\mathrm{DADC_{Nutr}} = 1 - (\mathrm{Y_{diet}} \times \mathrm{Nutr_{feces}})/(\mathrm{Y_{feces}} \times \mathrm{Nutr_{diet}})$$

where $\mathrm{Y_{diet}}$ and $\mathrm{Y_{feces}}$ represent the yttrium content of the diet and feces, respectively, and $\mathrm{Nutr_{diet}}$ and $\mathrm{Nutr_{feces}}$ represent the nutritional parameter of concern (dry matter, protein, lipid, or energy) content of the diet and feces, respectively. The digestibility values for each of the test ingredients in the test diets examined in this study were calculated according to the formulae:

$$IADC_{ingredient} = \frac{(AD_{test} \times Nutr_{test} - (AD_{basal} \times Nutr_{basal} \times 0.7))}{\left(0.3 \times Nutr_{ingredient}\right)}$$

where $Nutr \times AD_{ingredient}$ is the digestibility of a given nutrient from the test ingredient included in the test diet at 30%. AD_{test} is the apparent digestibility of the test diet. AD_{basal} is the apparent digestibility of the basal diet, which makes up 70% of the test diet. $Nutr_{Ingredient}$, $Nutr_{test}$, and $Nutr_{basal}$ are the level of the nutrient of interest in the ingredient, test diet, and basal diet, respectively. All raw material inclusion levels were adjusted to account for their dry matter content, and the potential impact of these adjustments on the actual ratio between the reference diet and the test ingredient was considered.

Ingredient digestibility greater than 100% were not corrected because we consider they are potentially indicative of interactive effects between the diet and test ingredient and should be stipulated as determined.

2.4. Statistical Analysis

Before the statistical analysis, all data were tested for normality and equality of variances by means of the Shapiro–Wilk and Levene's test, respectively. Normally distributed data were analyzed using one-way ANOVA, followed by Fisher's Least Significant Difference [31] test to compare significant differences between treatments. Data that failed the normality and equal variance tests were analyzed using the Kruskal–Wallis–H test and subsequently analyzed using Student–Newman–Keuls (NSK) test to compare significant differences ($p < 0.05$) between treatments. The software OriginPro 2020 (San Clemente, CA, USA) was used to employ all the statistical analyses.

3. Results

3.1. Diet Feed Intake Effects

No substantial variability in the feed intake (g/fish/day) among all the test diets was observed across the duration of the trial (Table 4). Although a lower ($p > 0.05$) feed intake was observed in fish fed diet D8, which included black soldier fly meal as the test ingredient. However, the other diets (D1–D7) containing different test ingredients did not show any significant differences compared to the D8 diet.

Table 4. Feed intake (palatability) assessment of basal (D1) and test diets (D2–D8) for Sobaity seabream *Sparidentex hasta* over 21 days trial.

Diets	Test Ingredients	Feed Intake (g/Fish/Day)
D1	Basal diet	3.2 ± 0.81 ^{a}
D2	Feather meal	3.3 ± 0.62 ^{a}
D3	Fermented feather meal	3.6 ± 0.69 ^{a}
D4	Mealworm meal	3.3 ± 0.72 ^{a}
D5	Chlorella mela	3.5 ± 0.61 ^{a}
D6	Poultry by-products meal	3.6 ± 0.24 ^{a}
D7	Corn meal	3.8 ± 0.93 ^{a}
D8	Black soldier fly meal	2.6 ± 0.58 ^{a}
	p value	$p = 0.557$

Values (means ± SEM, $N = 3$) within a column with a common superscript letter are not significantly different from the other dietary groups ($p > 0.05$).

3.2. Ingredient Digestibility

The IADC of the dry matter, crude protein, lipid, and energy of the test ingredients is provided in Table 5. The apparent digestibilities of dry matter (34.8–70.4%), crude protein (52.8–107.8%), crude lipid (67.7–112.9%), and energy (52.2–86.1%) were significantly affected among the different test ingredients ($p < 0.05$). The dry matter digestibility of mealworm meal was significantly higher (70.4%) compared to other ingredients. Feather meal (34.8%) and black soldier fly meal (40.4%) had significantly lower dry matter digestibility values. However, the dry matter digestibility of feather meal was significantly improved (50.1%) after the fermentation of this ingredient. Mealworm meal had the highest ($p < 0.05$) crude protein digestibility at 107.8%. In contrast, feather meal showed the lowest ($p < 0.05$) crude protein digestibility (52.8%) for this species when compared to the other ingredients. This value (52.8%) of non-fermented feather meal was significantly lower than that of fermented feather meal (85.6%). The dry matter and protein digestibility were not correlated ($r < 0.5$) with the ash, carbohydrate, and protein content of test ingredients ($p < 0.05$). Crude lipid was higher for the fermented feather meal than for the unfermented feather meal. The differences in energy digestibility reflected the differences in dry matter digestibility. The energy digestibility showed a significant positive correlation ($r = 0.870$) with the dry matter digestibility ($p < 0.05$). The energy digestibility of mealworm meal was significantly higher (86.1%, $p < 0.05$) than feather meal, Chlorella, and black soldier fly meal, whereas there was no significant difference when compared to fermented feather meal and poultry meal. Feather meal had the lowest energy digestibility ($p < 0.05$, 52.2%) than that of other ingredients.

Table 5. The ingredient apparent digestibility coefficients (IADC) values for the tested ingredients (D2–D8) fed to Sobaity seabream *Sparidentex hasta* over 21 days trial.

Test Diets	Test Ingredients	Dry Matter (IADC%)	Protein (IADC%)	Lipid (IADC%)	Energy (IADC%)
D2	Feather meal	34.8 ± 0.9 ^{c}	52.8 ± 0.7 ^{d}	91.0 ± 0.13 ^{b}	52.2 ± 0.4 ^{c}
D3	Fermented feather meal	50.1 ± 0.4 ^{b}	85.6 ± 0.6 ^{bc}	112.9 ± 0.17 ^{a}	77.1 ± 0.3 ^{ab}
D4	Mealworm meal	70.4 ± 0.9 ^{a}	107.8 ± 0.5 ^{a}	67.7 ± 0.7 ^{c}	86.1 ± 0.5 ^{a}
D5	Chlorella mela	36.8 ± 0.9 ^{c}	83.8 ± 0.6 ^{bc}	108.0 ± 0.17 ^{a}	66.6 ± 0.4 ^{bc}
D6	Poultry by-products meal	51.1 ± 0.8 ^{b}	88.9 ± 0.2 ^{b}	106.6 ± 0.5 ^{a}	79.4 ± 0.5 ^{ab}
D7	Corn meal	51.6 ± 0.7 ^{b}	79.7 ± 0.2 ^{c}	104.8 ± 0.3 ^{a}	68.1 ± 0.9 ^{bc}
D8	Black soldier fly meal	40.4 ± 0.06 ^{c}	72.1 ± 0.1 ^{c}	106.3 ± 0.1 ^{a}	59.9 ± 0.5 ^{c}
	p value	$p = 0.048$	$p = 0.023$	$p = 0.00008$	$p = 0.007$

Values (means ± SEM, $N = 3$) within a column with a common superscript letter are not significantly different from the other dietary groups ($p > 0.05$).

4. Discussion

4.1. Palatability and Digestibility

The palatability of feed ingredients can significantly influence feed intake, which is critical for evaluating the acceptability of alternative protein sources in aquafeeds. In this study, a lower relative feed intake ($p > 0.05$) in fish fed a diet containing insect meal indicates a lower palatability compared to other groups. This finding aligns with previous studies that have also noted a preference among some fish species for traditional ingredients over insect-based feeds [32]. The reduced palatability of insect meals has been attributed to factors such as fat oxidation susceptibility, high chitin content [12], unpleasant odors from essential oils, flavonoids, and terpenoids [33], and the presence of pupal hormones like ecdysone [34]. However, Moutinho et al. [35] found no adverse effects on the feed intake in *S. aurata* when fed diets containing defatted insect meal. Other researchers observed an improved voluntary feed intake in *Cynoglossus semilaevis* with a high inclusion of defatted insect meal [36]. It has been suggested that the inclusion of insect meal can influence diet acceptability due to differences in the nutrient content, which may act as feed stimulants [37,38].

The feather meal, fermented feather meal, mealworm meal, Chlorella, and poultry meal exhibited no significant differences in feed intake, indicating a similar palatability among these ingredients for this fish species. These results aid in refining ingredient selection and processing techniques to enhance feed acceptance in aquaculture diets, particularly when using less palatable ingredients. In this study, the apparent digestibility of the dry matter, crude protein, lipid, and energy of seven different feed ingredients when fed to Sobaity seabream was estimated and the results revealed that the apparent dry matter and crude protein digestibility of the tested ingredients varied from 34.8% to 70.4% and 52.8% to 107.8%, respectively, indicating the different utilization of protein ingredients by this Sobaity seabream.

4.2. Animal-Based Ingredients

4.2.1. Poultry By-Products Meal and Feather Meal

Poultry by-products meal and feather meal are two by-products of the poultry industry and cost-effective ingredients with the potential for use in fish feeds as a source of protein [39]. The results of this study showed that the dry matter digestibility of poultry by-products meal, feather meal, and fermented feather meal was 51.1, 34.8, and 50.1%, respectively. The dry matter ADC was moderate to high in all animal ingredients in earlier research conducted on European seabass (*Dicentrarchus labrax*) [18]. In that work, the authors reported a 69.5% dry matter digestibility for poultry by-product meal. Crude protein in the poultry by-products meal was found to be 88.9% digestible for this fish, which was similar to that reported for largemouth bass (*Micropterus salmoides*)—94% [40], sunshine bass (*Morone chrysops* $\times$ *M. saxatilis*)—75.16% [41], and cobia (*Rachycentron canadum*)—90.9% [42]. The apparent digestibility coefficients of gross energy for poultry by-products meal (79.4%) in this study were lower than those reported in cobia (91.8%) fed poultry by-products meal [43]. The apparent digestibility coefficients of lipids from poultry by-products meal was recorded to be 106.6%. However, the lipid ADC for cobia was reported at 82.9% for poultry by-products meals [43].

The earlier work reported a 73% dry matter digestibility for hydrolyzed feather meal [18]. In juvenile rockfish (*Sebastes schlegeli*), Lee [44] reported higher dry matter ADC values for steam-hydrolyzed feather meal (87%) compared to the Sobaity seabream in this study (35%). However, clear advantages were observed with the fermented feather meal when fed to the fish in this study. The fermentation treatment has been identified as an effective method to improve feather meal quality through biodegrading the keratin protein [45,46] and this may account for the elevated ADC of fermented feather meal. In the current research, the crude protein ADC values for fermented feather meal (85.6%) were found to be significantly higher compared to the steam-hydrolyzed feather meal (52.8%) for Sobaity seabream. This may be because of the higher (apparent) carbohydrate contents of the feather meal than the fermented feather meal, which may explain the different ADCs observed, with that carbohydrate content potentially acting like fiber and interfering with protein digestion as is often seen with other sources of non-starch polysaccharides [46,47]. The apparent digestibility coefficients of gross energy for feather meal and fermented feather meal in this study showed a significant difference from each other and were determined to be 52.2 and 77.1%, respectively. The lower values for feather meal and fermented feather meal were recorded in this study when compared to fermented feather meal (79%) and feather meal (77.5%) for cobia [43]. Another study reported a 51.3% gross energy digestibility of feather meal for speckled catfish *Pseudoplatystoma coruscans* [48]. The apparent digestibility coefficients of lipids from feather meal, and fermented feather meal in this study ranged between 91 and 112.9%. However, the lipid ADC for cobia was reported at 82.2% for fermented feather meal, and 71.1% for feather meal lipid digestibility [43]. The observed differences may be due to interspecies physiological differences in the digestive tract or to the ingredient composition, and, although Sobaity is similar to some species in its digestibility, how it compares to cobia is unknown.

4.2.2. Insects Meals

This study marks the first attempt to assess insect meal digestibility in Sobaity seabream. Insect meal (black soldier fly and mealworms) has shown promise as a potentially sustainable source of nutrients for aquafeeds, offering an alternative to expensive and ecologically undesirable ingredients, in the context of population growth and climate change. The dry matter digestibility of black soldier fly meal was found to be significantly low at 40.4%, which is lower ($p < 0.05$) than the mealworm meal digestibility (70.4%) when fed to this fish species. Similarly, the crude protein and energy digestibility values were significantly higher for mealworm meal than black soldier fly meal, indicating that mealworm meal may be a better ingredient in terms of nutrient digestibility over black soldier fly meal for this fish species. The traditional nitrogen-to-protein conversion factor of 6.25 may overestimate the insect protein content due to the presence of indigestible chitin in their exoskeletons. Recent studies suggest using species-specific factors to improve the accuracy in protein digestibility assessments [49,50]. Therefore, it is essential that we account for this when assessing the insect protein content in aquafeeds, as it could influence protein digestibility values and comparisons across different studies. The protein ADC (72.1%) observed for black soldier fly meal in this study was lower than an earlier study on barramundi *Lates calcarifer* (93.2%) [51], European seabream *Sparus aurata* 84.4% [35], and rainbow trout *Oncorhynchus mykiss* 85% [14], but higher than the values reported for turbot *Scophthalmus maximus* 63.1% [12].

4.3. Plant-Based Ingredients (Corn Meal and Chlorella)

Corn meal has a high carbohydrate content of 73.9%, and the ADC of energy for this ingredient was 68.1%, thus demonstrating some capability of Sobaity seabream to digest carbohydrates. The comparatively low digestion of energy of this ingredient was also reported (64.9%) for speckled catfish [48]. However, other researchers (Abimorad and Carneiro [52] and Abimorad, et al. [53]) have reported higher ADC values of energy in corn (75.6–86.7%) for pacu (*Piaractus brachypomus*) as compared with the results obtained in the present study. The crude protein digestibility was found to be 79.7% for the Sobaity seabream, which is lower than the earlier study conducted (85.8%) for the omnivorous pacu [53]. This result may be related to the feeding habits of fish. The digestibility of crude protein in the ingredients can vary depending on various factors, such as the source of the protein (animal-based or plant-based), the processing techniques used, and the specific amino acid profile of the ingredient, and other compositional features of the ingredient such as non-starch polysaccharide levels and types [54]. Furthermore, the quality of the protein source plays a significant role in determining its digestibility.

The Chlorella meal used in this study had a protein, lipid, carbohydrate, and ash content of 58.4, 9.2, 25.6, and 0.7%, respectively. The significantly lower energy IADC values of 66.6% than the mealworm meal obtained in this study may be due to the high fiber/non-starch polysaccharide content that might have inhibited the proteolytic enzymatic activity. In the present study, the Chlorella meal had an 83.8% crude protein ADC, which compares with the protein digestibility of Chlorella fed to European seabass reported at 85.5% [55]. Although the crude protein ADC of Chlorella for tilapia has been reported somewhat higher at 91% [56]. However, values recorded for the digestibility of *Chlorella* meal by other species have ranged from 68 to 80% [57]. In one study with Nile tilapia, Sarker et al. [58] have reported that the protein ADC was 80% for *Chlorella* sp., which is lower than the value reported in the current study for the processed Chlorella (83.8%). The differences in the nutrient digestibility of the ingredients are likely to be significantly impacted due to the technological processing of this algae, and further work needs to examine those effects.

In this study, we observed that ingredient digestibility values (IADCs) for certain test ingredients occasionally exceeded 100%. Several potential sources of error exist when calculating ingredient digestibility. These include inaccuracies in determining the composition of the test ingredient, errors in calculating the diet digestibility, and challenges in accurately assessing the inclusion or dilution of the test ingredient within the diet. The digestibility of an ingredient is calculated by comparing the digestibility of two diets (one containing the reference ingredient and one containing the test ingredient), which introduces additional complexity [59]. These calculations often assume that the inclusion of a test ingredient does not impact the digestibility of other diet components, such as those in the reference diet. However, this assumption is not always valid, as some ingredients can either enhance or reduce the digestibility of other components [17]. For instance, certain ingredients may contain compounds that enhance enzymatic activity, thereby improving the digestibility of the entire diet, leading to apparent digestibility values that exceed 100%. Conversely, some ingredients may contain anti-nutritional factors or indigestible components that can hinder the digestibility of other nutrients in the diet. This potential for interaction between ingredients and diet components makes the calculation of ingredient digestibility non-linear and prone to variability [59].

5. Conclusions

Since the apparent digestibility of feed ingredients is essential in formulating extruded mixed feed for aquaculture species as it helps determine how efficiently the species can utilize nutrients, by assessing the digestibility of nutrients, formulators can select ingredients that provide optimal nutrition while reducing waste. This ensures the feed is both cost-effective and energy-efficient, promoting better growth and health for the species while minimizing the environmental impact through reduced nutrient excretion. In this study, most of the tested ingredients were generally well-digested, with the exception of feather meal, which showed lower digestibility values. Mealworm meal demonstrated superior protein and energy digestibility and is, therefore, highly recommended as a key ingredient. Poultry meal and fermented feather meal also showed good protein digestibility and can be included in practical diets; however, feather meal should be used cautiously due to its comparatively poor digestibility in this species. These findings contribute to identifying nutritionally efficient and environmentally sustainable ingredients for formulating cost-effective diets for *S. hasta*, a species of growing importance in marine aquaculture.

Author Contributions: S.Z. was involved in the data curation, visualization, writing—original draft, writing—review and editing, analyzing and interpreting the data, and preparing and publishing manuscript. J.L., A.S.A.K., R.S., M.A.A. and P.D.M. was involved in the validation, interpretation of the data, and writing—review and editing. A.A. and Y.A. were part of MEWA that fully financed the project, and were involved in the data curation, conceptualization, and writing—review and editing. B.D.G. was involved in the conceptualization, methodology, and writing—review and editing. A.W.M. was responsible for the writing—review and editing, analyzing and interpreting the data, and project administration. All authors have read and agreed to the published version of the manuscript.

Funding: Ministry of Environment, Water, and Agriculture, Riyadh, Saudi Arabia (Applied Research Support for Enhancing Fisheries Production, Initiative No. 368).

Institutional Review Board Statement: The culture protocol and methodology of sampling in this study was approved by the Institutional Animal Care and Use Committee (IACUC) of the King Abdullah University of Science and Technology (KAUST) with IACUC no.: 17IACUC17. Approval date: 19 November 2020.

Informed Consent Statement: Not applicable.

Data Availability Statement: The data used in the outcomes of this research are included in this manuscript.

Acknowledgments: The authors are thankful to the King Abdullah University of Science and Technology, Thuwal, Jeddah, Saudi Arabia for providing necessary facilities. The authors are thankful to Nurhisham and Danial for their support during sampling and feeding.

Conflicts of Interest: The authors declare no competing interests.

References

1. Young, B.C.; Shaikhi, A.A. Sobaity Seabream Culture in High-Temperature Conditions. *N. Am. J. Aquac.* **2023**, *85*, 200–204. [CrossRef]
2. Agh, N.; Morshedi, V.; Noori, F.; Ghasemi, A.; Pagheh, E.; Rashidian, G. The effects of single and combined use of *Lactobacillus plantarum* and xylooligosacharide on growth, feed utilization, immune responses, and immune and growth related genes of sobaity (*Sparidentex hasta*) fingerlings. *Aquac. Rep.* **2022**, *25*, 101271. [CrossRef]
3. Aragão, C.; Gonçalves, A.T.; Costas, B.; Azeredo, R.; Xavier, M.J.; Engrola, S. Alternative proteins for fish diets: Implications beyond growth. *Animals* **2022**, *12*, 1211. [CrossRef] [PubMed]
4. Hussain, S.M.; Bano, A.A.; Ali, S.; Rizwan, M.; Adrees, M.; Zahoor, A.F.; Sarker, P.K.; Hussain, M.; Arsalan, M.Z.-u.-H.; Yong, J.W.H. Substitution of fishmeal: Highlights of potential plant protein sources for aquaculture sustainability. *Heliyon* **2024**, *10*, e26573. [CrossRef]
5. Annamalai, S.N.; Das, P.; Thaher, M.I.; Abdul Quadir, M.; Khan, S.; Mahata, C.; Al Jabri, H. Nutrients and energy digestibility of microalgal biomass for fish feed applications. *Sustainability* **2021**, *13*, 13211. [CrossRef]
6. Kamalam, B.S.; Medale, F.; Panserat, S. Utilisation of dietary carbohydrates in farmed fishes: New insights on influencing factors, biological limitations and future strategies. *Aquaculture* **2017**, *467*, 3–27. [CrossRef]
7. Woodgate, S.L.; Wan, A.H.; Hartnett, F.; Wilkinson, R.G.; Davies, S.J. The utilisation of European processed animal proteins as safe, sustainable and circular ingredients for global aquafeeds. *Rev. Aquac.* **2022**, *14*, 1572–1596. [CrossRef]
8. Belghit, I.; Liland, N.S.; Lundebye, A.-K.; Tibon, J.; Sindre, H.; Nilsen, H.; Hagemann, A.; Sele, V. Aquaculture sludge as feed for black soldier fly: Transfer of chemical and biological contaminants and nutrients. *Waste Manag.* **2024**, *187*, 39–49. [CrossRef]
9. Wang, X.; Luo, H.; Zheng, Y.; Wang, D.; Wang, Y.; Zhang, W.; Chen, Z.; Chen, X.; Shao, J. Effects of poultry by-product meal replacing fish meal on growth performance, feed utilization, intestinal morphology and microbiota communities in juvenile large yellow croaker (*Larimichthys crocea*). *Aquac. Rep.* **2023**, *30*, 101547. [CrossRef]
10. Randazzo, B.; Zarantoniello, M.; Cardinaletti, G.; Cerri, R.; Giorgini, E.; Belloni, A.; Contò, M.; Tibaldi, E.; Olivotto, I. *Hermetia illucens* and poultry by-product meals as alternatives to plant protein sources in gilthead seabream (*Sparus aurata*) diet: A multidisciplinary study on fish gut status. *Animals* **2021**, *11*, 677. [CrossRef]
11. Agbohessou, P.S.; Mandiki, S.N.; Gougbédji, A.; Megido, R.C.; Hossain, M.S.; De Jaeger, P.; Larondelle, Y.; Francis, F.; Lalèyè, P.A.; Kestemont, P. Total replacement of fish meal by enriched-fatty acid *Hermetia illucens* meal did not substantially affect growth parameters or innate immune status and improved whole body biochemical quality of Nile tilapia juveniles. *Aquac. Nutr.* **2021**, *27*, 880–896. [CrossRef]
12. Kroeckel, S.; Harjes, A.-G.; Roth, I.; Katz, H.; Wuertz, S.; Susenbeth, A.; Schulz, C. When a turbot catches a fly: Evaluation of a pre-pupae meal of the Black Soldier Fly (*Hermetia illucens*) as fish meal substitute—Growth performance and chitin degradation in juvenile turbot (*Psetta maxima*). *Aquaculture* **2012**, *364*, 345–352. [CrossRef]
13. Lock, E.; Arsiwalla, T.; Waagbø, R. Insect larvae meal as an alternative source of nutrients in the diet of Atlantic salmon (*Salmo salar*) postsmolt. *Aquac. Nutr.* **2016**, *22*, 1202–1213. [CrossRef]

14. Dumas, A.; Raggi, T.; Barkhouse, J.; Lewis, E.; Weltzien, E. The oil fraction and partially defatted meal of black soldier fly larvae (*Hermetia illucens*) affect differently growth performance, feed efficiency, nutrient deposition, blood glucose and lipid digestibility of rainbow trout (*Oncorhynchus mykiss*). *Aquaculture* **2018**, *492*, 24–34. [CrossRef]
15. Hancz, C. Feed efficiency, nutrient sensing and feeding stimulation in aquaculture: A review. *Acta Agrar. Kaposváriensis* **2020**, *24*, 35–54. [CrossRef]
16. Eriegha, O.J.; Ekokotu, P.A. Factors affecting feed intake in cultured fish species: A review. *Anim. Res. Int.* **2017**, *14*, 2697–2709.
17. Glencross, B.D.; Booth, M.; Allan, G.L. A feed is only as good as its ingredients—A review of ingredient evaluation strategies for aquaculture feeds. *Aquac. Nutr.* **2007**, *13*, 17–34. [CrossRef]
18. Campos, I.; Matos, E.; Aragão, C.; Pintado, M.; Valente, L. Apparent digestibility coefficients of processed agro-food by-products in European seabass (*Dicentrarchus labrax*) juveniles. *Aquac. Nutr.* **2018**, *24*, 1274–1286. [CrossRef]
19. NRC. *Nutrient Requirements of Fish and Shrimp*; National Academies Press: Washington, DC, USA, 2011.
20. Zehra, S.; Laranja, J.L.Q.; Abulkasim, A.S.; Saleh, R.; De Mello, P.H.; Pantanella, E.; Alarcon, J.; Al-Suwailem, A.M.; Al Shaikhi, A.; Glencross, B.D. Nutrient and Energy Apparent Digestibility of Protein-Based Feed Ingredients and Effect of the Dietary Factors on Growth Performance and Feed Utilization of Sobaity Seabream, *Sparidentex hasta*. *Animals* **2024**, *14*, 933. [CrossRef]
21. Glencross, B.D.; Carter, C.G.; Duijster, N.; Evans, D.R.; Dods, K.; McCafferty, P.; Hawkins, W.E.; Maas, R.; Sipsas, S. A comparison of the digestibility of a range of lupin and soybean protein products when fed to either Atlantic salmon (*Salmo salar*) or rainbow trout (*Oncorhynchus mykiss*). *Aquaculture* **2004**, *237*, 333–346. [CrossRef]
22. Mohamed, A.H.; Laranja, J.; Saleh, R.; Zehra, S.; De Mello, P.H.; Kasim, A.S.A.; Alarcon, J.; Alshaikhi, A.M.; Al-Suwailem, A.M.; Glencross, B.D. The digestible nutrient and energy values of diets across seven marine aquaculture species demonstrate the potential and limitations for cross utility of digestibility data. *Aquaculture* **2024**, *583*, 740586. [CrossRef]
23. Barreto, A.; Arenas, M.; Álvarez-González, A.; Suárez-Bautista, J.; Sánchez, A.; Maldonado, C.; Cuzon, G.; Gaxiola, G. Evaluation of in vitro and in vivo digestibility of potential feed ingredients for juvenile Yellowtail Snapper. *N. Am. J. Aquac.* **2024**, *86*, 179–192. [CrossRef]
24. Torfi Mozanzadeh, M.; Marammazi, J.G.; Yaghoubi, M.; Agh, N.; Pagheh, E.; Gisbert, E. Macronutrient requirements of silvery-black porgy (*Sparidentex hasta*): A comparison with other farmed sparid species. *Fishes* **2017**, *2*, 5. [CrossRef]
25. APHA. *Standard Methods for the Examination of Dairy Products*; American Public Health Association: Washington, DC, USA, 1992.
26. Blyth, D.; Tabrett, S.; Bourne, N.; Glencross, B. Comparison of faecal collection methods and diet acclimation times for the measurement of digestibility coefficients in barramundi (*Lates calcarifer*). *Aquac. Nutr.* **2015**, *21*, 248–255. [CrossRef]
27. Austreng, E. Digestibility determination in fish using chromic oxide marking and analysis of contents from different segments of the gastrointestinal tract. *Aquaculture* **1978**, *13*, 265–272. [CrossRef]
28. AOAC. Association of Official Analytical Chemists. 2005. Available online: https://www.aoac.org/ (accessed on 29 April 2025).
29. McQuaker, N.R.; Brown, D.F.; Kluckner, P.D. Digestion of environmental materials for analysis by inductively coupled plasma-atomic emission spectrometry. *Anal. Chem.* **1979**, *51*, 1082–1084. [CrossRef]
30. Folch, J.; Lees, M.; Stanley, G.S. A simple method for the isolation and purification of total lipides from animal tissues. *J. Biol. Chem.* **1957**, *226*, 497–509. [CrossRef]
31. Elsdon, T.S.; Gillanders, B.M. Interactive effects of temperature and salinity on otolith chemistry: Challenges for determining environmental histories of fish. *Can. J. Fish. Aquat. Sci.* **2002**, *59*, 1796–1808. [CrossRef]
32. Henry, M.; Gasco, L.; Piccolo, G.; Fountoulaki, E. Review on the use of insects in the diet of farmed fish: Past and future. *Anim. Feed Sci. Technol.* **2015**, *203*, 1–22. [CrossRef]
33. Finke, M.D. Complete nutrient composition of commercially raised invertebrates used as food for insectivores. *Zoo Biol. Publ. Affil. Am. Zoo Aquar. Assoc.* **2002**, *21*, 269–285. [CrossRef]
34. Rao, P.U. Chemical composition and nutritional evaluation of spent silk worm pupae. *J. Agric. Food Chem.* **1994**, *42*, 2201–2203. [CrossRef]
35. Moutinho, S.; Oliva-Teles, A.; Martínez-Llorens, S.; Monroig, Ó.; Peres, H. Total fishmeal replacement by defatted *Hermetia illucens* larvae meal in diets for gilthead seabream (*Sparus aurata*) juveniles. *J. Insects Food Feed* **2022**, *8*, 1455–1468. [CrossRef]
36. Li, X.; Qin, C.; Fang, Z.; Sun, X.; Shi, H.; Wang, Q.; Zhao, H. Replacing dietary fish meal with defatted black soldier fly (*Hermetia illucens*) larvae meal affected growth, digestive physiology and muscle quality of tongue sole (*Cynoglossus semilaevis*). *Front. Physiol.* **2022**, *13*, 855957. [CrossRef] [PubMed]
37. Oteri, M.; Di Rosa, A.R.; Lo Presti, V.; Giarratana, F.; Toscano, G.; Chiofalo, B. Black soldier fly larvae meal as alternative to fish meal for aquaculture feed. *Sustainability* **2021**, *13*, 5447. [CrossRef]
38. Karapanagiotidis, I.T.; Neofytou, M.C.; Asimaki, A.; Daskalopoulou, E.; Psofakis, P.; Mente, E.; Rumbos, C.I.; Athanassiou, C.G. Fishmeal replacement by full-fat and defatted *Hermetia illucens* prepupae meal in the diet of gilthead seabream (*Sparus aurata*). *Sustainability* **2023**, *15*, 786. [CrossRef]
39. Delgado, E.; Reyes-Jaquez, D. Extruded aquaculture feed: A review. In *Extrusion of Metals, Polymers and Food Products*; InTechOpen: London, UK, 2018; pp. 145–163.

40. Qiu, H.; Dai, M.; Chen, N.; Li, S. Apparent digestibility of ten protein ingredients for largemouth bass (*Micropterus salmoides*). *Aquac. Res.* **2022**, *53*, 6846–6854. [CrossRef]
41. Thompson, K.R.; Rawles, S.D.; Metts, L.S.; Smith, R.G.; Wimsatt, A.; Gannam, A.L.; Twibell, R.G.; Johnson, R.B.; Brady, Y.J.; Webster, C.D. Digestibility of dry matter, protein, lipid, and organic matter of two fish meals, two poultry by-product meals, soybean meal, and distiller's dried grains with solubles in practical diets for Sunshine bass, *Morone chrysops* × *M. saxatilis*. *J. World Aquac. Soc.* **2008**, *39*, 352–363. [CrossRef]
42. Zhou, Q.-C.; Tan, B.-P.; Mai, K.-S.; Liu, Y.-J. Apparent digestibility of selected feed ingredients for juvenile cobia *Rachycentron canadum*. *Aquaculture* **2004**, *241*, 441–451. [CrossRef]
43. Chi, S.; Wang, W.; Tan, B.; Dong, X.; Yang, Q.; Liu, H.; Zhang, S. The apparent digestibility coefficients of 13 selected animal feedstuff for Cobia, *Rachycentron canadum*. *J. World Aquac. Soc.* **2017**, *48*, 280–289. [CrossRef]
44. Lee, S.-M. Apparent digestibility coefficients of various feed ingredients for juvenile and grower rockfish (*Sebastes schlegeli*). *Aquaculture* **2002**, *207*, 79–95. [CrossRef]
45. Adelina, A.; Feliatra, F.; Siregar, Y.; Suharman, I.; Pamukas, N. Fermented chicken feathers using *Bacillus subtilis* to improve the quality of nutrition as a fish feed material. In Proceedings of the IOP Conference Series: Earth and Environmental Science, Ho Chi Minh City, Vietnam, 25–28 February 2019; p. 012008.
46. Safari, H.; Mohit, A.; Mohiti-Asli, M. Feather meal processing methods impact the production parameters, blood biochemical indices, gut function, and hepatic enzyme activity in broilers. *J. Anim. Sci.* **2024**, *102*, skae068. [CrossRef] [PubMed]
47. Yeh, R.-H.; Hsieh, C.-W.; Chen, K.-L. Two-stage fermented feather meal enhances growth performance and amino acid digestibility in broilers. *Fermentation* **2023**, *9*, 128. [CrossRef]
48. Gonçalves, E.G.; Carneiro, D.J. Coeficientes de digestibilidade aparente da proteína e energia de alguns ingredientes utilizados em dietas para o pintado (*Pseudoplatystoma coruscans*). *Rev. Bras. De Zootec.* **2003**, *32*, 779–786. [CrossRef]
49. Hasnan, F.F.B.; Feng, Y.; Sun, T.; Parraga, K.; Schwarz, M.; Zarei, M. Insects as Valuable Sources of Protein and Peptides: Production, Functional Properties, and Challenges. *Foods* **2023**, *12*, 4243. [CrossRef]
50. Ritvanen, T.; Pastell, H.; Welling, A.; Raatikainen, M. The nitrogen-to-protein conversion factor of two cricket species-*Acheta domesticus* and *Gryllus bimaculatus*. *Agric. Food Sci.* **2020**, *29*, 1–5. [CrossRef]
51. Le Boucher, R.; Chung, W.; Ng Kai Lin, J.; Tan, L.S.E.; Lee, C.S. Black Soldier Fly Larvae Meal vs. Soy Protein Concentrate Meal: A Comparative Digestibility Study in Barramundi (*Lates calcarifer*). *Aquac. Nutr.* **2024**, *2024*, 3237898. [CrossRef]
52. Abimorad, E.G.; Carneiro, D.J. Fecal collection methods and determination of crude protein and of gross energy digestibility coefficients of feedstuffs for pacu, *Piaractus mesopotamicus* (Holmberg, 1887). *Rev. Bras. De Zootec.* **2004**, *33*, 1101–1109. [CrossRef]
53. Abimorad, E.G.; Squassoni, G.; Carneiro, D. Apparent digestibility of protein, energy, and amino acids in some selected feed ingredients for pacu *Piaractus mesopotamicus*. *Aquac. Nutr.* **2008**, *14*, 374–380. [CrossRef]
54. Akharume, F.U.; Aluko, R.E.; Adedeji, A.A. Modification of plant proteins for improved functionality: A review. *Compr. Rev. Food Sci. Food Saf.* **2021**, *20*, 198–224. [CrossRef]
55. Batista, S.; Pintado, M.; Marques, A.; Abreu, H.; Silva, J.L.; Jessen, F.; Tulli, F.; Valente, L.M. Use of technological processing of seaweed and microalgae as strategy to improve their apparent digestibility coefficients in European seabass (*Dicentrarchus labrax*) juveniles. *J. Appl. Phycol.* **2020**, *32*, 3429–3446. [CrossRef]
56. Barone, R.S.C.; Sonoda, D.Y.; Lorenz, E.K.; Cyrino, J.E.P. Digestibility and pricing of *Chlorella sorokiniana* meal for use in tilapia feeds. *Sci. Agric.* **2018**, *75*, 184–190. [CrossRef]
57. Kotrbáček, V.; Doubek, J.; Doucha, J. The chlorococcalean alga Chlorella in animal nutrition: A review. *J. Appl. Phycol.* **2015**, *27*, 2173–2180. [CrossRef]
58. Sarker, P.; Gamble, M.; Kelson, S.; Kapuscinski, A. Nile tilapia (*Oreochromis niloticus*) show high digestibility of lipid and fatty acids from marine *Schizochytrium* sp. and of protein and essential amino acids from freshwater *Spirulina* sp. feed ingredients. *Aquac. Nutr.* **2016**, *22*, 109–119. [CrossRef]
59. Glencross, B.D. A feed is still only as good as its ingredients: An update on the nutritional research strategies for the optimal evaluation of ingredients for aquaculture feeds. *Aquac. Nutr.* **2020**, *26*, 1871–1883. [CrossRef]

fishes

Article

Dietary Supplementation of *Astragalus* Polysaccharides Modulates Growth Physiology, Metabolic Homeostasis, and Innate Immune Responses in Rice Field Eels (*Monopterus albus*)

Chengcheng Wu [1,2,†], Hang Yang [3,4,†], Yutong Yang [1,2], Quan Yuan [3,4], Weiwei Lv [3,4], Gelana Urgesa Ayana [1,2], Mingyou Li [1,2], Di Su [5], Wenzong Zhou [3,4,*] and Qinghua Zhang [1,2,*]

1 Key Laboratory of Exploration and Utilization of Aquatic Genetic Resources, Ministry of Education, Shanghai Ocean University, Shanghai 201306, China; chengchengwu2024@163.com (C.W.)
2 China National Pathogen Collection Center for Aquatic Animals, Shanghai Ocean University, Shanghai 201306, China
3 Eco-environmental Protection Research Institute, Shanghai Academy of Agricultural Sciences, Shanghai 201403, China; yhangqu2024@163.com (H.Y.)
4 Key Laboratory of Integrated Rice-Fish Farming Ecosystem, Ministry of Agriculture and Rural Affairs, Shanghai Academy of Agricultural Sciences, Shanghai 201403, China
5 Teaching Experimental Centre for Pharmacy, School of Pharmaceutical Sciences, Shanghai Jiao Tong University, 800 Dongchuan Road, Shanghai 200240, China
* Correspondence: zhouwz001@163.com (W.Z.); qhzhang@shou.edu.cn (Q.Z.)
† These authors contributed equally to this work.

Abstract: To investigate the dietary effects of *Astragalus* polysaccharides (APSs) on the growth performance, lipid metabolism, antioxidant activity, and non-specific immunity of Asian swamp eel (*Monopterus albus*) during the domestication stage, fish were randomly allocated into quadruplicate groups receiving *Tenebrio molitor*-based diets supplemented with *Astragalus* polysaccharides (APSs) at graded concentrations of 0 (CON), 700 (APS1), 1400 (APS2), and 2100 (APS3) mg/kg body weight for 28 days. The results showed that dietary APSs at 700–1400 mg/kg·bw significantly enhanced the weight gain rate (WG) and decreased the feed conversion ratio (FCR) of *M. albus* ($p < 0.05$). Concurrently, hematological analysis revealed that hemoglobin levels increased by 19.9% and 23.0% in the 700 and 1400 mg/kg APS groups, respectively ($p < 0.05$). In terms of lipid metabolism, supplementation with APSs significantly increased the serum high-density lipoprotein (HDL) content in all treatment groups ($p < 0.05$). Lower serum triglyceride (TG) levels were found in the APS2 group ($p < 0.05$), and decreased triglyceride (TG), cholesterol (CHO), and low-density lipoprotein (LDL) levels were displayed in the APS3 group ($p < 0.05$). Among the antioxidant parameters, the supplementation with 700 mg/kg·bw APSs significantly increased the glutathione peroxidase (GSH-Px) and catalase (CAT) activity levels of *M. albus* ($p < 0.05$). The APS2 group had a significantly increased total antioxidant capacity (T-AOC) and CAT activity levels ($p < 0.05$), and the APS3 group had significantly increased CAT activity levels ($p < 0.05$). In addition, the APS1 and APS3 groups had significantly reduced malondialdehyde (MDA) levels ($p < 0.05$). In terms of non-specific immunity, the APS1 and APS2 groups showed significantly increased superoxide dismutase (SOD) and lysozyme (LZM) activity levels of *M. albus* ($p < 0.05$), and the addition of 700 mg/kg·bw APSs significantly increased the levels of alkaline phosphatase (AKP) activity ($p < 0.05$). Furthermore, the levels of acid phosphatase (ACP) activity were significantly increased in all experimental groups ($p < 0.05$). In conclusion, the optimal APS addition for *T. molitor* as biocarrier bait is 700 mg/kg, corresponding to 352 mg/kg, which elicits improvements in the growth parameters, lipid homeostasis regulation, oxidative stress mitigation, and innate immune potentiation of *M. albus* during the domestication stage.

Keywords: *Monopterus albus*; *Astragalus* polysaccharides; growth; lipid metabolism; antioxidant capacity; innate immunity

Key Contribution: Dietary addition supplementation with *Astragalus* polysaccharides elicited improvements in the growth parameters, lipid homeostasis regulation, oxidative stress mitigation, and innate immune potentiation of *M. albus* during the domestication stage.

1. Introduction

The rice field eel (*Monopterus albus*), a protogynous hermaphroditic teleost species within the order Synbranchiformes, has emerged as a pivotal species in China's aquaculture industry owing to its substantial nutritional profile and pharmacological properties. As documented in the China Fishery Statistical Yearbook (2024), the national production of *M. albus* achieved a record yield exceeding 330,000 tons in 2024 [1]. However, the intensification of farming systems has precipitated physiological stress responses in this species, manifested through immunosuppression and increased disease susceptibility. Current prophylactic measures predominantly employ broad-spectrum antimicrobial agents, a practice that has engendered multidrug-resistant bacterial strains while exacerbating environmental contamination and residual toxicity in aquatic products. This paradigm underscores the urgent need to develop natural immunostimulants as sustainable alternatives within precision aquaculture frameworks. Phytogenic feed additives, particularly those derived from traditional Chinese medicine, have gained prominence due to their pleiotropic bioactivities, environmental compatibility, and absence of drug residues, positioning them as viable candidates for immunonutrition strategies in aquatic species [2].

Astragalus membranaceus, a leguminous plant with a 2000-year history in Chinese pharmacopeia, synthesizes *Astragalus* polysaccharides (APSs) as its principal bioactive constituents, comprising heteropolysaccharides with α-(1→4)-glucan backbones and arabinogalactan side chains [3]. Studies have demonstrated that APSs exhibit diverse biological functions, including lipid regulation, growth promotion, antioxidant activity, anti-inflammatory effects, and immunomodulation [4]. In recent years, dietary supplementation with APSs in livestock and poultry has been extensively documented for improving nutritional status and physiological conditions [5,6]. Beyond terrestrial applications, *Astragalus* polysaccharides exhibit cross-species applicability in aquaculture systems. Studies have reported that the addition of APSs significantly enhanced the growth potentiation of bluegill sunfish *Lepomis macrochirus* [7], koi *Cyprinus carpio* [8], and Nile tilapia *Oreochromis niloticus* [9]; improved the antioxidant capacity of largemouth bass *Micropterus salmoides* [10], large yellow croaker *Larimichthys crocea* [11], and turbot *Scophthalmus maximus* [12]; and enhanced the immunity of Nile tilapia [9], crucian carp *Carassius auratus* [8], and Amur catfish *Silurus asotus* [13]. Also, dietary APSs improved growth, immunity, and disease resistance in crustaceans, as well as enhancing growth performance and immune parameters in red swamp crayfish *Procambarus clarkii* [14] and upregulating immune-related gene expression in Chinese mitten crab *Eriocheir sinensis* [15]. A prior study evaluated the effects of a compound herbal formula (containing 0.05% andrographolide, 0.05% tributyrin, 0.02% taurine, and 0.05% APSs) on *M. albus* growth and physiology [16]. However, no research has yet investigated the isolated impact of APSs on the growth performance and immune function of *M. albus* during the domestication stage.

At present, most farmed *M. albus* originate from wild-caught juveniles, and they need to be domesticated during the feeding process, wherein *T. molitor* emerges as a superior functional carrier matrix for bioactive compound delivery during transitional husbandry protocols. Artificial cultivation requires a 28-day initial feeding acclimation period, during which mortality rates reach 40–80%. Therefore, this study evaluated the effects of APSs on growth, lipid metabolism, antioxidant capacity, and non-specific immunity in *M. albus* during the domestication stage, aiming to identify effective additives for specialized feeds.

2. Materials and Methods

2.1. Experimental Materials

Experimental specimens of *M. albus* (25.26 ± 0.16 g) and colony-reared *T. molitor* larvae were sourced from the Zhuanghang Comprehensive Experiment Station (Shanghai Academy of Agricultural Sciences, Shanghai, China). *Astragalus* polysaccharides (APSs > 70% purity, Xi'an Ruibo BioTech, Xi'an, China) were integrated into *T. molitor* biocarriers, whose nutritional profile comprised 52.0% crude protein, 32.0% lipids, 5.5% fiber, and 5.0% ash on a dry weight basis.

2.2. Experimental Design and Feeding Management

A total of 720 *M. albus* were stratified into four experimental groups (n = 3 replicates) receiving APS-fortified *T. molitor* formulations at incremental doses of 0 (CON), 700 (APS1), 1400 (APS2), and 2100 (APS3) mg/kg·bw. The method of addition involved evenly sprinkling CGA on hungry *T. molitor*, which were fed to the fish within 30 min after eating. HPLC quantification revealed actual APS retention levels of 0.5, 352, 781, and 1120 mg/kg in the respective biocarriers. The 28-day trial was conducted in recirculating aquaculture systems (RASs) equipped with concrete tanks (1.2 × 1.5 × 0.8 m) with continuous water treatment protocols. Water parameters included dissolved oxygen (6.0–7.5 mg/L), temperature (22–28 °C), and a natural photoperiod. *M. albus* were fed once daily at 16:00 h, with a daily feeding rate of 2–3% body weight. Residual feed was removed 20 min post-feeding, and feed intake was recorded. Mortalities were counted, and *M. albus* were weighed daily. Final body weights were measured for each group at trial termination.

2.3. Sample Collection and Analysis

At the end of the feeding experiment, the *M. albus* were fasted for 24 h. Three individuals per group were anesthetized with MS-222 (100 mg/L) for blood collection via decapitation. Whole blood was incubated at 4 °C overnight, followed by centrifugation at 3000 rpm for 10 min to isolate serum. Livers were excised immediately on ice, rinsed with 0.9% saline to remove surface blood, blotted with filter paper, weighed, and then stored for antioxidant enzyme activity and non-specific immune enzyme activity analyses.

2.3.1. Growth Indicators

Post-trial biometric analysis employed standardized aquacultural metrics for physiological evaluation. Vitality metrics were quantified as follows: Survival Rate (%) = (N_final/N_initial) × 100. Growth dynamics were assessed through the Weight Gain Ratio (%) = $[(W_t - W_0)/W_0] \times 100$, where W_0 and W_t represent the initial/final mean body mass (g). Nutritional efficiency was determined by the formula Feed Conversion Ratio = ΣFeed intake (g)/ΔBiomass (g).

2.3.2. Hematological Analysis

Whole blood parameters were quantified using a Mindray BC-2800vet veterinary hematology analyzer (Shenzhen, China), including white blood cell (WBC) count, red blood cell (RBC) count, hemoglobin (HGB) concentration, hematocrit (HCT), platelet count (PLT), mean platelet volume (MPV), and platelet distribution width (PDW).

2.3.3. Lipid Metabolism Analysis

Serum lipid profiling was conducted using commercial assay kits (Nanjing Jiancheng Institute of Biotechnology, Nanjing, China) through spectrophotometric quantification. Analytical parameters included the following: (1) triglyceride (TG) concentration and (2) cholesterol homeostasis markers comprising total cholesterol (TC), high-density lipoprotein cholesterol (HDL-C), and low-density lipoprotein cholesterol (LDL-C) subfractions. All biochemical analyses strictly adhered to manufacturer-certified protocols.

2.3.4. Antioxidant Capacity Assessment and Non-Specific Immunity Evaluation

Liver tissues were homogenized (1:9 *w*/*v*) in ice-cold 0.9% saline and centrifuged at 2500 rpm for 10 min, and the supernatant was collected for determination. The activity of total antioxidant capacity (T-AOC), catalase (CAT), malondialdehyde (MDA), glutathione peroxidase (GSH-PX), lysozyme (LZM), superoxide dismutase (SOD), alkaline phosphatase (AKP), and acid phosphatase (ACP) in the liver was determined using reagent kits (Jiancheng, Nanjing, China), and an assay was performed by referring to the kit's instructions. Simply, the liver was homogenized to obtain the supernatant, and the total protein (TP) content was measured for error calibration. Lastly, the supernatant and reagents were combined and used for measuring the abovementioned parameters using a microplate reader or colorimetry methods.

2.4. Statistical Analysis

All experimental data were recorded in Excel, with results expressed as mean ± standard deviation (SD). Statistical analyses were performed using SPSS 22.0. One-way analysis of variance (ANOVA) followed by Duncan's multiple range test was applied for multiple comparisons. Differences were considered statistically significant at $p < 0.05$.

3. Results

3.1. Growth Performance

Compared to the control group, the final weight and WG were significantly increased in the APS1 and APS2 groups ($p < 0.05$), while the FCR of the APS1 and APS2 groups was significantly lower than that of the control group ($p < 0.05$). Moreover, no significant differences ($p > 0.05$) in the SR were observed among all groups (Table 1).

Table 1. Influences of different doses of APSs on growth of *M. albus*.

Item	CON	APS1	APS2	APS3
Initial weight (g)	25.28 ± 0.09^{a}	25.44 ± 0.10^{a}	25.28 ± 0.10^{a}	25.06 ± 0.10^{b}
Final weight (g)	34.07 ± 0.42^{b}	39.61 ± 0.42^{a}	40.11 ± 0.26^{a}	34.28 ± 0.81^{b}
WG (%)	34.80 ± 1.24^{b}	55.68 ± 1.12^{a}	58.70 ± 0.52^{a}	36.82 ± 3.62^{b}
FCR	2.80 ± 0.20^{a}	2.01 ± 0.08^{b}	2.01 ± 0.04^{b}	2.92 ± 0.17^{a}
SR (%)	95.56 ± 1.93	97.78 ± 1.92	97.78 ± 0.96	95.00 ± 1.67

The means in the same column with different superscript letters are significantly different ($p < 0.05$). CON, 0 mg/kg·bw; APS1, 700 mg/kg·bw; APS2, 1400 mg/kg·bw; APS3, 2100 mg/kg·bw. Abbreviations: weight gain rate (WG), feed conversion ratio (FCR), survival rate (SR).

3.2. Hematological Parameters

AS shown in Table 2, no significant differences ($p > 0.05$) were observed in hematological parameters, including WBC, RBC, HCT, PLT, MPV, and PDW, among all experimental groups. However, HGB exhibited a dose-dependent trend, increasing initially and then decreasing with rising APS supplementation. Specifically, HGB concentrations significantly increased in the APS1 and APS2 groups ($p < 0.05$), while the APS3 group showed no significant change compared to the control.

Table 2. Influences of different doses of APSs on whole blood indices of *M. albus*.

Item	CON	APS1	APS2	APS3
WBC (10^9/L)	165.33 ± 16.19	168.07 ± 5.63	162.90 ± 7.11	166.77 ± 6.21
RBC (10^{12}/L)	1.74 ± 0.09	1.76 ± 0.11	1.71 ± 0.04	1.76 ± 0.34
HGB (g/L)	140.67 ± 2.08 ^{b}	168.67 ± 10.26 ^{a}	173.00 ± 8.72 ^{a}	146.33 ± 4.93 ^{b}
HCT (%)	21.70 ± 1.35	21.53 ± 2.07	22.70 ± 0.66	22.43 ± 0.61
PLT (10^9/L)	48.33 ± 1.53	51.67 ± 3.22	53.00 ± 4.36	50.33 ± 7.37
MPV (fL)	6.10 ± 0.26	6.23 ± 0.68	6.40 ± 0.36	6.13 ± 0.68
PDW (%)	19.33 ± 0.50	19.20 ± 0.46	19.23 ± 0.35	19.40 ± 0.36

Alphabetic superscripts denote significant intergroup variations ($p < 0.05$) within hematological indices. CON, 0 mg/kg·bw; APS1, 700 mg/kg·bw; APS2, 1400 mg/kg·bw; APS3, 2100 mg/kg·bw. Abbreviations: WBC (white blood cell), RBC (erythrocytes), HGB (hemoglobin), HCT (hematocrit), PLT (thrombocytes), MPV (platelet volume), PDW (platelet size heterogeneity).

3.3. Lipid Metabolism Parameters

In Table 3, the HDL content in all APS groups was significantly higher ($p < 0.05$), and the TG levels in APS2 and APS3 were lower than those of the control group ($p < 0.05$). In addition, APS supplementation with 2100 mg/kg resulted in significantly decreased CHO and LOD levels ($p < 0.05$).

Table 3. Influences of different doses of APSs on lipid metabolism of *M. albus*.

Item	CON	APS1	APS2	APS3
TG (mmol/L)	3.53 ± 1.25 ^{a}	2.75 ± 0.08 ^{ab}	2.00 ± 0.34 ^{bc}	1.20 ± 0.05 ^{c}
CHO (mmol/L)	3.69 ± 0.26 ^{a}	3.41 ± 0.21 ^{a}	3.47 ± 0.22 ^{a}	2.53 ± 0.26 ^{b}
HDL (mmol/L)	0.58 ± 0.06 ^{c}	1.47 ± 0.11 ^{a}	1.62 ± 0.16 ^{a}	0.93 ± 0.15 ^{b}
LDL (mmol/L)	1.56 ± 0.38 ^{a}	1.49 ± 0.18 ^{a}	1.13 ± 0.04 ^{ab}	0.73 ± 0.09 ^{b}

Alphabetic superscripts denote significant intergroup variations ($p < 0.05$) within hematological indices. CON, 0 mg/kg·bw; APS1, 700 mg/kg·bw; APS2, 1400 mg/kg·bw; APS3, 2100 mg/kg·bw. Abbreviations: TG (triglyceride), CHO (cholesterol), HDL (high-density lipoprotein), LDL (low-density lipoprotein).

3.4. Antioxidant Capacity Parameters

In Figure 1, the antioxidant capacity parameters were influenced by the supplementation of APSs. T-AOC exhibited a quadratic response to increasing APS doses, peaking in the APS2 group with significantly higher activity levels compared to the control ($p < 0.05$). Similarly, GSH-Px activity followed a dose-dependent pattern, reaching the maximum levels in the APS1 group ($p < 0.05$). All experimental groups showed significantly higher CAT activity levels than the control group ($p < 0.05$). Furthermore, MDA content was significantly reduced in the APS1 and APS3 groups ($p < 0.05$).

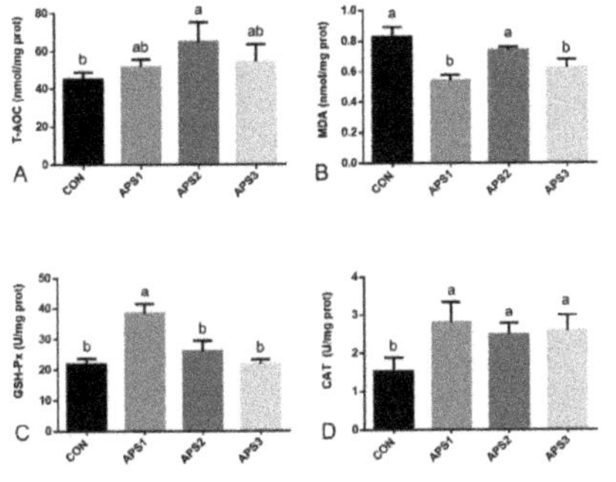

Figure 1. Antioxidant enzyme activities of *M. albus* livers in different experimental groups. Different letters indicate statistical significance ($p < 0.05$) between dietary regimens. (**A**): total antioxidant capacity (T-AOC), (**B**): malondialdehyde (MDA), (**C**): glutathione peroxidase (GSH-PX), (**D**): catalase (CAT). CON, 0 mg/kg·bw; APS1, 700 mg/kg·bw; APS2, 1400 mg/kg·bw; APS3, 2100 mg/kg·bw.

3.5. Non-Specific Immunity Parameters

In Figure 2, all experimental groups exhibited elevated hepatic non-specific immune enzyme activities when compared to the control group. The activities of SOD, LZM, and AKP displayed a quadratic dose–response pattern, peaking in the APS1 group (SOD and AKP, $p < 0.05$) and APS2 group (LZM, $p < 0.05$). In contrast, ACP activity showed a linear increase, with all APS-treated groups demonstrating significantly higher activity levels than the control ($p < 0.05$).

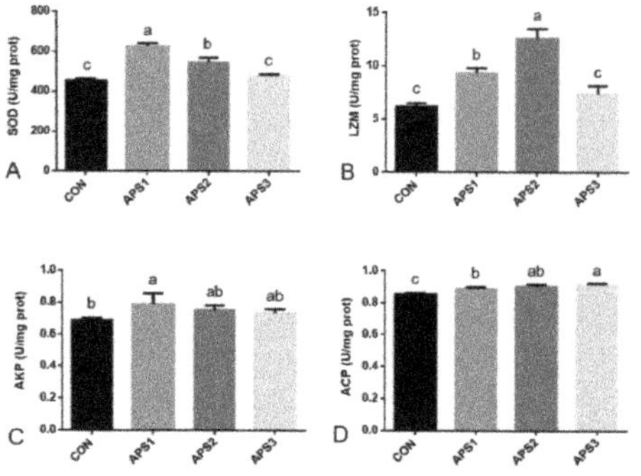

Figure 2. Non-specific immune enzyme activities of *M. albus* livers in different experimental groups. Different letters represent significant differences at $p < 0.05$; same or no letter represents no significant difference ($p > 0.05$). (**A**): superoxide dismutase (SOD), (**B**): lysozyme (LZM), (**C**): alkaline phosphatase (AKP), (**D**): acid phosphatase (ACP). CON, 0 mg/kg·bw; APS1, 700 mg/kg·bw; APS2, 1400 mg/kg·bw; APS3, 2100 mg/kg·bw.

4. Discussion

As the primary bioactive component of *Astragalus membranaceus*, APSs have been reported to enhance growth performance in various farmed fish species [17]. In this study, APS supplementation with 700 mg/kg and 1400 mg/kg in biocarrier bait significantly increased the WG and reduced the FCR in *M. albus* during the domestication stage. Similar findings were observed in large yellow croaker larvae, where dietary APSs with 500–1000 mg/kg improved growth performance [18], and a largemouth bass diet supplemented with 2000 mg/kg APS enhanced growth and feed utilization [10]. Comparable growth-promoting effects of APSs have also been documented in Nile tilapia [9] and bluegill sunfish [7]. Notably, however, the highest APS level (2100 mg/kg·bw) in this study resulted in no significant improvement in the FCR. This aligns with observations in crucian carp, where excessive APS supplementation failed to positively influence feed efficiency [19]. These discrepancies may arise from inter-specific variations, developmental stages, feeding methods, experimental designs, or trial durations. The present study employed a bioencapsulation strategy using *T. molitor* as carriers. Current farming practices rely heavily on wild-caught juveniles that resist formulated feeds during early acclimation. Live feeds, such as *T. molitor*, *Hermetia illucens*, and *Musca demestica*, are thus critical for transitional training. Moreover, live feeds provide comprehensive nutrition, particularly advantageous for juvenile and broodstock stages due to their high protein content and palatability. Additionally, the bioactive compounds encapsulated in live carriers remain structurally intact, avoiding the thermal denaturation and functional deactivation that often occur during the high-temperature processing of pelleted feeds. This method also minimizes nutrient leaching into water, which is a critical limitation of conventional feed delivery systems. By replicating the natural feeding ecology of *M. albus* during the initial acclimation phase, our experimental design provides actionable insights for commercial farming practices. These findings collectively suggested that moderate APS supplementation enhances growth performance, likely through the upregulation of digestive enzyme activities and the stimulation of nutrient absorption pathways [18].

Blood composition in fish is influenced by metabolism, nutritional status, and immune responses. As the primary bioactive component of Astragalus membranaceus, APSs exhibit potent hematopoietic activity by stimulating colony-forming unit-erythrocyte (CFU-E) and burst-forming unit-erythrocyte (BFU-E), thereby enhancing hemoglobin synthesis [20]. Our results demonstrated that dietary supplementation with 700 mg/kg and 1400 mg/kg APSs significantly elevated HGB levels in *M. albus* compared to the control group. Hemoglobin, primarily responsible for oxygen transport and storage, also regulates pH homeostasis, modulates carbon monoxide levels, and participates in immune responses [21]. Similar findings were reported in *Catla catla*, where HGB levels increased following APS supplementation at 200–300 mg/kg [22]. These results suggest that APSs may improve oxygen acquisition in *M. albus*, enhancing survival under hypoxic conditions (e.g., high-density farming or transportation stress) while simultaneously meeting the elevated oxygen demands of physiological metabolic processes.

As a critical metabolic transport system, blood regulates lipid metabolism in fish, with serum TG and total CHO levels serving as key indicators [23]. Our results revealed a dose-dependent reduction in serum TG and CHO levels in *M. albus* supplemented with APSs. Specifically, higher APS supplementation resulted in significant decreases ($p < 0.05$) in these parameters. This aligns with findings in Nile tilapia, where 0.1% Astragalus extract supplementation for 40 days significantly reduced serum TG and CHO [24], and mice (*Mus musculus*) demonstrated improved lipid metabolism following APS administration [25]. Notably, APS supplementation induced a quadratic response in serum HDL levels, peaking at intermediate doses while remaining significantly elevated compared to the control across all treatment groups. Conversely, LDL levels exhibited a gradual decline, with a significant reduction observed only at the highest APS dose (2100 mg/kg) ($p < 0.05$). These findings correlate with prior reports that APSs modulate lipid profiles in Nile tilapia exposed to heavy metal thallium [26]. Mechanistically, HDL facilitates reverse cholesterol transport from peripheral tissues to the liver for catabolism, whereas excessive LDL promotes arterial lipid deposition, increasing the risks of atherosclerosis and vascular injury [27]. We hypothesize that APSs delivered via mealworm carriers (1400–2100 mg/kg bw) regulate lipid metabolism in M. albus by modulating lipoprotein concentrations, thereby enhancing lipid transport efficiency. However, the precise molecular mechanisms underlying this require further investigation.

SOD, LZM, AKP, and ACP are key non-specific humoral immune factors in fish, playing critical roles in immunological responses [28]. SOD activity is closely associated with immune function, while LZM serves as a sensitive biomarker for detecting foreign invasions. Additionally, ACP and AKP hydrolyze phosphate groups and are recognized as vital indicators for evaluating the immune status of aquatic animals [29]. Extensive studies have demonstrated the broad application of APSs as immunostimulants in aquaculture. In this study, the activities of SOD, LZM, AKP, and ACP in *M. albus* were significantly elevated following dietary APS supplementation. Specifically, the addition of 700 mg/kg and 1400 mg/kg APSs markedly enhanced SOD, LZM, and ACP activities. Similar findings have been reported in large yellow croaker, in which 1000 mg/kg APSs significantly increased SOD, LZM, AKP, and ACP activities [30]. APS supplementation at 1200 mg/kg also improved SOD, LZM, AKP, and ACP activities in red swamp crayfish [14] and Chinese mitten crab [15], reducing disease susceptibility. Furthermore, the dietary inclusion of 0.15% and 0.30% APSs has been shown to enhance immunity and significantly elevate LZM activity in Nile tilapia [31]. Mechanistically, APSs may regulate host immunity by activating the TLR4-mediated MyD88-dependent signaling pathway [32]. Collectively, this study demonstrates that APS supplementation enhances the immunity of *M. albus* by upregulating SOD, AKP, LZM, and ACP activities.

5. Conclusions

In summary, dietary supplementation with APSs elicited improvements in the growth parameters, lipid homeostasis regulation, oxidative stress mitigation, and innate immune potentiation of *M. albus* during the domestication stage. Based on the findings of this study, an optimal APS dosage of 700 mg/kg, corresponding to 352 mg/kg, is recommended when administered via *T. molitor* as a biocarrier bait.

Author Contributions: Conceptualization, Q.Y.; methodology, W.L.; software, G.U.A.; validation, W.L.; formal analysis, H.Y. and W.L.; investigation, H.Y.; resources, M.L.; data curation, D.S.; writing—original draft, C.W. and Y.Y.; writing—review and editing, H.Y.; visualization, G.U.A. and M.L.; supervision, W.Z. and Q.Z.; project administration, W.Z. and Q.Z.; funding acquisition, H.Y. All authors have read and agreed to the published version of the manuscript.

Funding: This work was supported by the National Key Research and Development Program of China (2024YFD2401505); China Agriculture Research System of MOF and MARA (CARS-46); the Science and Technology Innovation Action Plan project of the Shanghai Municipal Commission of Science and Technology (No. 22YF1416600); and the Science and Technology Commission of Shanghai Municipality (22DZ2291200).

Institutional Review Board Statement: The ethical approval research scheme was approved by the Animal Care and Use Committee of Shanghai Ocean University (SHOU-DW-2023-006).

Informed Consent Statement: Written informed consent has been obtained from the patient(s) to publish this paper.

Data Availability Statement: The datasets generated for this study are available on request to the corresponding author.

Conflicts of Interest: The authors declare no conflicts of interest.

References

1. Ministry of Agriculture and Rural Affairs. *China Fishery Statistical Yearbook*; China Agriculture Press: Beijing, China, 2024; p. 25.
2. Abdel-Latif, H.M.; Abdel-Daim, M.M.; Shukry, M.; Nowosad, J.; Kucharczyk, D. Benefits and applications of *Moringa oleifera* as a plant protein source in aquafeed: A review. *Aquaculture* **2022**, *547*, 737369. [CrossRef]
3. Shao, B.M.; Xu, W.; Dai, H.; Tu, P.; Li, Z.; Gao, X.M. A study on the immune receptors for polysaccharides from the roots of *Astragalus membranaceus*, a Chinese medicinal herb. *Biochem. Biophys. Res. Commun.* **2004**, *320*, 1103–1111. [CrossRef] [PubMed]
4. Zhang, W.; Ma, W.; Zhang, J.; Song, X.; Sun, W.; Fan, Y. The immunoregulatory activities of astragalus polysaccharide liposome on macrophages and dendritic cells. *Int. J. Biol. Macromol.* **2017**, *105*, 852–861. [CrossRef] [PubMed]
5. Wang, H.F.; Yang, W.R.; Yang, H.W.; Wang, Y.; Yang, Z.B.; Jiang, S.Z.; Zhang, G.G. Effects of *Astragalus membranaceus* on growth performance, carcass characteristics, and antioxidant status of broiler chickens. *Acta Agric. Scand. Sect. A* **2010**, *60*, 151–158.
6. Gu, C.; Zeng, Y.; Tang, Z.; Wang, C.; He, Y.; Feng, X.; Zhou, L. *Astragalus* polysaccharides affect insulin resistance by regulating the hepatic SIRT1-PGC-1α/PPARα-FGF21 signaling pathway in male Sprague Dawley rats undergoing catch-up growth. *Mol. Med. Rep.* **2015**, *12*, 6451–6460. [CrossRef]
7. Elabd, H.; Wang, H.P.; Shaheen, A.; Yao, H.; Abbass, A. *Astragalus membranaceus* (AM) enhances growth performance and antioxidant stress profiles in bluegill sunfish (*Lepomis macrochirus*). *Fish Physiol. Biochem.* **2016**, *42*, 955–966. [CrossRef]
8. Wu, S. Dietary *Astragalus membranaceus* polysaccharide ameliorates the growth performance and innate immunity of juvenile crucian carp (*Carassius auratus*). *Int. J. Biol. Macromol.* **2020**, *149*, 877–881. [CrossRef]
9. Zahran, E.; Risha, E.; AbdelHamid, F.; Mahgoub, H.A.; Ibrahim, T. Effects of dietary *Astragalus* polysaccharides (APS) on growth performance, immunological parameters, digestive enzymes, and intestinal morphology of Nile tilapia (*Oreochromis niloticus*). *Fish Shellfish Immun.* **2014**, *38*, 149–157. [CrossRef]
10. Lin, S.M.; Jiang, Y.; Chen, Y.J.; Luo, L.; Doolgindachbaporn, S.; Yuangsoi, B. Effects of *Astragalus* polysaccharides (APS) and chitooligosaccharides (COS) on growth, immune response and disease resistance of juvenile largemouth bass, *Micropterus salmoides*. *Fish Shellfish Immun.* **2017**, *70*, 40–47. [CrossRef]

11. Liu, Y.; Miao, Y.; Xu, N.; Ding, T.; Cui, K.; Chen, Q.; Zhang, J.; Fang, W.; Mai, K.; Ai, Q. Effects of dietary *Astragalus* polysaccharides (APS) on survival, growth performance, activities of digestive enzyme, antioxidant responses and intestinal development of large yellow croaker (*Larimichthys crocea*) larvae. *Aquaculture* **2020**, *517*, 734752. [CrossRef]
12. Sun, Y.; Wang, X.; Zhou, H.; Mai, K.; He, G. Dietary *Astragalus* polysaccharides ameliorates the growth performance, antioxidant capacity and immune responses in turbot (*Scophthalmus maximus* L.). *Fish Shellfish Immun.* **2020**, *99*, 603–608. [CrossRef] [PubMed]
13. Abdel-Latif, H.M.; Ahmed, H.A.; Shukry, M.; Chaklader, M.R.; Saleh, R.M.; Khallaf, M.A. *Astragalus membranaceus* extract (AME) enhances growth, digestive enzymes, antioxidant capacity, and immunity of *Pangasianodon hypophthalmus* juveniles. *Fishes* **2022**, *7*, 319. [CrossRef]
14. Chang, Z.Q.; Ge, Q.Q.; Sun, M.; Wang, Q.; Lv, H.Y.; Li, J. Immune responses by dietary supplement with *Astragalus* polysaccharides in the Pacific white shrimp, *Litopenaeus vannamei*. *Aquacult. Nutr.* **2018**, *24*, 702–711. [CrossRef]
15. Fu, L.; Zhou, G.; Pan, J.; Li, Y.; Lu, Q.; Zhou, J.; Li, X. Effects of *Astragalus* polysaccharides on antioxidant abilities and non-specific immune responses of Chinese mitten crab, *Eriocheir sinensis*. *Aquacult. Int.* **2017**, *25*, 1333–1343. [CrossRef]
16. Che, C.B.; Shi, Y.; Hu, Y. Effects of two kinds of composite immunopotentiators on growth, immunity and physiological and biochemical indexes of rice field eel (*Monopterus albus*). *Feed Ind.* **2020**, *41*, 43–48.
17. Jin, M.; Zhao, K.; Huang, Q.; Shang, P. Structural features and biological activities of the polysaccharides from *Astragalus* membranaceus. *Int. J. Biol. Macromol.* **2014**, *64*, 257–266. [CrossRef]
18. Liu, H.; Liu, K.; Fu, H.; Ji, R.; Qu, X. Sunlight mediated cadmium release from colored microplastics containing cadmium pigment in aqueous phase. *Environ. Pollut.* **2020**, *263*, 114484. [CrossRef]
19. Liu, J.; Zhang, P.; Wang, B.; Lu, Y.; Li, L.; Li, Y.; Liu, S. Evaluation of the effects of *Astragalus* polysaccharides as immunostimulants on the immune response of crucian carp and against SVCV in vitro and in vivo. *Comp. Biochem. Phys. C* **2022**, *253*, 109249. [CrossRef]
20. Ren, F.; Qian, X.H.; Qian, X.L. *Astragalus* polysaccharide upregulates hepcidin and reduces iron overload in mice via activation of p38 mitogen-activated protein kinase. *Biophys. Res. Commun.* **2016**, *472*, 163–168. [CrossRef]
21. Giardina, B. Hemoglobin: Multiple molecular interactions and multiple functions. An example of energy optimization and global molecular organization. *Mol. Asp. Med.* **2022**, *84*, 101040. [CrossRef]
22. Harikrishnan, R.; Devi, G.; Van Doan, H.; Tapingkae, W.; Balasundaram, C.; Arockiaraj, J.; Ringø, E. Changes in immune genes expression, immune response, digestive enzymes-antioxidant status, and growth of catla (*Catla catla*) fed with *Astragalus* polysaccharides against edwardsiellosis disease. *Fish Shellfish Immun.* **2022**, *121*, 418–436. [CrossRef] [PubMed]
23. Qiang, J.; Tao, Y.F.; Bao, J.W.; Chen, D.J.; Li, H.X.; He, J.; Xu, P. High fat diet-induced miR-122 regulates lipid metabolism and fat deposition in genetically improved farmed tilapia (GIFT, *Oreochromis niloticus*) liver. *Front. Physiol.* **2018**, *9*, 1422. [CrossRef] [PubMed]
24. Wu, F.; Yang, C.; Wen, H.; Zhang, C.; Jiang, M.; Liu, W.; Tian, J.; Yu, L.; Lu, X. Improving low-temperature stress tolerance of tilapia, Oreochromis niloticus: A functional analysis of *Astragalus membranaceus*. *J. World Aquacult. Soc.* **2019**, *50*, 749–762. [CrossRef]
25. Li, B.; Hong, Y.; Gu, Y.; Ye, S.; Hu, K.; Yao, J.; Ding, K.; Zhao, A.; Jia, W.; Li, H. Functional metabolomics reveals that *Astragalus* polysaccharides improve lipids metabolism through microbial metabolite 2-hydroxybutyric acid in obese mice. *Engineering* **2022**, *9*, 111–122. [CrossRef]
26. Farag, M.R.; Alagawany, M.; Khalil, S.R.; Moustafa, A.A.; Mahmoud, H.K.; Abdel-Latif, H.M. *Astragalus membranaceus* polysaccharides modulate growth, hemato-biochemical indices, hepatic antioxidants, and expression of HSP70 and apoptosis-related genes in *Oreochromis niloticus* exposed to sub-lethal thallium toxicity. *Fish Shellfish Immun.* **2021**, *118*, 251–260. [CrossRef]
27. Li, M.; Qiang, J.; Zhu, X.; Bao, J.; Tao, Y.; Zhu, H. Effect of Siberian ginseng water extract as a dietary additive on growth performance, blood biochemical indexes, lipid metabolism, and expression of PPARs pathway-related genes in genetically improved farmed tilapia (*Oreochromis niloticus*). *Fishes* **2022**, *7*, 149. [CrossRef]
28. Xu, D.; Han, P.; Xia, L.; Gan, J.; Xu, Q. A comparative transcriptome analysis focusing on immune responses of Asian swamp eel following infection with *Aeromonas hydrophila*. *Aquaculture* **2021**, *539*, 736655. [CrossRef]
29. Hu, X.; Yang, H.L.; Yan, Y.Y.; Zhang, C.X.; Ye, J.D.; Lu, K.L.; Hu, L.H.; Zhang, J.J.; Ruan, L.; Sun, Y.Z. Effects of fructooligosaccharide on growth, immunity and intestinal microbiota of shrimp (*Litopenaeus vannamei*) fed diets with fish meal partially replaced by soybean meal. *Aquacult. Nutr.* **2019**, *25*, 194–204. [CrossRef]
30. Zhang, W.; Zhang, M.; Cheng, A.; Hao, E.; Huang, X.; Chen, X. Immunomodulatory and antioxidant effects of *Astragalus* polysaccharide liposome in large yellow croaker (*Larimichthys crocea*). *Fish Shellfish Immun.* **2020**, *100*, 126–136. [CrossRef]
31. Farag, M.R.; Alagawany, M.; Khalil, S.R.; El-Hady, E.W.; Elhady, W.M.; Ismail, T.A.; Marini, C.; Cerbo, A.; Abdel-Latif, H.M. Immunosuppressive effects of thallium toxicity in Nile Tilapia fingerlings: Elucidating the rescue role of *Astragalus membranaceus* polysaccharides. *Front. Vet. Sci.* **2022**, *9*, 843031. [CrossRef]
32. Zhou, L.; Liu, Z.; Wang, Z.; Yu, S.; Long, T.; Zhou, X.; Bao, Y. *Astragalus* polysaccharides exerts immunomodulatory effects via TLR4-mediated MyD88-dependent signaling pathway in vitro and in vivo. *Sci. Rep.* **2017**, *7*, 44822.

fishes

Article

Influence of the Silkworm-Derived (*Bombyx mori*) Functional Substance (Silkrose-BM) on the Fish Meat Quality of Yellowtail (*Seriola quinqueradiata*)

Athira Athira [1,2], Haruki Nishiguchi [2], Daichi Hayashi [2], Yuki Otsu [2], Chiemi Miura [2], Ibnu Bangkit Bioshina Suryadi [2,3], Muhammad Fariz Zahir Ali [2,4] and Takeshi Miura [2,*]

1 Department of Fisheries, Faculty of Marine Science and Fisheries, Universitas Hasanuddin, Jl. Perintis Kemerdekaan km 10, Makassar 90245, Indonesia; athiraahyar1999@gmail.com

2 Graduate School of Agriculture, Ehime University, 3-5-7 Tarumi, Matsuyama 790-8566, Japan; haruwest314@icloud.com (H.N.); hayashi@hayashi-suisan.com (D.H.); ohtsu2525@gmail.com (Y.O.); chiemi8@agr.ehime-u.ac.jp (C.M.); ibnu.bangkit@unpad.ac.id (I.B.B.S.); muhammad.fariz.zahir.ali@brin.go.id (M.F.Z.A.)

3 Department of Fisheries, Faculty of Fisheries and Marine Sciences, Universitas Padjadjaran, Jl. Ir. Soekarno km 21, Sumedang 45363, Indonesia

4 Research Center for Marine and Land Bio Industry, National Research and Innovation Agency, Jl. Raya Senggigi, Kodek Bay, Pemenang 83352, Indonesia

* Correspondence: miutake@agr.ehime-u.ac.jp; Tel.: +81-89-946-3684; Fax: +81-050-37378455

Abstract: Popular foods such as sushi and sashimi depend on the quality of raw fish meat to maintain consumer satisfaction. Recently, dietary insect meal and insect-derived substances have been extensively studied for application in aquaculture as a protein alternative or immunostimulant. However, the impact of insect functional substances on the fish meat quality of teleosts remains unclear. Here, we investigated the influence of dietary inclusion of silkrose-BM, a novel bioactive polysaccharide derived from the silkworm, *Bombyx mori*, on the meat quality of yellowtail (*Seriola quinqueradiata*). This study was conducted by comparing two groups given different feeds, commercial EP and feeds containing Silkrose-BM (0.1%), after a culture period of six months in separate floating-net cages. The yellowtail were specifically cut into loins and several meat quality parameters were observed, including proximate, meat color changes, total collagen, drip loss, muscle histology, and gene expression (qRT-PCR). The results of the color-change analysis showed that discoloration of red muscle in the EP feed group occurred faster than in the silkrose-BM group, indicating an antioxidant property of silkrose-BM. Dietary silkrose-BM also significantly reduced drip loss and increased the total collagen content of yellowtail meat. Furthermore, qRT-PCR analysis showed that genes related to lipid and protein degradation were downregulated in the muscles of fish fed on silkrose-BM. In contrast, proximate analysis indicated no significant change in the nutritional composition of the meat between the groups. Taken together, our results suggest that dietary silkrose-BM could improve fish meat quality by minimizing protein denaturation and inhibiting lipid oxidation during fish meat storage.

Keywords: fish meat quality; insects; silkworm; silkrose; polysaccharide

Key Contribution: Bioactive polysaccharide from silkworm: oral administration of silkrose-BM was shown to be effective in improving the meat quality of cultured yellowtail.

1. Introduction

Maintaining the quality of fish meat is crucial, as it is a perishable food. Quality indicators of fish meat can have a major impact on consumer preferences, so it is essential to maintain its quality at the highest possible level [1]. Raw fish meat is highly valued worldwide, for example in popular foods such as sushi and sashimi. In Japan, these dishes traditionally use yellowtail (*Seriola quinqueradiata*), specifically 'hamachi' (young yellowtail) and 'buri' (mature yellowtail) as the main ingredients. Fish meat is highly nutritious, containing macronutrients and micronutrients that are important for humans, such as protein and fat, as well as some carbohydrates, vitamins, and minerals. However, the overall quality of fish meat, including its nutritional composition, can be influenced by various factors such as the type of feed and the farming methods used to rear the fish. These factors also impact texture, flavor, and freshness—key attributes that determine the desirability of fish meat that is to be consumed raw. Quality parameters of fish meat that are commonly considered critical in assessing the suitability of raw fish for sushi or sashimi include its color, overall appearance, aroma, wateriness, fattiness, taste, flavor, and texture [2,3].

The use of functional compounds or substances from insects are being increasingly widely studied, and insects are considered to be a promising source of various beneficial compounds to optimize fish health and growth during cultivation and culture [4–7]. The pupae of silkworm (*Bombyx mori*) contain many potentially useful compounds, including a bioactive polysaccharide known as silkrose-BM [8]. The use of silkrose-BM in the aquaculture industry has recently been investigated due to its unique advantage in terms of availability as a by-product of silk production. This suggests there may be potential synergies between the aquaculture and sericulture industries in the future [8,9].

Silkrose-BM was initially studied as a feed supplement to improve the resistance of farmed fish to bacterial diseases and *Ectoparasites* by stimulating their immunity. However, further analysis revealed that its effects go beyond immune stimulation, inducing changes to muscle fiber structure that affect meat quality and enhance stress tolerance, including tolerance to high water temperatures. Silkrose-BM consists of the monosaccharides 1-rhamnose, 1-fucose, 1-arabinosa, d-glucuronic acid, d-mannose, d-glucose, d-galactosa, N-acetyl-D-glucosamine (GlcNAc), and N-acetyl-D-galactosamine (GalNAc) [10,11]. GlcNAc and GalNAc are bioactive monosaccharides that belong to the group collectively known as glycosaminoglycans (GAGs). GAGs are important components of connective tissue structure, including in collagen, which forms the structure of muscle tissue [12,13]. It is likely that GAGs support collagen, which influences the quality of fish meat, particularly in terms of texture and muscle tissue structure.

The application of silkroses (including silkrose-BM and silkrose-AY, the latter derived from the Japanese oak silkmoth, *Antheraea yamamai*) as feed additives in aquaculture has proven effective in enhancing the immune systems and survival rates of various fish species, such as white trevally (*Pseudocaranx dentex*) [5], paneids (*Litopenaeus vannamei* and *Marsupenaeus japonicus*) [8], and Japanese medaka fish (*Oryzias latipes*) [14]. However, studies of the effects of a silkrose-BM diet on fish meat quality remain limited. In this study, we aim to investigate the impact of a silkrose-BM feeding treatment on fish meat quality indicators in yellowtail (*Seriola quinqueradiata*), such as nutritional value, appearance, texture, and gene expression. This study provides important insights into the potential of silkworm-derived functional substances to enhance the quality and market value of aquaculture products.

2. Materials and Methods

2.1. Experimental Animals

One-year-old yellowtail (*S. quinqueradiata*) were used in this study. They were fed on commercial extruded pellets (EP) until they reached an average body weight of 1.4–1.9 kg (young yellowtail), before being subjected to the silkrose-BM diet treatment.

2.2. Preparation of Experimental Diets

The silkrose-BM was provided by Shintoa Co., Ltd. (Tokyo, Japan). The basal feed used in this experiment was commercial extruded pellets (EP) produced by Sakamoto Feed Co., Ltd. (Choshi, Japan). For the treated group, silkrose-BM was added to the EP feed at a concentration of 0.1% at the time of pellet preparation on consignment by Sakamoto Feed Co., Ltd. The concentration of silkrose-BM at 0.1% was chosen based on the previous studies [5,8]. The feed composition and nutritional profiles are shown in Table 1.

Table 1. Feed composition and proximate analysis of diets fed to yellowtail during the feeding period.

Ingredients (%)	Diet		
	EP Feed		Silkrose-BM diet
Animal meal (fish meal, krill meal)	50.0		50.0
Grain (flour, starch)	15.0		14.9
Vegetable meal (soybean meal, corn gluten meal)	9.0		9.0
Vegetable oil	1.0		1.0
Others (fish meal oil, calcium phosphate, marigold petal extract, licorice extract, rice germ extract)	25.0		25.0
Silkrose-BM			0.1
Proximate analysis (%)			
Crude protein		±40.0 or more	
Crude fat		±28.0 or more	
Crude fiber		±3.0 or less	
Ash		±13.0 or less	
Calcium		±1.0 or more	
Phosphate		±1.0 or more	
Total feed intake (kg)	11,726		12,709

The feed composition and nutritional profile of the diet are provided by Sakamoto Feed Co., Ltd.

2.3. Fish Culture

The fish were cultured in two separate floating-net cages (10 × 10 × 6 m), in Uwajima, Japan. The number of fish in each cage ranged from approximately 3080 to 3100 individuals. The fish were maintained for six months (March to September), with an average sea water temperature 14 °C in the spring and 24 °C in the summer [15]. The fish were fed twice daily, once in the morning and once in the afternoon, until satisfied.

2.4. Sampling Techniques

At the end of the feeding trial, five fish from each group were randomly selected for sampling. The sampled fish were all females to minimize bias in the interpretation of results due to mixing samples from both sexes, especially for molecular analyses such as qRT-PCR. Different sexes of fish can express distinct genes, which made the determination of the sex of the fish crucial for this study.

Yellowtail are large fish that are typically divided into several sections for consumption. Their ordinary muscle (OM) is categorized by type into dorsal ordinary muscle (DOM), ventral ordinary muscle (VOM), and caudal ordinary muscle (COM) regions. These sections are further processed into loins and then distributed in the marketplace [16]. The sampling techniques used in this study were modified from the methodologies described by Ando et al. [17] and Shioya et al. [18]. The modifications were implemented to align with the specific observed parameters, as shown below (Figure 1).

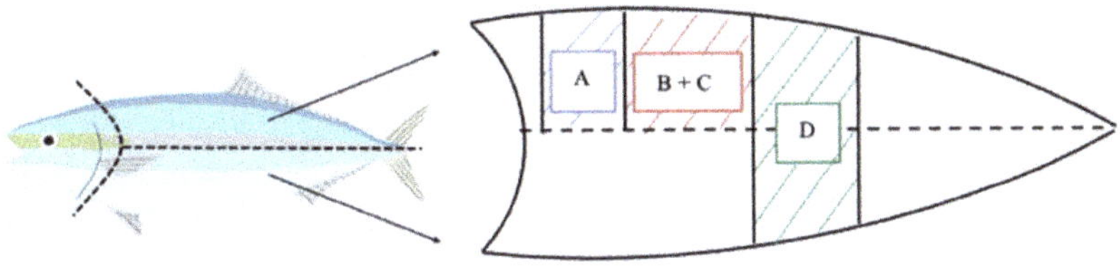

Figure 1. Muscle parts of yellowtail sampled for this study. (**A**) Proximate (kept in a sealed plastic bag); (**B**) histology (kept in fixative in a bottle); RNA (kept in RNAlater solution); (**C**) total collagen quantification (kept in a sealed plastic bag); (**D**) discoloration analysis (kept in a sealed plastic bag); and drip loss (dealt with immediately).

The proximate composition of yellowtail fish meat varies according to the specific part of the body. Therefore, to analyze the proximate composition of yellowtail fish meat, previous studies have sampled meat from the dorsal, ventral, and caudal sections of the fish and then combined them [19–21]. Shioya et al. [18] stated that although the proximate composition in different parts of farmed yellowtail fish varies, sampling the proximate composition from the 7th to the 10th internal vertebrae (musculus latero-dorsalis) provides a comprehensive overview of the overall proximate content. As nutrient content varies in fish meat obtained from different sections of the fish, this sampling technique was adapted to ensure more accurate results and minimize bias from sampling inconsistent or diverse parts of the fish.

2.5. Proximate Analysis

The proximate composition (moisture, crude fat, crude protein, and ash) was analyzed based on the Association of Analytical Communities (AOAC) method [22]. Briefly, the content of crude protein was analyzed with the Kjeldahl method [22]. "Kjeltab" (containing K_2SO_4) was added to the samples, and the samples were digested in a block heater (TecatorTM Digestion Systems 2520, FOSS, Hilleroed, Denmark). The nitrogen content was analyzed using an auto analyzer (KjeltecTM 8400, FOSS, Hilleroed, Denmark). The crude protein was obtained based on the calculation of the nitrogen content with the nitrogen–protein conversion factor, 6.25. Crude fat was analyzed with the Soxhlet extraction method [22]. Extraction from the samples was conducted with petroleum ether in an Automated extractor (SoxtecTM 8000, FOSS, Hilleroed, Denmark). The content of ash was analyzed with an electric furnace (MMF-1, AS ONE, Osaka, Japan).

2.6. Color Change Analysis

Color changes or discoloration in the red muscle of fish were measured using a chroma meter (CR-400, Konica Minolta Inc., Tokyo, Japan). Color changes were observed for 0, 24, 48 h after storage at 4 °C. Analysis of red muscle color changes in fish using a chromameter (CR-400, Konica Minolta) based on the CIELAB color space (L*a*b*), which is a color space defined by the International Commission on Illumination (abbreviated as CIE) in 1976. CIELAB expresses color as three values: L* for perceptual brightness, a* and b* for the unique colors of human vision: red, green, blue, and yellow. CIELAB is intended to be a perceptually uniform space, where a given numerical change corresponds to an observed color change. Images of each sample were taken using a mobile phone camera (iPhone 6 Plus type, Apple Inc., Zhengzhou, China) during measurement using a chromameter instrument as supporting data for statistical analysis of fish meat color change. Image imaging was performed using the remove-background feature on the hardware (MacBook

Air 13, Apple Inc., Zhengzhou, China). The data obtained from the chromameter instrument was then calculated using the formula below:

$$\Delta E^{*}ab = \sqrt{((L_1 - L_0)^2 + (a_1 - a_0)^2 + (b_1 - b_0)^2)}$$

note:

ΔE^*ab = total color difference (calculated for each observation 0, 24, and 48 h)
L_1 = mean L* value of silkrose-BM
L_0 = mean L* value of EP feed
a_1 = mean a* value silkrose-BM
a_0 = mean a* value EP feed
b_1 = mean b* value of silkrose-BM
b_0 = mean b* value of EP feed

2.7. Total Collagen Determination

Total collagen content testing was performed using the Total Collagen Assay kit (QuickZyme Biosciences, Leiden, The Netherlands), with the testing principle based on the detection of hydroxyproline. This assay kit measures the total amount of hydroxyproline formed covering all types of collagen present in the sample including pro-collagen, moisture collagen, and collagen degradation products. The absorbance of the assay is read using a microplate reader at 570 nm. The reading results of the assay kit plate were then calculated based on the manufacturer's protocol. Calculation of total collagen content was done by entering the absorbance data into the standard curve obtained as shown in the Figure 2 below:

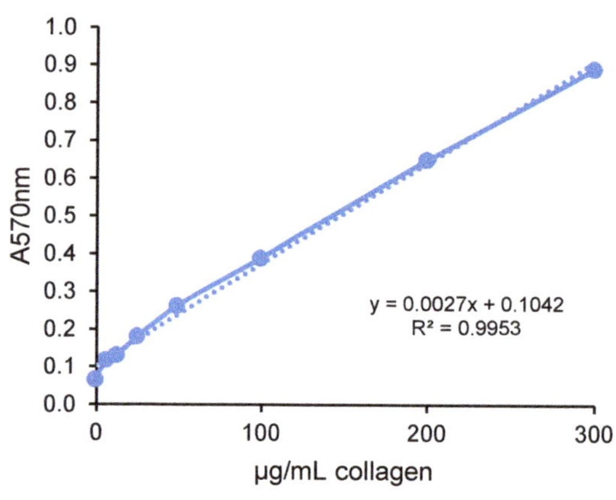

Result of serial dilution standard

No.	Colla concentration (µg/mL)	Absorbance (nm)
1	300	0.892
2	200	0.651
3	100	0.389
4	50	0.263
5	25	0.182
6	12.5	0.132
7	6.25	0.119
8	0	0.067

Figure 2. Absorbance value of total collagen standard solution using total collagen assay kit.

2.8. Drip Loss Analysis

Drip loss was assessed by placing the fish fillet between drip sheets and observing after 24 h storage at 4 °C. Before that, the sheets were weighed one by one and marked, and the filleted fish meat was weighed separately. After 24 h of storage time, the sheets were weighed again one by one, as well as the fillet. The calculation of drip loss used the formula below:

$$Drip\ loss\ (\%) = \frac{W_0 - W_1}{W_0} \times 100$$

note:
W_0 = weight of fillet before storage time (g)
W_1 = weight of fillet after storage time (g)

2.9. Histological Analysis

The samples for histological analysis were fixed immediately following their collection during the rigor mortis phase. The sample was fixed in a 100 mL bottle containing 70 mL of 12% formaldehyde. The fixation solution was refreshed daily for 3–5 days, after which it was replaced with 70% histol. Under these conditions, the samples could be stored and preserved for up to one year for further analysis.

The samples were embedded in paraffin blocks and, following fixation, sectioned to a thickness of 0.3 mm using a microtome. The specimens were stained using a modified Gomori trichrome method, incorporating aniline blue to highlight collagen fibers, which subsequently appeared stained blue in the fish meat. The staining method using Histological Method for CNS modified with Gomori's One Step Trichrome (S3). The histological samples were then observed under a microscope at 40× magnification.

2.10. Analysis of Gene Expression (Total RNA Isolation and qRT-PCR)

Total RNA was extracted from the yellowtail muscle using ISOGEN II (Nippon Gene, Toyama, Japan), according to the manufacturer's protocol. The total RNA yield was assessed spectrophotometrically using a Nanophotometer P330 (Implen, Munich, Germany). RNA from n = 5 individuals sampled from each group was used for qRT-PCR analysis following cDNA synthesis. First-strand cDNA synthesized from the total RNA was generated using a High-Capacity cDNA Reverse Transcription Kit (Applied Biosystems, Foster City, USA), according to the manufacturer's protocol. Gene-specific primers used for qRT-PCR were *Ampd1*, *hao1*, *mecr*, *acadm*, *mstn*, and *fbxo32* (Table S1). Melting curve analyses were performed following the amplification of each gene to confirm that only a single product had been amplified. The thermocycling process was conducted in a 96-well white Multiplate PCR plate (Bio-Rad Laboratories, Hercules, CA, USA) using a quantitative RT-PCR detection system (Bio-Rad Laboratories) with the following cycling conditions: 95 °C for 30 s, followed by 40 cycles of 95 °C for 5 s, and 55 °C for 5 s. The relative gene expression was calculated using the comparative threshold (CT) cycle method described by Livak and Schmittgen [23], with 18s rRNA as an endogenous reference.

2.11. Statistical Analysis

Each item measured was expressed as mean ± standard error (SE). Statistical analysis of proximate composition and total collagen was performed using Mann–Whitney U test, color changes data were analyzed using a two-way ANOVA followed by a Duncan post-hoc test ($p < 0.05$). Statistical analysis of drip loss, histology, and gene expression was performed using the Shapiro–Wilk for normality test and Levene's test of variance, and continued using a t-test to compare the differences between each group. The level of significance was 0.05 (95%). The statistical analyses were performed using Rstudio (R 4.3.3 GUI 1.80 Big Sur ARM build (8340)).

3. Results and Discussion

The concentration of silkrose-BM (0.1%) was tested at the most effective concentration in accordance with our previous papers [5,8]. Dietary silkrose-BM at a concentration of 0.1% effectively prevents vibriosis in penaeid prawns (*Litopenaeus vannamei* and *Marsupenaeus japonicus*) and significantly reduced skin parasitism in yellowtail and white trevally (*Pseudocaranx dentex*). Furthermore, silkrose-BM also significantly reduced cortisol level and altered expression of various genes, including those involved in immunity, stress response, wound healing, and heat responses, in yellowtail, white trevally, and Japanese medaka (*Oryzias latipes*) [4,5,8]. In the present study, we used yellowtail fed with 0.1% silkrose-BM for six months to reveal the effect of silkworm-derived polysaccharides on the fish meat quality. Below, we highlighted several key constituents for yellowtail meat quality affected by dietary silkrose-BM.

3.1. Proximate Composition

The proximate composition of fish meat is influenced by various factors such as diet, breeding season, and environmental conditions. Previous studies have shown that the proximate composition, particularly lipid content, of yellowtail fish meat is related to gonadal maturity [24], as well as to the water temperature and the type of feed provided [18]. The water quality data during the feeding period of yellowtail are displayed in Table 2.

Table 2. Overall water quality measurements during the feeding period of yellowtail.

Time	Parameter				
	Salinity (%)	Temperature (°C)	DO [1] (%)	DOM [2] (mg/L C)	Turbidity (ntu [3])
April	34.47 ± 0.02	16.74 ± 0.02	108.55 ± 2.05	8.68 ± 0.09	0.36 ± 0.03
May	34.48 ± 0.04	16.69 ± 0.01	113.68 ± 0.17	8.96 ± 0.01	0.32 ± 0.01
June	34.21 ± 0.04	18.20 ± 0.01	104.63 ± 0.49	8.02 ± 0.04	0.45 ± 0.03
July	33.87 ± 0.02	20.28 ± 0.11	104.11 ± 0.89	7.70 ± 0.06	0.32 ± 0.02

[1] DO = dissolved oxygen; [2] DOM = dissolved organic matter; [3] ntu = nephelometric turbidity units.

In this study, we determined the sampling method from specific parts of the fish meat, such as the proximate samples taken from the inner part of the *Musculus latero-dorsalis*. The results of the proximate composition analysis are shown in Table 3.

Table 3. The proximate values in yellowtail meat after six months of aquaculture.

Parameters	Compositions Level (%)		*p*-Value
	EP Feed Group	Silkrose-BM Diet Group	
Moisture	58.39 ± 1.81	60.60 ± 4.39	1
Crude fat	20.18 ± 2.08	20.59 ± 3.12	0.84
Crude protein	20.07 ± 0.46	19.56 ± 1.90	0.31
Crude ash	1.63 ± 0.04	1.38 ± 0.13	0.09

Data are presented as mean ± S.E., (n = 5) fish from each group.

The proximate composition of yellowtail meat was determined in both feeding treatment groups. The mean values for the EP feed group for moisture, crude fat, crude protein, and crude ash were different; fish in the silkrose-BM diet group had higher crude fat and moisture values, but statistical analysis revealed no significant differences between the groups.

3.2. Discoloration in Yellowtail Red Muscle

Color changes in the red muscle of fish are one of the indicators of a decline in meat quality and have a considerable impact on consumer preferences. The color of red muscle in fish from each treatment group (n = 5) was measured using a CR-400 instrument (Konika Minolta Inc., Tokyo, Japan) before being stored for 48 h at 4 °C, with further color measurements taken every 24 h. The total color difference before storage began was ΔE^*ab = 0.84, with no significant differences observed between the feeding treatment groups.

Differences in L* values indicate variations in the lightness level of an object. (−L*) value indicates a darker color, while a (+L*) value indicates a lighter color. The a* value indicates differences between redness (+a*) and greenness (−a*), while the b* value indicates differences between yellowness (+b*) and blueness (−b*) in an object. The higher the value of the color index, the more colorful the observed object.

In this study, the L* values in the EP feed group increased drastically ($p < 0.05$) after 48 h of storage, but the same phenomenon was not observed in the silkrose-BM diet treatment group. A significant decrease ($p < 0.01$) in the a* value occurred every 24 h in the EP feed group; meanwhile, in the silkrose-BM diet group, the decrease in a* value was not significant at 24 h, and the changes seemed to slow even after the 24 h mark. The changes in b* values in the EP feed group were significantly different every 24 h ($p < 0.01$), while the b* values in the silkrose-BM diet group only changed after storage for 48 h ($p < 0.05$). The mean values of the results of this analysis are shown in Table S2. The conclusions we drew from the color changes in yellowtail fish meat are summarized in Table 4.

Table 4. Color changes in yellowtail fish meat stored at 4 °C, observed after 0, 24, and 48 h.

Color	EP Feed Group	Silkrose-BM Diet Group
L*	Increased drastically over 48 h of storage	Decreased over 48 h of storage
a*	Decreased drastically every 24 h during the 48 h storage period	Decreased drastically after 24 h of storage, but the decrease was less drastic after 48 h
b*	Increased drastically every 24 h during the 48 h storage period	Increased drastically after 48 h of storage.

EP feed, commercial extruded pellets feed; *value of L*, a*, b* indicates lightness, redness, and yellowness, respectively.

The extended storage of fish (48 h) from both groups led to considerable color changes of fish meat in values of lightness (L*), decreases in redness (a*), and an increase in yellowness (b*). Color changes in the EP feed group occurred more rapidly, indicating that the oxidation and degradation of pigments and lipids were more intense. Color changes happened more slowly in the silkrose-BM diet group. Overall, the extended storage period led to lipid oxidation and pigment degradation, which caused the fish meat to become darker/more dull (decreases/increases in L*); less red (decreased a*), indicating the degradation of myoglobin; and more yellow (increase in b*), suggesting a loss of quality in terms of pigment and chemical composition due to lipid oxidation.

The lower intensity of the red a* value found in the EP feed treatment group was probably associated with the denaturation of sarcoplasmic proteins, including myoglobin. This denaturation causes increased light scattering and a greater amount of oxidized myoglobin in the muscle. These findings align with those of previous studies by Langer & Zhang [25,26]. Figure 3 shows a comparison of the appearance of yellowtail red muscle between groups.

In addition to the chemical composition of yellowtail fish meat being able to influence color changes, especially in the red muscle, microorganisms that proliferate at specific pH levels are also suspected to have an indirect influence by contributing to the degradation of fish meat quality and altering its color [27]. The nutrients contained in fish meat provide an ideal medium for microbial growth, so the presence of microorganisms in fish meat accelerates the deterioration and spoilage process through biochemical reactions once the fish have been killed, which can directly affect the quality of fish meat.

Lipid oxidation and myoglobin oxidation in meat lead to the development of off-flavor and discoloration, respectively, and they are influenced by both external environmental factors and endogenous substances [28]. These processes often appear to be linked, and the oxidation of one can lead to the formation of chemical species that can exacerbate the oxidation of the other [29]. Oxidation can generate harmful substances that pose health risks to humans. Lipid oxidation is thus the primary factor in quality degradation, a process mediated by myoglobin [30,31].

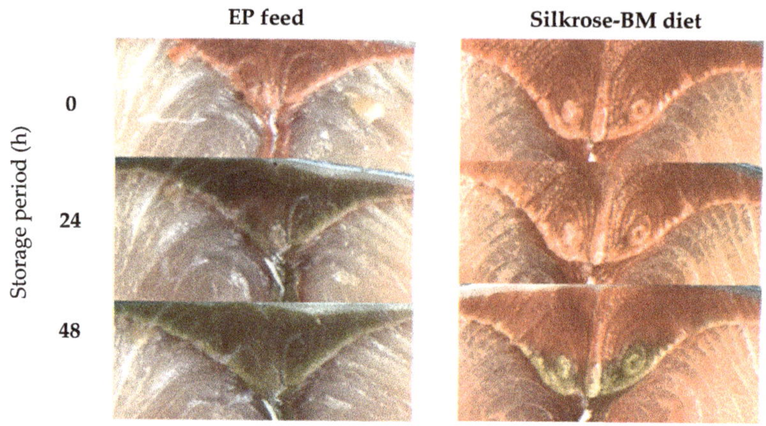

Figure 3. Changes in the appearance of sliced meat prepared from farmed yellowtail during storage at 4 °C. Sliced meat was prepared from fish from both groups and stored on ice for approximately 3 to 5 h after filleting the fish meat.

When alive, fatty fish have an antioxidant system that stabilizes their high content of unsaturated lipids. This endogenous antioxidant system includes compounds that can capture free radicals and enzymes that eliminate reactive oxygen species, such as superoxide radicals, hydrogen peroxide, and lipid peroxides. Following death, these endogenous antioxidants are sequentially consumed, and several studies have linked the depletion of these antioxidants to the development of oxidation. Antioxidants are molecules that can inhibit damage in food products by minimizing lipid oxidation. They can perform various actions in food, including capturing free radicals, binding metal ions, and neutralizing singlet oxygen. Biological tissues have a highly efficient endogenous antioxidant system, which arises from the need to protect aerobic organisms from oxygen and its reactive species [30,32,33]. Several investigators have reported the preservation of color in fresh meat following the inclusion of antioxidant ingredients [29].

The bioactive monosaccharides present in silkrose-BM mainly play a role in biological activities related to enhanced antioxidant activity, via protein and cell modification. These bioactive compounds, including GlcNAc and GalNAc, are involved in modifying proteins via glycation, thereby affecting cellular activity and responses to oxidative stress. According to Chatham et al. [34], protein modification by sugars is one of the most common post-translational modifications of protein. In addition, L-fucose, a unique monosaccharide used by cells in a fucosylation process, is also present in silkrose-BM. Fucosylation is a process used to post-translationally modify and regulate the behavior and function of proteins. It plays an important role in various cellular processes, including cell signaling, immune responses, and the regulation of protein interactions. Adding fucose residues to proteins enables cells to alter the stability, localization, and activity of proteins, influencing various biological functions, including antioxidant defense and stress responses [35].

The mechanisms via which bioactive polysaccharides from silkrose-BM affect color changes in fish meat, and their molecular pathways, remain unclear. However, feeding fish a silkrose-BM diet likely plays a role in slowing the color changes seen in the red muscle compared with the speed of the color changes seen in the EP feed group. This phenomenon, which leads to the production of antibacterial substances, is promoted by silkrose-BM and helps to prevent microbial growth and oxidative stress in the muscle tissue. The effects may occur because the innate immunity of fish fed with silkrose-BM diet was activated. As a result, the oxidative processes that typically accelerate color change and spoilage in fish muscle, such as the oxidation of myoglobin and lipids, may be slowed down, so the color and quality of the fish meat can be preserved during longer storage periods [9]. Thus, the process of quality deterioration in fish meat post-mortem, such as lipid oxidation and protein degradation, slows down.

3.3. *Quantification of Total Collagen*

Collagen is the main protein in the extracellular matrix. Collagens are a family of fibrous proteins found in all multicellular animals. The primary types of collagen found in connective tissues are types I, II, III, V, and XI; however, in fish meat, types I and V collagen are more dominant. Other key components of connective tissue include proteoglycans. These are complex multifunctional molecules consisting of a core protein with a molecular weight ranging from 40 to 350 kDa, covalently linked to several dozen GAG chains [36]. GAGs are complex polysaccharides that play an important role in growth, differentiation, morphogenesis, cell migration, and bacterial/viral infections [13]. The main GAG chains in vertebrates consist of amino sugar units and include GlcNAc and GalNAc, which are also present in silkrose-BM. The total collagen content in yellowtail in our study is shown in Figure 4.

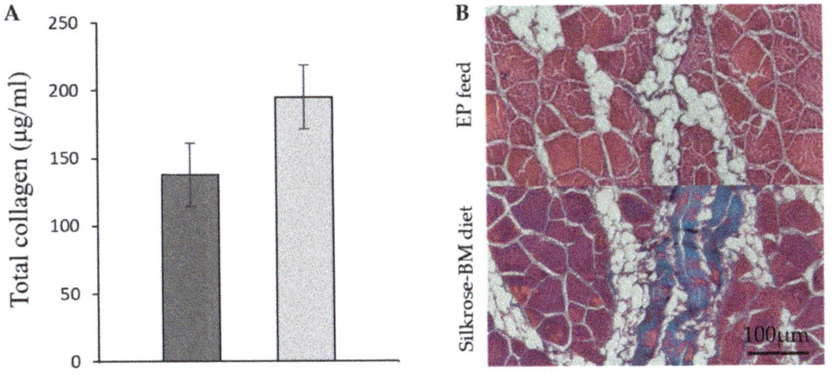

Figure 4. (**A**) Total collagen in yellowtail muscle from the EP group (dark grey) and the silkrose-BM group (pale grey). Data are presented as the mean ± SE (n = 5). (**B**) Photomicrographs showing collagen appearance in the perimysium area of yellowtail muscle, 200× magnification.

Our quantitative analysis of total collagen revealed values of 137.8 µg/mL for the EP feed group and 195.0 µg/mL for the silkrose-BM diet group (Figure 4). Although no statistically significant differences were observed, the collagen content in the fish muscle of the silkrose-BM diet group was notably higher than that of the EP feed group. This difference suggests a potential trend worth further investigation. In addition, distinct visual changes were observed in the muscle fibers of the silkrose-BM diet group. The muscle tissue exhibited a noticeable blue-to-purple hue, with prominent blue lines around the perimysium area, which may reflect alterations in the structural organization or optical properties of the collagen fibers (Figure 4B). The collagen content in fish meat is very important and contributes to the toughness of sliced raw meat [37].

The silkrose-BM diet includes bioactive polysaccharides, so we assume that the provision of this feed has the potential to increase the total collagen content in fish muscle by enhancing protein functions related to the formation of muscle structure and extracellular matrix proteoglycan networks. This hypothesis is supported by the research of Listrat et al. [36], who stated that proteoglycans, one of the main components of connective tissue, are complex multifunctional molecules consisting of a core protein covalently linked to several dozen GAG chains. This structure is essential for the function of proteoglycans in maintaining tissue integrity and modulating various biological processes, including the formation and stabilization of muscle and connective tissue structures [38,39].

3.4. Drip Loss

The drip loss results presented the amount of fluid lost from the yellowtail fish meat that influence the meat weight after 24 h, as shown in Figure 5.

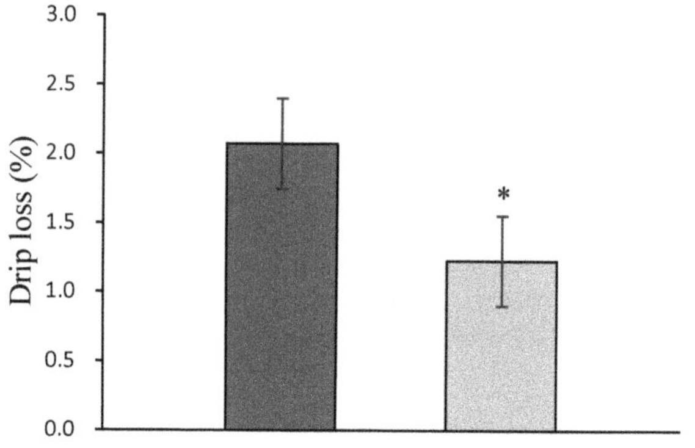

Figure 5. The percentage of drip loss from farmed yellowtail fish meat fed on EP feed (dark grey) and fed on a silkrose-BM diet (pale grey). The asterisk indicates a significant difference between the groups ($p < 0.05$).

The results of the drip loss analysis showed a significant difference between the groups, with average drip loss values of 2.07% and 1.23% for EP feed and the silkrose-BM diet, respectively. We assumed the reduction in drip loss value in the silkrose-BM diet group was due to the presence of functional compounds in silkrose-BM that may influence protein stability and lipid oxidation. These effects could indirectly contribute to the enhanced quality of yellowtail fish meat in this group compared with the EP feed group, as evidenced by the observed color changes and improved moisture retention.

Drip loss is positively correlated with water-holding capacity [40], and changes in water-holding capacity correlate with lipid oxidation. Wada and Ogawa [41] highlighted this, hypothesizing that water-binding molecules are broken down during lipid oxidation. The denaturation of proteins such as myoglobin, which is related to lipid oxidation, alters the molecular structure and water-holding capacity of fish meat, particularly after it has been stored for two days.

In the EP feed treatment group, the degradation that occurred during storage affected the shrinkage of the sample, as drip loss is generally a continuous process involving the transfer of water from myofibrils in muscle tissue to the extracellular space. Thus, differences in drip loss are associated with changes in protein function in the muscle after it is cut. Protein denaturation reduces the ability of the fish muscle to retain water, resulting in an open structure that increases the amount of drip released from the muscle. The amount of drip lost from the fish tissue can indicate a lack of protein functionality in the binding water. This leads to poor water-holding capacity and facilitates the loss of sarcoplasmic proteins from muscle cells into the extracellular space [40,42,43].

3.5. Histology

In terms of economic value as a food product, the texture of the muscle is important in many types of fish, including yellowtail, when the muscle is eaten raw, such as in sushi and sashimi [18,36]. Microstructural analysis not only allows for the assessment of quality but also enables the determination of the quantitative ratio of the components in a meat product.

Histological examination is one of the methods that enables the constituent components of meat products to be identified, including the degree of autolytic processes in muscle fiber [44]. The histology specimens were stained using the Gomori trichrome special staining method under the same time and conditions between the groups. The blue area in photomicrographs indicates the collagen levels (Figure 4B); cross-sectional areas of yellowtail muscle were observed under a microscope at a magnification of 40× (Figure 6). The muscle condition in the EP feed group indicated unhealthy muscle (atrophy) compared with that seen in the silkrose-BM diet group (Figure 6).

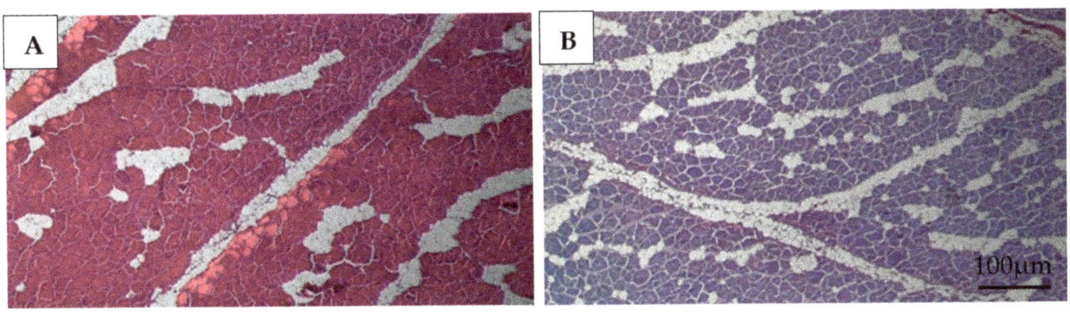

Figure 6. Cross-sectional areas of yellowtail muscle under a microscope at a magnification of 40×. The sections shown are from the perimysium area of yellowtail muscle from the (**A**) EP feed group and (**B**) the silkrose-BM group.

The histological analysis was used to compare the area of the muscle fibers in the perimysium area of the two groups. The area of the muscle fiber is shown in Figure 7.

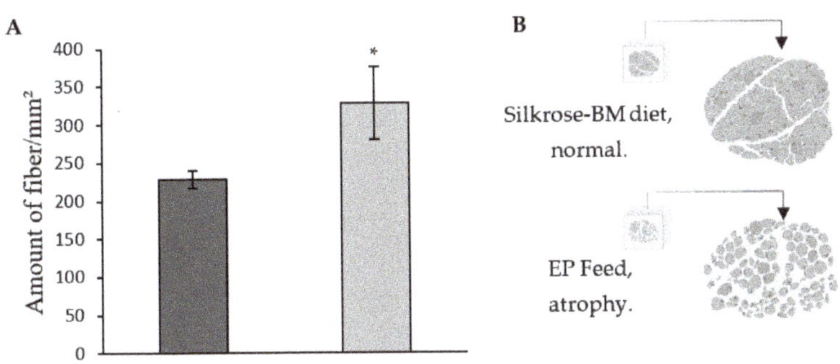

Figure 7. (**A**) The area of yellowtail muscle fiber (mm^2) in cultured yellowtail in the EP feed group (dark grey) and the silkrose-BM diet group (pale grey). The asterisk indicates a significant difference between the groups (* $p < 0.05$). (**B**) Images of muscle condition in the two groups.

A significant difference was observed, with a mean muscle fiber area of 228.63/mm^2 and 327.48/mm^2 for the EP feed and silkrose-BM diet groups, respectively. The greater area observed in the silkrose-BM diet group indicates that the fish possess greater numbers of fiber cells binding the muscle tissue structure, making the muscle in these fish more compact than that of the fish in the EP feed group. This finding suggests that feeding fish with silkrose-BM can enhance the functionality of proteins in their meat, thereby optimizing their role in muscle development and contributing to improving the texture profile as a quality indicator.

3.6. Gene Expression Analysis (qRT-PCR)

Gene expression analysis was conducted using qRT-PCR to detect differences in gene expression between the groups to provide additional data to support our overall analysis. The results are shown in Figure 8.

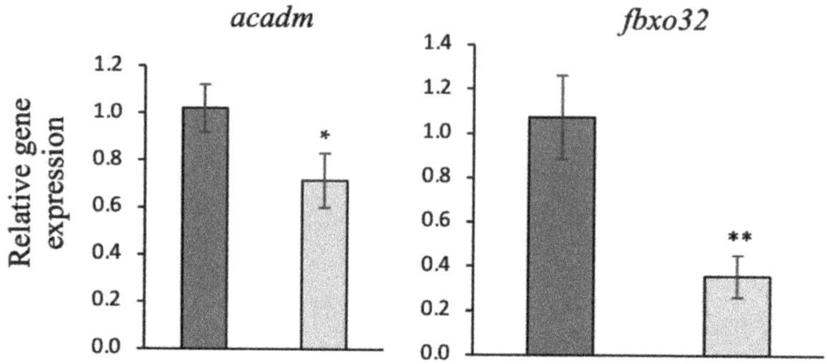

Figure 8. Relative expression of genes in yellowtail muscle of the EP feed (dark grey) and silkrose-BM diet (pale grey) groups. Data are presented as mean ± SE (n = 5). Asterisks indicate a statistically significant difference between the groups (* $p < 0.05$), (** $p < 0.01$).

Figure 8 shows that significant differences in the expression of the *acadm and fbxo32* genes were observed between the groups. The average relative gene expression values for *acadm* were 1.02 and 0.71 and for *fbxo32* were 1.07 and 0.36 for the EP feed and silkrose-BM diet groups, respectively. Both genes showed higher relative expression in the EP feed treatment group compared with the silkrose-BM diet group.

An increase in the capacity of the *F-box protein atrogin-1/fbxo32* to polyubiquitinate proteins via increases in the expression of ubiquitination machinery is largely responsible for faster rates of protein degradation associated with atrophy [45,46]. Silkrose-BM has pro-inflammatory properties, stimulating the activation of the innate immune system in crustaceans and mammals (as seen in mice RAW264 cells) through its function as a pathogen-associated molecular pattern (PAMP) that is recognized by pathogen recognition receptors (PRRs) [8]. Polysaccharides containing PAMPs affect various types of cells in various animal species, including both vertebrates and invertebrates, as has been reported for lipopolysaccharide (LPS) and β-glucans. According to Tacchi et al. [47], the expression of *atrogin-1 (fbxo32)* increases following stimulation by LPS, indicating a catabolic process occurs during the inflammatory response. It is possible that feeding silkrose-BM to yellowtail stimulates the formation of cells due to maximal protein catabolism, which then halts, leading to a decrease in the expression of the *fbxo32* gene.

The elevated relative expression of these genes in the EP feed treatment group suggested increased lipid metabolism (*acadm*) [48], and protein degradation (*fbxo32*) in fish meat [47], which contributed to the reduced overall meat quality compared with the silkrose-BM diet group. The higher expression of the *fbxo32* gene in the EP feed group indicated a greater extent of muscle protein breakdown that potentially causes fish muscle to atrophy. This finding aligns with the observed differences in muscle condition, further emphasizing the potential advantages of the silkrose-BM diet in preserving muscle quality vis the minimization of muscle atrophy (Figure 7).

4. Conclusions

A silkrose-BM diet containing bioactive polysaccharides may enhance the quality of fish meat by optimizing protein performance and inhibiting the lipid metabolism that occurs in fish meat during storage. Feeding a silkrose-BM diet to yellowtail had an impact on changes in fish meat color during storage, the total collagen content, the quantity of drip loss, the histological structure, and gene expression. Optimizing protein performance in fish muscle may help to minimize protein denaturation, thereby preserving the functionality of proteins in muscle tissue by promoting the production of muscle cells, as evidenced by the higher muscle density observed in the silkrose-BM diet treatment group. However, the molecular pathways underlying this mechanism remain unclear. Further research is needed to determine relevant RNA sequences and to elucidate the mechanisms and molecular pathways underlying the changes in fish meat quality induced by silkrose-BM.

Supplementary Materials: The following supporting information can be downloaded at: https://www.mdpi.com/article/10.3390/fishes10030130/s1, Table S1. Gene-specific primers for qRT-PCR of yellowtail muscle (EP feed and silkrose-BM diet treatment groups) [49–55]; Table S2. Differences in the color changes (ΔE*ab) in the dark flesh of yellowtail fish during storage at 4 °C, observed every 24 h. Table S3. Modified Gomori Thricrome staining method of yellowtail muscle.

Author Contributions: Conceptualization, Y.O., D.H., C.M. and T.M.; methodology, A.A., H.N., D.H., Y.O., M.F.Z.A., C.M., I.B.B.S. and T.M.; formal analysis, A.A., H.N. and D.H.; investigation, D.H., Y.O. and T.M.; writing—original draft, A.A., M.F.Z.A.; writing—review and editing, C.M., I.B.B.S., M.F.Z.A. and T.M.; Supervision, T.M.; Project administration, T.M.; Funding acquisition, T.M. All authors have read and agreed to the published version of the manuscript.

Funding: This work was supported by the JSPS Kakenhi (20K20574).

Institutional Review Board Statement: Animal experiments were carried out following the guidelines of Ehime University. The study protocol was accepted by the Institutional Animal Care and Use Committee (IACUC) of Ehime University (Permit Number: 08K2-1). Approval date 4 December 2019.

Data Availability Statement: The data presented in this study are available on request from the corresponding author, upon reasonable request.

Acknowledgments: We are grateful to the staff of the Laboratory of Fish Reproductive Physiology, the Graduate School of Agriculture, Ehime University for their assistance in this study. This work was supported by JSPS Kakenhi (20K20574).

Conflicts of Interest: The authors declare no conflicts of interest. The funders had no role in the design of the study; in the collection, analyses, or interpretation of data; in the writing of the manuscript; or in the decision to publish the results.

References

1. Cheng, J.H.; Sun, D.W.; Han, Z.; Zeng, X.A. Texture and Structure Measurements and Analyses for Evaluation of Fish and Fillet Freshness Quality: A Review. *Compr. Rev. Food Sci. Food Saf.* **2014**, *13*, 52–61. [CrossRef] [PubMed]
2. Watanabe, M.; Suzuki, T.; Ichimaida, K.; Hattori, T.; Ueda, R. Do Consumers Actually Sense that Sashimi Made from Frozen-Thawed Fish Tastes Worse than Non-Frozen One? *Int. J. Refrig.* **2020**, *111*, 94–102. [CrossRef]
3. Kiessling, A.; Ruohonen, K.; Bjornevik, M. Muscle Fibre Growth and Quality in Fish. *Arch. Tierz.* **2006**, *49*, 137–146.
4. Ali, M.F.Z.; Nakahara, S.; Otsu, Y.; Ido, A.; Miura, C.; Miura, T. Effects of Functional Polysaccharide from Silkworm as an Immunostimulant on Transcriptional Profiling and Disease Resistance in Fish. *J. Insects Food Feed* **2021**, *8*, 1221–1233. [CrossRef]
5. Miura, T.; Nishikawa, M.; Otsu, Y.; Ali, M.F.Z.; Hashizume, A.; Miura, C. The Effects of Silkworm-Derived Polysaccharide (Silkrose) on Ectoparasitic Infestations in Yellowtail (*Seriola quinqueradiata*) and White Trevally (*Pseudocaranx dentex*). *Fishes* **2022**, *7*, 14. [CrossRef]

6. Zou, X.; Liu, M.; Li, X.; Pan, F.; Wu, X.; Fang, X.; Zhou, F.; Peng, W.; Tian, W. Applications of insect nutrition resources in animal production. *J. Agric. Food Res.* **2024**, *15*, 100966. [CrossRef]
7. Nishiguchi, H.; Suryadi, I.B.B.; Ali, M.F.Z.; Miura, C.; Miura, T. Dietary Black Soldier Fly (*Hermetia illucens*)-Dipterose-BSF-Enhanced Zebrafish Innate Immunity Gene Expression and Resistance to *Edwardsiella tarda* Infection. *Insects* **2024**, *326*, 326. [CrossRef]
8. Ali, M.F.Z.; Yasin, I.A.; Ohta, T.; Hashizume, A.; Ido, A.; Takahashi, T.; Miura, C.; Miura, T. The Silkrose of *Bombyx mori* Effectively Prevents Vibriosis in Penaeid Prawns via The Activation of Innate Immunity. *Sci. Rep.* **2018**, *8*, 8836. [CrossRef]
9. Bharath, K.B.; Chandrashekaer, S.; Pallavi. Silkworm Pupae: A Goldmine Waste. *J. Enomology Zool. Stud.* **2024**, *12*, 231–236. [CrossRef]
10. Ohta, T.; Ido, A.; Kusano, K.; Miura, C.; Miura, T. A Novel Polysaccharide in Insects Activities the Innate Immune System in Mouse Macrophage RAW264 Cells. *PLoS ONE* **2014**, *9*, e114823. [CrossRef]
11. Ohta, T.; Kusano, K.; Miura, C.; Miura, T. Silkrose: A Novel Acidic Polysaccharide from The Silkmoth that can Stimulate The Innate Immune Response. *Carbohydr. Polym.* **2016**, *136*, 995–1001. [CrossRef]
12. Yamada, S.; Sugahara, K.; Ozbeck, S. Evolution of Glycosaminoglycans, Comparative Biochemical Study. *Commun. Intergrative Biol.* **2011**, *4*, 150–158. [CrossRef] [PubMed]
13. Sahu, B.; Sharma, D.D.; Jayakumar, G.C.; Madhan, B.; Zameer, F. A Review on an Imperative By-Product: Glycosaminoglycans-A Holistic Approach. *Carbohydr. Polym. Technol. Appl.* **2023**, *5*, 100275. [CrossRef]
14. Ali, M.F.Z.; Kameda, K.; Kondo, F.; Iwai, T.; Kurniawan, R.A.; Ohta, T.; Ido, A.; Takahashi, T.; Miura, C.; Miura, T. Effects of Dietary Silkrose of *Antheraea yamamai* on Gene Expression Profiling and Disease Resistance to *Edwardsiella tarda* in Japanese Medaka (*Oryzias latipes*). *Fish Shellfish Immunol.* **2021**, *114*, 207–217. [CrossRef]
15. Climate and Average Weather Year Round in Uwajima, Japan. 2024. Available online: https://weatherspark.com/y/143164/Average-Weather-in-Uwajima-Japan-Year-Round (accessed on 25 December 2024).
16. Mabuchi, R.; Adachi, M.; Kikutani, H.; Tanimoto, S. Discriminants Analysis of Muscle Tissue Type in Yellowtail *Seriola quinqueradiata* Muscle Based on Metabolic Component Profiles. *Food Sci. Technol. Res.* **2018**, *24*, 883–891. [CrossRef]
17. Ando, M.; Mok, W.J.; Maeda, Y.; Miki, R.; Fukuda, T.; Tsukamasa, Y. Quality Assessment of Yellowtail (*Seriola quinqueradiata*) Meat Cultured in an Offshore Floating Flexible Facility. *Food Sci. Nutr.* **2022**, *10*, 3024–3033. [CrossRef]
18. Shioya, I.; Takemura, S.; Ishizuka, R.; Yamaguchi, T. Variations in the Proximate Composition of Mucle in Cultured Yellowtail *Seriola quinqueradiata* at Different Anatomical Portions. *Food Sci. Technol.* **2012**, *78*, 725–733. [CrossRef]
19. Date, K.; Yamamoto, Y. Seasonal Variations with Growth in Nutritive Components in Meat of Cultured Yellowtail *Seriola quinqueradiata*. *Nippon Suisan Gakkaishi* **1988**, *54*, 1041–1047. [CrossRef]
20. Saeki, K.; Kumagai, H. Muscle Components of Wild and Cultured yellowtail. *Shokuhin Eiseigaku Zasshi* **1979**, *20*, 101–105. [CrossRef]
21. Shimizu, Y.; Tada, M.; Endo, K. Seasonal Variations in Chemical Constituents of Yellowtail Muscle 1. Water, Lipid and Crude Protein. *Nippon Suisan Gakkaishi* **1973**, *39*, 993–999. [CrossRef]
22. Association of Official Analytical Chemist (AOAC). *Official Methods of Analysis*, 18th ed.; Association of Official Analytical Chemist Inc.: Rockville, MD, USA, 2005.
23. Livak, K.J.; Schmittgen, T.D. Analysis of Relative Gene Expression Data Using real-time quantitative PCR and the 2 (-Delta Delta C(T)) Method. *Methods* **2001**, *25*, 402–408. [CrossRef] [PubMed]
24. Ochiai, A.; Nabeshima, H.; Umeda, S.; Hasegawa, I. Gonad Maturation and Dynamics of Crude Lipid Content in The Body of Yellowtail During the Spawning Season. *Nippon Suisan Gakkaishi* **1980**, *46*, 407–412. [CrossRef]
25. Langer, R.O.S.; Simoes, G.S.; Soares, A.L.; Oba, A.; Rossa, A.; Shimokomaki, M.; Ida, E.I. Broiler Transportation Conditions in A Brazilian Commercial Line and The Accurence of Breast PSE (Pale, Soft, Exudative) Meat and DFD-Like (Dark, Firm, Dry) Meat. *Braz. Arch. Biol. Technol.* **2010**, *53*, 1161–1167. [CrossRef]
26. Zhang, Z.Y.; Jia, G.Q.; Zhang, Y.; Lei, J.; Ren, L.; Feng, D.Y. Effects of Constant and Cyclic Heat Stress on Muscle Metabolism and Meat Quality of Broiler Breast Fillet and Thigh Meat. *Poult. Sci.* **2012**, *91*, 2931–2937. [CrossRef]
27. Gong, H.; Wang, B.; Jiang, H.; Li, W.; Zhang, Y.; Ji, C.; Lin, X.; Zhang, S. Quality Assessment and Microbial Community Succession of Finely Chopped Common Carp (*Cyprinus Carpio*) Under Low-Temperature Storage: Implication for Nutritional Composition and Spoilage Indicators. *Food Control* **2025**, 171. [CrossRef]
28. Wu, H.; Tatiyaborwortham, N.; Hajimohammad, M.; Decker, E.A.; Richards, M.P.; Undeland, I. Model Systems for Studying Lipid Oxidation Associated with Muscle Foods: Methods, Challenges, and Prospects. *Crit. Rev. Food Sci. Nutr.* **2024**, *64*, 153–171. [CrossRef]
29. Faustman, C.; Sun, Q.; Mancini, R.; Suman, S.P. Myoglobin and Lipid Oxidation Interactions: Mechanistic Bases and Control. *Meat Sci.* **2010**, *86*, 86–94. [CrossRef]

30. Bayram, I.; Decker, E.A. Underlying Mechanisms of Synergistic Antioxidant Iteractions During Lipid Oxidation. *Trends Food Sci. Technol.* **2023**, *133*, 219–230. [CrossRef]
31. Cao, J.; Yan, H.; Ye, B.; Shen, Y.; Liu, L. Effects of Mailard Reaction Products on Myoglobin-Mediated Lipid Oxidation During Refrigerated Storage of Carp. *Food Chem.* **2024**, *434*, 137465. [CrossRef]
32. Choe, E.; Min, D.B. Mechanisms of Antioxidants in the Oxidation of Foods. *Compr. Rev. Food Sci. Food Saf.* **2009**, *8*, 345–358. [CrossRef]
33. Pazos, M.; Gonzalez, M.J.; Gallardo, J.M.; Torres, J.L.; Medina, I. Preservation of The Endogeneous Antioxidant System of Fish Muscle by Grape Polyphenols During Frozen Storage. *Eur. Food Res. Technol.* **2005**, *220*, 514–519. [CrossRef]
34. Chatham, J.C.; Zhang, J.; Wende, A.R. Role of O-linked N-acetylglucosamine (O-glcNAc) Protein Modification in Cellular (patho)physiology. *Am. Physiol. Soc. Physiol. Rev.* **2021**, *101*, 427–493. [CrossRef]
35. Adhikari, E.; Liu, Q.; Burton, C.; Mockabee-Macias, A.; Lester, D.K. L-fucose, A Sugary Regulator of Antitumor Immunity and Immunotherapies. *Mol. Carciogenesis* **2022**, *61*, 439–453. [CrossRef]
36. Listrat, A.; Lebret, B.; Louveau, I.; Astruc, T.; Bonnet, M.; Lefaucheur, L.; Picard, B.; Bugeon, J. How Muscle Structure and Composition Influence Meat and Flesh Quality. *Sci. World J.* **2016**, *2016*, 3182746. [CrossRef]
37. Sato, K.; Yoshinaka, R.; Sato, M.; Shimizu, Y. Collagen content in the muscle of fishes in association with their swimming movement and meat texture. *Bull. Jpn. Soc. Sci. Fish.* **1986**, *52*, 1595–1600. [CrossRef]
38. Puri, S.; Coulson-Thomas, Y.M.; Gesteira, T.F.; Coulson-Thomas, V.J. Distribution and Function of Glycosaminoglycans and Proteoglycans in the Development Homeostasis and Pathology of the Ocular Surface. *Front. Cell Dev. Biol.* **2020**, *8*, 731. [CrossRef]
39. Brooke, L.F.; Melrose, J. The Glycosaminoglycan Side Chains and Modular Core Proteins of Heparan Sulphate Proteoglycans and the Varied Ways They Provide Tissue Protection by Regulating Physiological Processes and Cellular Behaviour. *Int. J. Mol. Sci.* **2023**, *24*, 14101. [CrossRef]
40. Rini, S.R.; Sugiharo; Mahfudz, L.D. The Effect of Different Rearing Temperatures on the Physical Quality of Broiler Meat in the Finisher Period (in Bahasa Indonesia). *J. Sains Peternak. Indones.* **2019**, *14*, 387–395. [CrossRef]
41. Wada, S.; Ogawa, Y. High Pressure Effects on Fish Lipid Degradation: Myoglobin Change and Water Holding Capacity. *Prog. Biotechnol.* **1996**, *13*, 351–356. [CrossRef]
42. Nkukwana, T.T.; Muchenje, V.; Masika, P.J.; Pieterse, E.; Hoffman, L.C.; Dzama, K. Proximate composition and variation in colour, drip loss and pH of breast meat from broilers supplemented with *Moringa oleifera* leaf meal over time. *Anim. Prod. Sci.* **2015**, *56*, 1208–1216. [CrossRef]
43. Kale, L.D.; Eikevik, T.M.; Rustad, T.; Nordtvedt, T.S. Changes in Water Holding Capacity and Driploss of Atlantic Salmon (*Salmo salar*) Muscle during Superchilled Storage. *LWT-Food Sci. Technol.* **2014**, *55*, 528–535. [CrossRef]
44. Mkrtchyan, M.; Safronov, D.; Tokarev, A.; Makavchik, S.; Orlova, D. Determination The Quality of Meat, Manufactured Meat, and Meat Products Via The Histological Method. *Int. Trans. J. Eng. Manag. Appl. Sci. Technol.* **2020**, *11*, 1115.
45. Bodine, S.C.; Latres, E.; Baumhueter, S.; Lai, V.K.; Nunez, L.; Clarke, B.A.; Poueymirou, W.T.; Panaro, F.J.; Na, E.; Dharmarajan, K.; et al. Identification of Ubiquitin Ligases Required for Skeletal Muscle Atrophy. *Science* **2001**, *294*, 1704–1708. [CrossRef] [PubMed]
46. Nakashima, K.; Komatsu, T.; Yamazaki, M.; Abe, H. Effects of Fasting and Refeeding on Expression of Proteolytic-Related Genes in Skeletal Muscle of Chicks. *J. Nutr. Sci. Vitaminol.* **2005**, *51*, 248–253. [CrossRef]
47. Tacchi, L.; Bickerdike, R.; Scombes, C.J.; Pooley, N.J.; Urquhart, K.L.; Collet, B.; Martin, S.A.M. Ubiquitin E3 ligase atrogin-1 (Fbox-32) in Atlantic salmon (*Salmo salar*): Sequence Analysis, Genomic Structure and Modulation of Expression. *Comp. Biochem. Physiol. Part B* **2010**, *157*, 364–373. [CrossRef]
48. Zhang, J.; Xu, X.; Zhu, H.; Wang, Y.; Hou, Y.; Liu, Y. Dietary Fish Oil Supplementation Alters Liver Gene Expressions to Protect Against LPS-Induced Liver Injury in Weanling Piglets. *Innate Immun.* **2019**, *25*, 60–72. [CrossRef]
49. Ranjithkumar, K.; Sudhan, C.; Roy, U.; Rao, M.B. Ribosomal RNA and Their Applications in Species Identification. *J. Aquac. Trop.* **2018**, *33*, 91–99. [CrossRef]
50. He, L.; Shi, X.; Han, K.; Huang, W.; Chen, D.; Lian, Z.; Ruan, S. Molecular Characterization of Adenosine Monophosphate Deaminase 1 and the Correlation Analysis between its mRNA Expression Levels and Inosine Monophosphate Content in Large Yellow Croaker (*Larimichthys crocea*). *Comp. Biochem. Physiol. Part B Biochem. Mol. Biol.* **2024**, *272*. [CrossRef]
51. Pinel, K.; Heraud, C.; Morin, G.; Dias, K.; Marcé, A.; Beauclair, L.; Fontagné-Dicharry, S.; Masagounder, K.; Klünemann, M.; Seiliez, I.; et al. Are the Main Methionine Sources Equivalent? A Focus on DL-Methionine and DL-Methionine Hydroxy Analog Reveals Differences on Rainbow Trout Hepatic Cell Lines Functions. *I. J. Mol. Sci.* **2022**, *23*, 2935. [CrossRef]
52. Persson, B.; Hedlund, J.; Jornvall, H. Medium- and short-chain Dehydrogenase/Reductase Gene and Protein Families: The MDR Superfamily. *Cell. Mol. Life Sci.* **2008**, *65*, 3879–3894. [CrossRef]
53. Mason, E.; Hindmarch, C.C.T.; Dunham-Snary, K.J. Medium-chain Acyl-CoA Dehydrogenase Deficiency: Pathogenesis, Diagnosis, and Treatment. *Endocrinol. Diabetes Metab.* **2022**, *6*, 1–11. [CrossRef]

54. Xu, C.; Wu, G.; Zohar, Y.; Du, S. Analysis of Myostatin Gene Structure, Expression and Function in Zebrafish. *J. Exp. Biol.* **2003**, *206*, 4067–4079. [CrossRef] [PubMed]
55. Cleveland, B.M.; Evenhuis, J.P. Molecular Characterization of Atrogin-1/F-box protein-32 (FBXO32) and F-box protein-25 (FBXO25) in Rainbow Trout (Oncorhynchus mykiss): Expression Across Tissues in Response to Feed Deprivation. *Comp. Biochem. Physiol. Part B Biochem. Mol. Biol.* **2010**, *157*, 248–257. [CrossRef] [PubMed]

fishes

MDPI

Article

Effects of Blending *Curcuma longa* Hydrolate and *Lactobacillus plantarum* on the Growth and Health of Nile Tilapia

Adolfo Jatobá [1,*], Marina de Oliveira Pereira [1], Gabriel Fernandes Alves Jesus [2], Scheila Anelise Pereira Dutra [3], José Luiz Pedreira Mouriño [4], Marco Shizuo Owatari [5] and Delano Dias Schleder [1]

1 Aquaculture Laboratory, Federal Institute Catarinense, Campus Araquari, BR-280 km 27, Araquari 89245-000, SC, Brazil; marinapereira104@gmail.com (M.d.O.P.); delano.schleder@ifc.edu.br (D.D.S.)
2 Biohall Pesquisa & Inovação, Rua Pedro Rangel, 165, Itajaí 88304-430, SC, Brazil; marinapereira@gmail.com
3 Marine Shrimp Laboratory—LCM, Aquaculture Department, Federal University of Santa Catarina (CCA/UFSC), Beco dos Coroas, 503, Florianópolis 88061-600, SC, Brazil; scheila.anelise@ufsc.br
4 Aquatic Organisms Health Laboratory—AQUOS, Aquaculture Department, Federal University of Santa Catarina (UFSC), Rodovia Admar Gonzaga 1346, Florianópolis 88040-900, SC, Brazil; jose.mourino@ufsc.br
5 Laboratory of Algae Culture—LCA, Aquaculture Department, Federal University of Santa Catarina (UFSC) Beco dos Coroas, 503, Barra da Lagoa, Florianópolis 88061-600, SC, Brazil; marco.owatari@ufsc.br
* Correspondence: adolfo.bezerra@ifc.edu.br; Tel.: +55-(47)-3803-7271

Abstract: In the present study, *Curcuma longa* (CL) hydrolate and the probiotic *Lactobacillus plantarum* (LP) were provided as dietary supplements to Nile tilapia. One hundred ninety-two juvenile tilapias (2.25 ± 0.14 g and 4.5 ± 0.10 cm) were distributed in sixteen experimental units, and four experimental groups were established: CL [CUR]: fish fed a diet supplemented with CL hydrolate at 2.5%; probiotic [PRO]: a diet supplemented with LP; LP + CL [COMB]: diet supplemented with the LP strain cultivated in media supplemented with 2.5% CL hydrolate; and control [CTRL]: diet without supplementation. After 70 days, the final average weight was significantly greater in the PRO group (33.26 ± 1.12 g) than in the CTRL and CUR groups, whereas the specific growth rate was significantly greater in the PRO and COMB groups than in the CTRL and CUR groups. Feed conversion decreased significantly in the PRO group (1.03 ± 0.11). Dietary supplementation did not change the body composition of tilapia. Leukocyte and lymphocyte counts were greater in the PRO treatment than in the CTRL group. Compared with those in the CTRL group, total serum protein was significantly increased in the PRO group. Immunoglobulins were higher in the COMB and PRO groups. In the experimental challenge, all the fish in the treated groups presented lower cumulative mortality rates. The combination of LP and CL improved the growth parameters of Nile tilapia.

Keywords: aquaculture; curcumin; hematology; immunity; phytotherapy; tilapia fish

Key Contribution: *Lactobacillus plantarum* + Curcuma longa hydrolate reduced feed conversion. *L. plantarum* + *C. longa* hydrolate improved immunoglobulin levels. *L. plantarum* increased total plasma protein.

1. Introduction

Phytotherapeutics, such as essential oils and their bioactive compounds, are products obtained from plants and are used for curative, prophylactic, or adjuvant nutraceutical purposes in the prevention and treatment of diseases [1–3]. Historically, plants and their derivatives have been used to treat different diseases in humans and animals [4] and control bacterial and parasitic infections in the aquaculture industry [1,5].

In aquaculture, many products derived from plants have already been identified as authentic therapeutic agents for use in fish farming owing to their medicinal, antioxidant, antifungal, antibacterial, and immunostimulant properties [6,7]. Among them, *Lippia origanoides* stands out as an antiparasitic agent for *Colossoma macropomum* [5]. *Psidium guajava* leaf extract minimized the toxicity effects of cypermethrin on Nile tilapia (*Oreochromis niloticus*) [8]. *Punica granatum* peel mitigated the adverse effects of silver nanoparticles on Nile tilapia welfare [9]. *Aloe vera* induced immune gene expression in *Oncorhynchus mykiss* [10]. *Allium sativum* was found to prevent monogenean infection [11]. *Zingiber officinale* improved the immune response of the Asian sea bass *Lates calcarifer* against *Vibrio harveyi* [12] and enhanced the growth and immune response of fish against *Streptococcus agalactiae* [13]. *Ocimum gratissimum* increases the growth and immune response of *Lophiosilurus alexandri* and serves as an anesthetic agent [14]. *Silybum marianum* acted as a hepatic protector and immunomodulator in Nile tilapia during *S. agalactiae* infection [15], and *Curcuma longa* hydrolate was found to improve the immune response and survival of fish [16,17].

The hydrolate is a byproduct of the distillation process in the extraction of essential oils. During this process, some volatile and bioactive elements of the original plant not present in the oil are captured by the water vapor that passes through the vegetable mass. This occurs because some plant compounds are hydrophilic and only have an affinity for water [16].

Curcuma longa L. (Zingiberaceae), known in Brazil as *açafrão-da-terra* [18], is a perennial plant widely cultivated in tropical regions of Asia [4]. It is a plant with a small stem and large and oblong leaves with oval, elliptical, or piriform rhizomes that are occasionally branched and brownish-yellow. Turmeric is the component from which the predominant yellow color of dry *C. longa* powder originates. Indian Turmeric or Haridra, Yellow Ginger, Kyoo, or Ukon are just a few names. Turmeric is commonly used in Indian culture as a home therapy in the treatment of human diseases [4,19,20].

Nile tilapia was the top four species produced in 2022, with around 5.3 million tonnes [21]. In tilapia farming, bacterial diseases, one of the main weaknesses of the production sector, which are caused, for example, by the bacteria of the genus *Streptococcus* [22] and *Aeromonas* [23], are preventable with the use of probiotics [24]. In addition, probiotics, which are beneficial microorganisms that cohabit with the diverse microbial community of the host's digestive tract [25], can also improve fish appetite, leading to greater growth performance and optimizing feed conversion [25–27]. Among probiotic microorganisms, *Lactobacillus plantarum* has already demonstrated beneficial effects on the immunocompetence and growth parameters of Nile tilapia [28–30].

According to Yue et al. [31], polysaccharides contained in bioactive plant ingredients can be degraded into absorbable metabolites by bacteria that colonize the intestine. Consequently, the intestinal microbiota plays a vital role in the catabolism of these polysaccharides. In addition, polysaccharides can modulate the composition and activities of the intestinal microbiota, promoting the growth of beneficial bacteria and inhibiting the colonization of pathogenic bacteria.

In recent decades, the growing individualized use of phytotherapeutics and probiotics, driven by scientific discovery, has become an alternative to the chemical products and antibiotics frequently used in aquaculture, aiming at sustainability in fish production [24,27,32–34]. However, little is known about the crosstalk between phytotherapeutics and probiotics when combined with dietary supplementation and the possible effects of such cross-actions on Nile tilapia.

Therefore, this study aimed to investigate the effects of the phytobiotic *Curcuma longa* hydrolate and the probiotic *Lactobacillus plantarum*, which are provided as dietary supplements to Nile tilapia.

2. Materials and Methods

This research was approved by the National Council for the Control of Animal Experimentation (CONCEA) under protocol number 263/2018. All the fish used in the biological analyses were previously anesthetized with eugenol (50 mg L^{-1}) and euthanized by spinal cord transection. Following the guidelines of the Ethics Committee on Animal Use (CEUA) and aiming at the principles of "reduction, replacement and refinement" in the use of animals in scientific and academic centers, the present research used a minimal, but adequate, number of animals per experimental procedure to minimize animal discomfort.

2.1. In Vitro Assay to Test the Dosage of Curcuma longa Hydrolate in Probiotic Culture Medium

C. longa hydrolate was included in the final growth phase of *L. plantarum*. For this purpose, Man Rogosa Sharpe (MRS) culture media were prepared with five different concentrations of vehicle (Table 1). *L. plantarum* was sown in an MRS tube and incubated at 35 °C for 24 h. After incubation, the number of colony-forming units per mL (CFU mL^{-1}) of each concentration was estimated through a serial dilution (factor of 1:10) in MRS agar culture medium, which was performed in triplicate.

Table 1. Vehicle volume used to cultivate *Lactobacillus plantarum*.

Treatments	Vehicle (mL)	
	***Curcuma longa* Hydrolate (%)**	**Distilled Water (%)**
100	10.00	0.00
50	5.00	5.00
25	2.50	7.50
12.5	1.25	8.75
0	0.00	10.00

The inhibitory capacity against *Pseudomonas* sp. *Streptococcus* sp., *Aeromonas hydrophila*, and *A. veronii* was verified. For this purpose, *L. plantarum* was cultivated on MRS agar, as shown in Table 1. The different treatments were evaluated for their inhibitory effects via the disc agar diffusion technique, as described by Jatobá et al. [28]. The bacteria grown in different vehicles were seeded in a Petri dish containing MRS agar and incubated at 35 °C for 24 h. After this period, discs 1 cm in diameter were removed from these MRS plates with *L. plantarum*, which were superimposed on the surface of the same newly sown pathogenic bacteria and incubated at 30 °C for 24 h. Antagonistic activity was expressed as the diameter (mm) of the zone of inhibition around the agar discs superimposed in triplicate.

2.2. In Vivo Experimental Design

In total, 192 Nile tilapia with initial average weight and length of 2.25 ± 0.14 g and 4.5 ± 0.10 cm were distributed in 16 experimental units (12 fish per unit), with a useful volume of 800 L, which were arranged in a recirculation aquaculture system equipped with physical and biological filtration, ultraviolet sterilizers, constant aeration, and central water heating.

The experiment was a completely randomized design with four groups in quadruplicate. The groups were fish fed (1) *Curcuma longa* (CUR), a diet supplemented with 2.5% *C. longa* hydrolate according to Pereira et al. (2020); (2) probiotic (PRO), a diet supplemented with *L. plantarum*; (3) *L. plantarum* + *C. longa* (COMB), a diet supplemented with *L. plantarum* cultivated with 2.5% *C. longa* hydrolate as a vehicle; or (4) control (CTRL), a diet without supplementation.

The experimental diet (Supplementary File S1) was sprayed with PRO, CUR, COMB, or distilled water (CTRL). PRO and COMB were previously grown in MRS culture media at a rate of 100 mL kg $feed^{-1}$. Subsequently, it was stored in a drying oven at 24 to 35 °C. Next, the feed was dried in an oven for 24 h at 35 °C. For the CUR group, the commercial diet was supplemented with 2.5% *C. longa* hydrolate, according to the protocols established by Pereira et al. [16], and the control feed was sprayed with distilled water (100 mL kg $feed^{-1}$). To quantify the lactic acid bacteria content in the feed, serial dilutions (1:10) were carried out. The final count of lactic acid bacteria in the PRO and COMB groups was greater than 1×10^8 CFU g $feed^{-1}$.

The fish were fed four times a day with 6% biomass, and weekly biometrics were performed to monitor growth and adjust feeding management. After 70 days, at the end of the growth performance period, the final weight, specific growth rate (SGR) = ($_{\text{ln}}$Final weight − $_{\text{ln}}$Initial weight)/(cultivation days) × 100), feed conversion rate (FCR) = (consumed feed/weight gain), survival rate (%) = [(number of animals at the end/number of animals at the beginning) × 100], yield (g m^{-3}) = [(final biomass − initial biomass)/(experimental unit volume)], hepatosomatic index (HIS %) = [(liver weight/fish body weight) × 100], and viscerosomatic index (VSI %) = [(viscera weight/fish body weight) × 100] were evaluated. In addition, the carcasses of the fish were used to determine moisture, ash, lipid, and crude protein contents following the methodologies of the AOAC [35].

Water quality variables were measured throughout the experimental period and included dissolved oxygen at 5.12 ± 0.76 mg L^{-1} and a temperature of 29.01 ± 2.59 °C (YSI PRO20 Oximeter), ammonia at 0.11 ± 0.09 mg L^{-1}, nitrite below 0.10 mg L^{-1}, nitrate below 0.20 ± 0.12 mg L^{-1}, alkalinity at 101.2 mg $CaCO_2$ L^{-1} and pH at 6.86 ± 0.21. Dissolved oxygen and temperature were checked twice daily, and the other variables were checked once weekly. The experimental units were siphoned twice weekly to remove excess organic solids accumulated at the bottom of the boxes.

2.3. Hematoimmunological Analysis

For hematological analysis, blood from 20 fish from each treatment (five fish per experimental unit) was collected via puncture of the caudal vessel with insulin syringes coated with the anticoagulant solution HEMSTB (EDTA K2 15 g dL^{-1}) for the preparation of blood smears in duplicate, and the following hematological analyses were performed: determination of hematocrit via the standard microhematocrit method [36]; glucose (G-TECH free®, Accumed-Glicomed, Duque de Caxias, RJ, Brazil); total hemocyte count via a Neubauer hemocytometer; and hemoglobin concentration [37]. Hematimetric absolute rates of the mean corpuscular volume (MCV), mean corpuscular hemoglobin (MCH), and mean corpuscular hemoglobin concentration (MCHC) were also obtained [38]. Blood smear slides were stained with May Grunwald–Giemsa stain (MGG) [39] for total and differential leukocyte counts according to Ishikawa et al. [40].

For immunological analyses, blood was collected without anticoagulant for serum withdrawal. The blood was kept in Eppendorf tubes for 1 h at 25 °C until coagulation and subsequently centrifuged at 1400× *g* for 15 min for serum withdrawal and stored at −20 °C for immunological analyses. The serum antimicrobial titer was determined against *A. hydrophila* in a 96-well flat-bottom microplate [41]. The inoculum of *A. hydrophila* was grown in brain heart infusion (BHI) broth at 28 °C for 24 h and prepared in poor broth (PB) culture medium at a concentration of 1 × 10^5 CFU mL^{-1}. The serum was diluted 1:3 in PB in the first well (50 µL of plasma:100 µL of PB solution) and was serially diluted by a factor of 1:2 for the other wells. For the positive and white controls, saline solution was diluted in PB in a manner similar to that used for the serum. Finally, 20 µL of *A. hydrophila* was added to the wells with diluted serum and a positive control. The microplates were incubated at 28 °C for 24 h. The growth of the microorganisms was determined in a microplate reader at a wavelength of 550 nm. The plasma antimicrobial titer was the reciprocal of the last dilution that showed bactericidal activity, i.e., evidence of total inhibition of microbial growth.

Blood serum protein was measured via a commercial total protein kit. The total immunoglobulin (Ig) concentration was measured according to the method described by Amar et al. [42]. This protocol involves mixing 50 µL of the serum with 50 µL of 12% polyethylene glycol solution (PEG, 10,000 MW, Sigma Chemical, St. Louis, MO, USA) and incubating the mixture at 25 °C for two hours to precipitate the immunoglobulin molecules. The immunoglobulin precipitate was removed by centrifugation (5000× *g* at 4 °C for 10 min), and the supernatant was removed. The amount of total protein was also measured with a commercial kit, in which bovine serum albumin was used to construct a standard curve. The concentration of total immunoglobulin was expressed in mg mL^{-1} and calculated as GI (mg mL^{-1}) = total whey protein − PEG-treated protein.

2.4. Experimental Infection of Aeromonas

After 70 days of growth (fish weight above 28.03 ± 3.44 g), two experimental infections were carried out against *A. hydrophila* and *A. veronii*, two pathogens that cause large losses in tilapia farming [23]. Both infections were completely randomized designs with the same four groups described above in triplicate.

For the *A. hydrophila challenge*, 60 juvenile Nile tilapia were distributed in 15 polyethylene boxes (40 L) equipped with a biological filter and thermostat. These parameters were measured daily: dissolved oxygen above 5.52 mg L^{-1} and a temperature of 28.54 ± 0.23 °C (YSI PRO20 Oximeter); ammonia, 0.04 NH_3 mg L^{-1}; and pH, 6.93 ± 0.06.

For the challenge, 1×10^7 CFU mL^{-1}, approximately 1.5×10^5 CFU g^{-1}, was administered by intraperitoneal (i.p.) injection such that each fish received 100 µL of the bacterial mixture. The inoculum cultivated in BHI for 24 h at 30 °C was centrifuged for 30 min at 1800× *g*. The supernatant was discarded, and the sediment was resuspended in a 0.65% sterile saline solution at a concentration of 1×10^7 CFU mL^{-1}.

For infection of *A. veronii*, another batch of 60 fish was used, and the process described above was repeated. All the parameters were measured daily: dissolved oxygen above 5.64 mg L^{-1} and a temperature of 28.23 ± 0.19 °C (YSI PRO20 Oximeter); ammonia below 0.04 NH_3 mg L^{-1}; and a pH of 6.94 ± 0.07. After both challenges, the fish were observed for nine days to obtain the mortality rate, and the percentage of deaths and cumulative mortality were calculated. Mortality was checked every 8 h, and dead fish were immediately removed from the tank.

2.5. Statistical Analysis

The data were subjected to the Kolmogorov-Smirnov test to determine normality, and Levene's test was used to verify homoscedasticity. The data obtained met the prerequisites of normality and homoscedasticity, and one-way ANOVA and significant differences among treatments were analyzed via the Student-Newman-Keuls (SNK) test. The vehicle concentrations were subjected to second-order polynomial regression. All tests were conducted at a 5% level of significance.

3. Results

3.1. In Vitro Assay

The in vitro results revealed greater growth of *L. plantarum* when cultivated in 2.5% *C. longa* hydrolate as a vehicle (Figure 1). The inhibition halo was significantly ($p < 0.05$) reduced against *S. agalactiae* when *L. plantarum* was cultivated in culture media supplemented with 10.0% *C. longa* hydrolate (Table 2).

Table 2. Inhibition halo (mm) of *Lactobacillus plantarum* cultivated under different concentrations of *Curcuma longa* hydrolate as a vehicle against pathogenic bacteria.

Treatments	*Aeromonas hydrophila*	*A. veronii*	*Pseudomonas* sp.	*Streptococcus agalactiae*
0	10.5 ± 1.00 ^{a}	12.0 ± 0.50 ^{a}	17.0 ± 0.00 ^{a}	15.0 ± 0.00 ^{a}
12.5	11.5 ± 1.50 ^{a}	12.5 ± 1.00 ^{a}	17.0 ± 1.00 ^{a}	15.0 ± 1.00 ^{a}
25	11.0 ± 1.00 ^{a}	11.5 ± 1.50 ^{a}	17.0 ± 1.00 ^{a}	15.0 ± 0.00 ^{a}
50	10.5 ± 1.00 ^{a}	13.0 ± 0.50 ^{a}	16.0 ± 0.00 ^{a}	14.0 ± 1.00 ^{a}
100	11.0 ± 0.50 ^{a}	12.5 ± 0.00 ^{a}	15.0 ± 1.00 ^{a}	12.0 ± 0.00 ^{b}

Different letters indicate significant differences in ANOVA and SNK ($p > 0.05$).

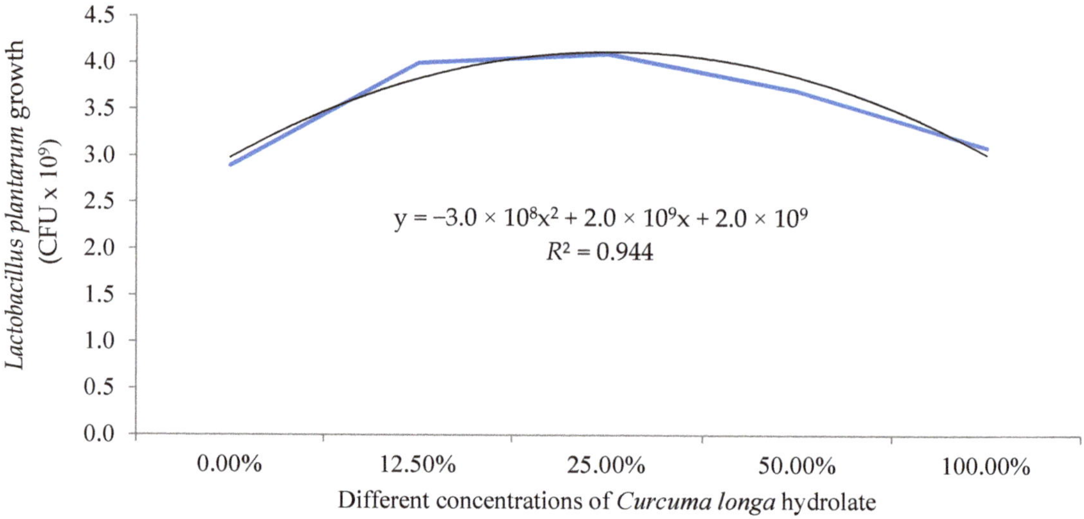

Figure 1. Growth curve of *Lactobacillus plantarum* cultivated with different concentrations of *C. longa* hydrolate as a vehicle. The vehicle concentrations were subjected to second-order polynomial regression.

3.2. *In Vivo Assay*

3.2.1. Growth Performance

The average final weight of the fish in the PRO treatment group was greater ($p < 0.05$) (33.26 ± 1.12 g), whereas those in the CUR and COMB groups did not differ from those in the control group. The specific growth rate was significantly ($p < 0.05$) greater in the fish from the PRO (1.64 ± 0.01% day^{-1}) and COMB (1.61 ± 0.04% day^{-1}) groups than in those from the CUR and CTRL groups. Feed conversion was significantly ($p < 0.05$) lower for the fish in the PRO and COMB groups. The lowest survival rate (83.33 ± 5.56%) and lowest yield (345.42 ± 24.44 g m^{-3}) were observed in the CTRL group (Table 3). The body indices did not significantly differ ($p > 0.05$) among the treatments (Table 4).

Table 3. Growth performance (mean ± standard deviation) for Nile tilapia (*O. niloticus*) reared in RAS, fed a diet supplemented with *Curcuma longa* hydrolate and probiotic *Lactobacillus plantarum* in combination or alone.

Growth Indexes	CTRL	CUR	PRO	COMB
Final average weight (g)	28.03 ± 3.44 ^{b}	30.81 ± 1.02 ^{b}	33.26 ± 1.12 ^{a}	32.62 ± 1.96 ^{ab}
SGR (% day^{-1})	1.51 ± 0.10 ^{b}	1.58 ± 0.02 ^{b}	1.64 ± 0.01 ^{a}	1.61 ± 0.04 ^{a}
FCR	1.50 ± 0.19 ^{a}	1.22 ±0.11 ^{ab}	1.03 ±0.11 ^{c}	1.11 ±0.08 ^{bc}
Survival (%)	83.33 ± 5.56 ^{b}	100.00 ± 0.00 ^{a}	97.22 ± 3.70 ^{a}	97.22 ± 3.70 ^{a}
Yield (g m^{-3})	345.42 ± 24.44 ^{b}	462.15 ± 15.23 ^{a}	484.35 ± 10.83 ^{a}	474.51 ± 11.34 ^{a}

CTRL = control group; CUR = *C. longa* hydrolate in an isolated way; PRO = *L. plantarum* probiotic in an isolated way; and COMB = *C. longa* hydrolate and probiotic *L. plantarum* in a combined way. SGR = specific growth rate; FCR = feed conversion rate. Different letters in the lines indicate statistical differences in ANOVA and SNK ($p < 0.05$). The fish had an initial average weight and length of 2.5 ± 0.10 g and 4.9 ± 0.10 cm.

Table 4. Whole-body composition analysis of Nile tilapia (*O. niloticus*) reared in RAS, fed a diet supplemented with *Curcuma longa* hydrolate and probiotic *Lactobacillus plantarum* in a combined or isolated way.

Body Indexes	Treatments			
	CTRL	CUR	PRO	COMB
Hepatosomatic Index (%)	2.25 ± 0.27	2.57 ± 0.25	2.25 ± 0.27	2.57 ± 0.25
Viscerosomatic Index (%)	6.36 ± 0.56	6.32 ± 0.26	6.36 ± 0.56	6.32 ± 0.26
Moisture (%)	68.33 ± 0.94	67.29 ± 1.41	67.21 ± 0.76	66.47 ± 2.02
Crude Protein (%)	16.82 ± 1.84	16.06 ± 2.32	16.44 ± 1.23	17.00 ± 1.93
Crude lipids (%)	11.03 ± 1.49	12.64 ± 2.32	10.99 ± 1.78	12.11 ± 2.04
Ashes (%)	2.62 ± 0.12	2.63 ± 0.09	2.65 ± 0.09	2.65 ± 0.08

CTRL = control group; CUR = *C. longa* hydrolate in an isolated way; PRO = *L. plantarum* probiotic in an isolated way; and COMB = *C. longa* hydrolate and probiotic *L. plantarum* in a combined way.

3.2.2. Hematoimmunological Analyses

According to the hemogram, the number of total leukocytes and lymphocytes in the fish from the PRO group was significantly greater ($p < 0.05$) than that in the CTRL group; however, this increase was equal to that in the other treatment groups. The other hematological indices did not significantly differ ($p > 0.05$) among the treatments (Table 5).

Table 5. Hematological parameters (mean ± standard deviation) for Nile tilapia (*O. niloticus*) reared in RAS, fed a diet supplemented with *Curcuma longa* hydrolate and probiotic *Lactobacillus plantarum* in a combined or isolated way.

Total and Differential Count	CTRL	CUR	PRO	COMB
Thrombocytes ($\times 10^3$ μL^{-1})	21.08 ± 8.12	20.90 ± 7.32	36.74 ± 6.26	24.34 ± 10.97
Total leukocytes ($\times 10^3$ μL^{-1})	25.47 ± 6.76 ^{b}	30.73 ± 9.58 ^{ab}	40.92 ± 6.04 ^{a}	33.56 ± 10.29 ^{ab}
Lymphocytes ($\times 10^3$ μL^{-1})	20.73 ± 7.04 ^{b}	25.29 ± 8.76 ^{ab}	36.95 ± 7.12 ^{a}	28.62 ± 11.09 ^{ab}
Monocytes ($\times 10^3$ μL^{-1})	2.23 ± 1.85	3.45 ± 0.97	2.54 ± 0.42	2.93 ± 1.60
Neutrophils ($\times 10^3$ μL^{-1})	2.51 ± 0.64	1.99 ± 0.51	1.43 ± 0.57	2.01 ± 0.69
Hematimetric Parameters				
Erythrocytes ($\times 10^6$ μL^{-1})	1.96 ± 0.64	1.97 ± 0.45	1.52 ± 0.43	1.58 ± 0.42
Hematocrit (%)	26.89 ± 3.19	25.89 ± 4.71	27.29 ± 3.60	26.00 ± 1.44
Hemoglobin (g dL^{-1})	6.51 ± 0.65	7.20 ± 0.68	6.76 ± 0.85	7.87 ± 0.98
MCV (fL)	154.81 ± 58.86	137.77 ± 38.36	192.81 ± 54.28	183.41 ± 75.36
MCH (10^{-5} pg)	40.57 ± 5.26	36.17 ± 6.98	37.19 ± 5.64	35.25 ± 5.27
MCHC (g dL^{-1})	24.44 ± 3.07	28.99 ± 7.29	27.56 ± 4.53	29.66 ± 3.78
Glucose (mg d L^{-1})	30.31 ± 4.71	31.29 ± 4.25	30.54 ± 5.06	34.37 ± 7.02
MIC (%)	5.0	5.0	5.33	5.66
Agglutination titer	5.01 ± 1.0	4.66 ± 0.58	5.00 ± 0.34	5.38 ± 0.34

CTRL = control group; CUR = *C. longa* hydrolate in an isolated way; PRO = *L. plantarum* probiotic in an isolated way; and COMB = *C. longa* hydrolate and probiotic *L. plantarum* in a combined way. MCV = mean corpuscular volume; MCH = mean corpuscular hemoglobin; MCHC = mean corpuscular hemoglobin concentration. MIC = minimum inhibitory concentration. Different letters in the lines indicate statistical differences in ANOVA and SNK ($p < 0.05$).

Compared with the fish in the CTRL group (48.12 mg mL^{-1}), the fish in the PRO treatment group presented a significant increase ($p < 0.05$) in total plasma protein (73.55 mg mL^{-1}), whereas the CUR (54.69 mg mL^{-1}) and COMB (64.76 mg mL^{-1}) groups did not differ from the PRO or the CTRL groups. However, significantly greater ($p < 0.05$) immunoglobulin concentrations were detected in the fish in the COMB (40.34 mg mL^{-1}) and PRO (39.64 mg mL^{-1}) groups than in the CUR (31.74 mg mL^{-1}) and CTRL (28.09 mg mL^{-1}) groups (Figure 2). The MIC and agglutination titer did not significantly differ ($p > 0.05$) among the treatments (Table 5).

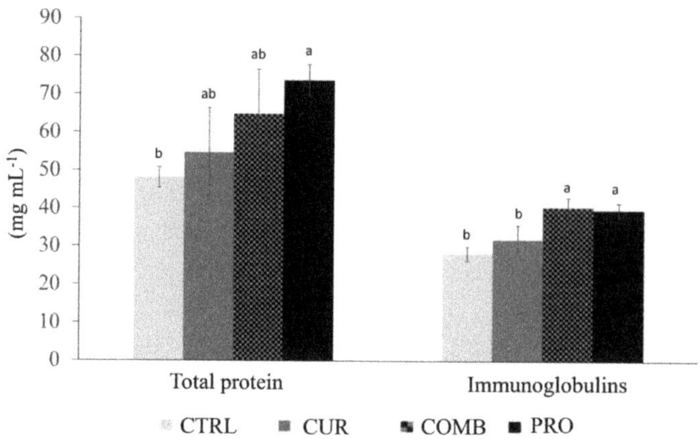

Figure 2. Total serum protein and immunoglobulins (mean ± standard deviation) of Nile tilapia (*O. niloticus*) reared in a recirculating aquaculture system and fed a diet supplemented with *Curcuma longa* hydrolate and the probiotic *Lactobacillus plantarum*, either in combination or alone. CTRL = control group; CUR = *C. longa* hydrolate in an isolated way; PRO = *L. plantarum* probiotic in an isolated way; and COMB = *C. longa* hydrolate and probiotic *L. plantarum* in a combined way. Different letters indicate significant differences according to ANOVA and SNK ($p < 0.05$).

3.2.3. Experimental Challenge

Among the evaluated groups, the PRO treatment group presented the lowest cumulative mortality, followed by the CUR and COMB groups, with the highest cumulative mortality recorded in the CTRL group (Figure 3).

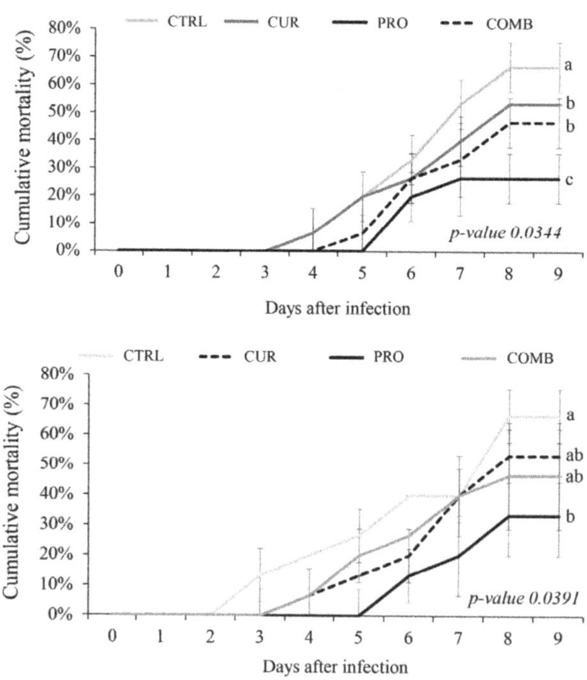

Figure 3. Cumulative mortality of Nile tilapia (*O. niloticus*) fed a diet supplemented with *Curcuma longa* hydrolate and the probiotic *Lactobacillus plantarum*, either in combination or alone: experimental

challenge with *Aeromonas hydrophila* (above); experimental challenge with *A. veronii* (below). Different letters indicate significant differences according to ANOVA and SNK ($p < 0.05$).

4. Discussion

Functional diets are increasingly gaining prominence in the aquafeed industry. For example, *Nigella sativa* exhibited hepatoprotective and antioxidant effects against silver nanoparticle toxicity [43], while papaya extract, *Carica papaya*, alleviated endocrine disruption and hepatic DNA damage and counteracted the subchronic toxicity of chlorpyrifos [44] in African catfish (*Clarias gariepinus*). This can be attributed to their ability to promote health and prevent disease owing to the wide range of scientifically tested bioactive substances with proven efficacy for inclusion in fish diets [3,7,34,44].

Among supplemental feeds, phytotherapeutics [1] and probiotics [45] stand out because of their antioxidant, anti-inflammatory and immunomodulatory properties [24,32]. In the present study, a novel dietary plan for Nile tilapia was formulated based on diets enriched with bioactive components derived from *Curcuma longa*, highlighting the combined dietary effects of *L. plantarum* + *C. longa*.

C. longa is a plant that has bioactive compounds such as curcumin, a powerful antioxidant that fights free radicals, safeguarding cells and tissues from oxidative harm. It also possesses anti-inflammatory characteristics that can lower cytokine production [18,20]. Similarly, *L. plantarum* is recognized for its antioxidant properties and other advantageous substances that promote intestinal health in aquacultured fish [24].

Phytotherapeutics can influence the microbiota by helping to strengthen and balance the intestinal flora [31]. Some compounds found in plants, such as the polyphenols ar-turmerone, α-tumerone, and α-curcumene, present in *C. longa* hydrolate [17], have shown beneficial effects [16–20,46–48].

Some herbal products are natural sources of prebiotics, which are substances used as substrates for probiotic bacteria, favoring their multiplication in the intestine to increase the diversity of beneficial bacteria and protect against the growth of harmful bacteria [1]. In addition, a clean, pathogen-free, and more stable water environment, such as the recirculation system used in this research, is a fundamental strategy for improving fish health.

Su et al. [49] reported the potential for combined therapy with herbal medicines and probiotics in microbial infections caused by *Staphylococcus aureus* and *Streptococcus pyogenes*. At that time, researchers reported that the combination of green tea extract and probiotics significantly reduced the viable count of both pathogens at 4 h and at 24 h, suggesting a possible synergistic effect between the two products. Indeed, in the present study, the addition of 25% and 100% *C. longa* hydrolate to the culture medium promoted the greatest in vitro growth of the probiotic bacterium *L. plantarum* and inhibited the in vitro growth of the pathogenic bacterium *S. agalactiae*, corroborating the studies cited above.

The effects of the probiotic *L. plantarum* have already been reported in other studies since it has been widely tested and used as a probiotic for Nile tilapia in different life stages and culture environments, promoting growth and improving innate immunity [28–30,50]. In the present study, the highest daily specific growth rates were observed in the fish from the groups in which the probiotic was present.

In clear water RAS, the specific growth rate (SGR) of Nile tilapia (initial weight of 0.85 ± 0.08 g) fed with feed containing 2.5% *C. longa* hydrolate was equal to that of the control group, reaching $2.21 \pm 0.05\%$ day^{-1} after 40 days of cultivation [17]. These results corroborate the findings of the present research, in which the SGRs of the CUR and CTRL groups remained the same.

Combining probiotics and phytotherapeutics results in a synergistic effect in which probiotics balance the intestinal microbiota, whereas phytotherapeutics provide bioactive compounds that promote the regulation of physiological functions that, in turn, improve the absorption of nutrients, improve growth, and strengthen the immune system [51].

Indeed, in the present study, improvements were observed in the zootechnical performance of Nile tilapia fed *L. plantarum* cultivated in a culture medium supplemented with *C. longa* (*L. plantarum* + *C. longa*). However, the effects were the same as those for the PRO group, and when used alone, *C. longa* hydrolate presented indices similar to those of the control group. The present study demonstrated that the probiotic *L. plantarum* was responsible for the improved growth performance of tilapia, given that probiotic supplementation alone increased the average final weight, SGR, and feed conversion compared with those of the CUR group.

Despite high variability in the microbiota of fish, lactic acid bacteria (LAB), such as *L. plantarum*, are sometimes abundant in the gut, especially in freshwater fish. Fortunately, most LAB are harmless, and some strains have been reported to have beneficial health effects by boosting the immune system in fish, along with pathogen antagonism, another key feature of candidate bacteria used as probiotics [52–54]. These characteristics were confirmed in the present study when we performed hematoimmunological analyses. A significant increase in plasma protein levels was observed in the blood of the fish in the groups treated with either probiotics or *C. longa* hydrolate alone, as well as *L. plantarum* + *C. longa*, indicating that both probiotics and phytotherapeutics caused these changes. Indeed, phytotherapeutics may also be used to improve hematoimmunological variables in fish. Abdel-Tawwab and Hamed [8] reported that dietary guava (*Psidium guajava*) leaf extract significantly enhanced hematobiochemical and immunity variables in cypermethrin-intoxicated Nile tilapia, significantly minimizing the negative impacts caused by synthetic pyrethroid insecticide. The increase in plasma protein in the present study is an important indicator of the health status of the fish, revealing that tilapia in these groups were better prepared to trigger an immune response to pathogens.

On the other hand, Santos et al. [45] reported significant reductions in immunoglobulin concentrations in Nile tilapia after an experimental challenge, demonstrating that infectious processes promote considerable reductions in this blood component. In contrast, Moraes et al. [27] did not observe changes in immunoglobulins in juvenile tilapia when the fish were fed diets supplemented with probiotic additives containing *Bacillus* spp. or *Saccharomyces cerevisiae* microencapsulated. In the present study, the level of immunoglobulins was altered in the PRO and *L. plantarum* + *C. longa* groups. However, in the group of fish receiving *C. longa* alone, the immunoglobulin concentration was the same as that in the control group, suggesting that the probiotic caused changes in immunoglobulin levels.

Leukocytes are fundamental cells for the immune defense of animals [38]. Even without coming into contact with the host's blood system, lactic acid bacteria can interact directly with immune cells through specialized cells of the intestinal epithelium, such as M cells, and induce the activation and multiplication of the cells responsible for defense [55]. In the present study, dietary supplementation with the probiotic *L. plantarum* may have induced greater production or release of total leukocytes and circulating lymphocytes, which may suggest a nonspecific response to infections [28].

Recently, Pereira et al. [16] tested different concentrations of *Curcuma longa* hydrolate (0.0%, 2.5%, 7.5%, and 10.0%) in the diet of juvenile Nile tilapia (7.45 $\pm$ 0.38 g). Pereira et al. [17] demonstrated that dietary supplementation with 2.5% *C. longa* hydrolate improved the immune response and survival and promoted beneficial changes in the intestinal microbial community of tilapia fingerlings (0.85 $\pm$ 0.08 g). Furthermore, the major compounds of *C. longa* hydrolate are ar-turmerone, α-tumerone, and α-curcumene, representing approximately 93.50% purity. Lee [47] reported that ar-turmerone and arachidonic acid effectively inhibited platelet aggregation induced by collagen (IC50, 14.4 μM and IC50, 43.6 μM, respectively). However, ar-turmerone had no effect on platelet activation factor or thrombin-induced platelet aggregation, suggesting that it could be useful as a lead compound to inhibit platelet aggregation. In the blood count of the present study, we observed that the group that received *C. longa* alone in the diet presented, in absolute number, a smaller number of thrombocytes, but the difference was not significant. Furthermore, the COMB group showed no significant improvement beyond that of either the probiotic

L. plantarum or *C. longa* alone since no possible synergistic effects were observed in the hematology analyses.

Experimental challenges that use the intraperitoneal route for infections can be aggressive, offering few opportunities for fish to recover after exposure to the pathogen and causing severe damage to organs responsible for vital functions, such as the brain, liver, and spleen [56]. When juvenile tilapias were challenged with *A. hydrophila*, Moraes et al. [27] reported that treatments supplemented with probiotics resulted in lower cumulative mortality during the experimental challenge. Similarly, Santos et al. [45] reported lower mortality rates for juvenile tilapia fed the probiotic *Bacillus* spp. or 0.1% benzoic acid alone. However, when *Bacillus* was combined with 0.1% benzoic acid, no synergistic effects were observed during the experimental challenge.

Many isoprenoids and phenolic compounds with antioxidant properties and molecular interactions that affect inflammatory processes have been discovered in *C. longa* [48]. Ferreira et al. [46] verified that a fraction consisting of ar-turmerone isolated from *C. longa* neutralized the hemorrhagic activity present in *Bothrops jararaca* venom and inhibited the proliferation and natural killer activity of human lymphocytes. Devi et al. [57] also verified the in vitro and in vivo efficacy of partially purified herbal extracts against ornamental fish pathogens. At that time, researchers revealed that turmeric extract exhibited significant antibacterial activity and greater antioxidant properties and phenol content, among which the highest presence of ar-turmerone (42.85%) was revealed. The authors also suggested that partially purified turmeric extract improved the hematobiochemical profile and survival rate of *Pseudochromis dilectus* against *Proteus mirabilis*.

These characteristics attributed to ar-turmerone may have influenced the survival of fish in the *C. longa* and *L. plantarum* + *C. longa* groups during experimental challenge with *A. hydrophila*. Since the genus *Aeromonas* causes severe hemorrhage, this condition may have been alleviated in the affected fish, considering the lower mortality rates observed relative to those in the control group. However, the lowest cumulative mortality in both challenges was attributed to the group that received the probiotic alone.

Probiotics are considered effective alternatives to chemotherapy [58,59]. However, the effectiveness of probiotics can be contradictory; therefore, it is important to search for alternatives to improve their effectiveness [51]. According to Yue et al. [31], some polysaccharides present in natural products derived from plants can be catabolized with difficulty, and the intestinal microbiota plays a fundamental role in the degradation of these polysaccharides. Therefore, it is necessary to further investigate the mode of combined action between *L. plantarum* and other plant-derived products that may exhibit synergistic action and thus enhance the efficacy of this probiotic. Nonetheless, it is worth mentioning that *C. longa* can be considered a medicine for the treatment of inflammation, microbial infections, biliary disorders, and liver disorders, including antioxidant, antibacterial, antifungal, hepatoprotective, and immunostimulant activities in animals [19].

5. Conclusions

This study evaluated the effects of supplementing Nile tilapia diets with *L. plantarum* and *C. longa* hydrolate. The results show that both probiotic *L. plantarum* and phytobiotic *C. longa* improved the growth performance of Nile tilapia in a recirculating aquaculture system. While *L. plantarum* alone showed better results, the combination of *L. plantarum* and *C. longa* was also effective. All additives enhanced the innate immune resistance of juvenile tilapia, reducing mortality rates under experimental conditions.

Supplementary Materials: The following supporting information can be downloaded at: https://www.mdpi.com/article/10.3390/fishes9120503/s1, Supplementary file S1 Composition of experimental diet before inclusion of feed additives.

Author Contributions: A.J.—experimental execution, conceptualization, methodology, validation, investigation, data curation, visualization; G.F.A.J.—conceptualization, project administration; S.A.P.D.—hematoimmunological analysis, data curation, microbiological count; M.d.O.P.—experimental execution, investigation, data curation; D.D.S.—experimental execution, investigation, data curation; J.L.P.M.—conceptualization, supervision; M.S.O.—experimental execution, data curation, manuscript review, final writing; supervision. All authors have read and agreed to the published version of the manuscript.

Funding: This research received no external funding.

Institutional Review Board Statement: This research was approved by the National Council for the Control of Animal Experimentation (CONCEA) under protocol number 263/2018, and it was performed in the Aquaculture Laboratory (LAq) of IFc-Araquari—IFCA. All the fish used in the biological analyses were previously anesthetized with eugenol (50 mg L^{-1}) and euthanized by spinal cord transection. Following the guidelines of the Ethics Committee on Animal Use (CEUA) and aiming at the principles of "reduction, replacement and refinement" in the use of animals in scientific and academic centers, the present research used a minimal, but adequate, number of animals per experimental procedure to minimize animal discomfort.

Informed Consent Statement: Not applicable.

Data Availability Statement: The data related to this research are available upon prior request.

Acknowledgments: The authors thank the National Council for Scientific and Technological Development (CNPq) for grant to Adolfo Jatobá (308661/2023-0); and the company Presence–Brazil for providing diets used in this research.

Conflicts of Interest: The authors declare no conflicts of interest.

References

1. Elumalai, P.; Kurian, A.; Lakshmi, S.; Faggio, C.; Esteban, M.A.; Ringø, E. Herbal immunomodulators in aquaculture. *Rev. Fish. Sci. Aquac.* **2020**, *29*, 33–57. [CrossRef]
2. Jeyavani, J.; Sibiya, A.; Sivakamavalli, J.; Divya, M.; Preetham, E.; Vaseeharan, B.; Faggio, C. Phytotherapy and combined nanoformulations as a promising disease management in aquaculture: A review. *Aquac. Int.* **2022**, *30*, 1071–1086. [CrossRef]
3. Abdel-Tawwab, M.; Hamed, H.S.; Monier, M.N.; Amen, R.M. The ameliorative effects of dietary rosemary (*Rosmarinus officinalis*) against growth retardation, oxidative stress, and immunosuppression induced by waterborne lead toxicity in *Nile tilapia* fingerlings. *Ann. Anim. Sci.* **2024**, *24*, 139–149. [CrossRef]
4. Ammon, H.P.; Wahl, M.A. Pharmacology of *Curcuma longa*. *Planta Med.* **1991**, *57*, 1–7. [CrossRef] [PubMed]
5. Soares, B.V.; Cardoso, A.C.F.; Campos, R.R.; Gonçalves, B.B.; Santos, G.G.; Chaves, F.C.M.; Chagas, E.C.; Tavares-Dias, M. Antiparasitic, physiological and histological effects of the essential oil of *Lippia origanoides* (Verbenaceae) in native freshwater fish *Colossoma macropomum*. *Aquaculture* **2017**, *469*, 72–78. [CrossRef]
6. Hamed, H.S.; El-Sayed, Y.S. Antioxidant activities of *Moringa oleifera* leaf extract against pendimethalin-induced oxidative stress and genotoxicity in Nile tilapia, *Oreochromis niloticus* (L.). *Fish Physiol. Biochem.* **2019**, *45*, 71–82. [CrossRef]
7. Abdel-Tawwab, M.; Elabd, H.; Mahboub, H.H.; Assayed, M.E.M.; Hamed, H.S.; Elsayyad, A.; Mohamed, E.M. The protective efficacy of dual dietary rosemary plus cinnamon mix against lead nitrate-induced immune suppression, genotoxicity, and oxidant/antioxidant status in Nile tilapia fingerlings. *Aquac. Int.* **2023**, *32*, 4009–4029. [CrossRef]
8. Abdel-Tawwab, M.; Hamed, H.S. Antagonistic effects of dietary guava (*Psidium guajava*) leaves extract on growth, hemato-biochemical, and immunity response of cypermethrin-intoxicated Nile tilapia, *Oreochromis niloticus*, fingerlings. *Aquaculture* **2020**, *529*, 735668. [CrossRef]
9. Hamed, H.S.; Abdel-Tawwab, M. Dietary pomegranate (*Punica granatum*) peel mitigated the adverse effects of silver nanoparticles on the performance, haemato-biochemical, antioxidant, and immune responses of Nile tilapia fingerlings. *Aquaculture* **2021**, *540*, 736742. [CrossRef]
10. Zanuzzo, F.S.; Urbinati, E.C.; Rise, M.L.; Hall, J.R.; Nash, G.W.; Gamperl, A.K. *Aeromonas salmonicida* induced immune gene expression in *Aloe vera* fed steelhead trout, *Oncorhynchus mykiss* (Walbaum). *Aquaculture* **2015**, *435*, 1–9. [CrossRef]
11. Militz, T.A.; Southgate, P.C.; Carton, A.G.; Hutson, K.S. Dietary supplementation of garlic (*Allium sativum*) to prevent monogenean infection in aquaculture. *Aquaculture* **2013**, *408*, 95–99. [CrossRef]
12. Talpur, A.D.; Ikhwanuddin, M.; Bolong, A.M.A. Nutritional effects of ginger (*Zingiber officinale* Roscoe) on immune response of Asian sea bass, *Lates calcarifer* (Bloch) and disease resistance against *Vibrio harveyi*. *Aquaculture* **2013**, *400*, 46–52. [CrossRef]
13. Brum, A.; Pereira, S.A.; Owatari, M.S.; Chagas, E.C.; Chaves, F.C.M.; Mouriño, J.L.P.; Martins, M.L. Effect of dietary essential oils of clove basil and ginger on Nile tilapia (*Oreochromis niloticus*) following challenge with *Streptococcus agalactiae*. *Aquaculture* **2017**, *468*, 235–243. [CrossRef]
14. Boaventura, T.P.; Souza, C.F.; Ferreira, A.L.; Favero, G.C.; Baldissera, M.D.; Heinzmann, B.M.; Baldisserotto, B.; Luz, R.K. Essential oil of *Ocimum gratissimum* (Linnaeus, 1753) as anesthetic for *Lophiosilurus alexandri*: Induction, recovery, hematology, biochemistry and oxidative stress. *Aquaculture* **2020**, *529*, 735676. [CrossRef]

15. Owatari, M.S.; Jesus GF, A.; Brum, A.; Pereira, S.A.; Lehmann, N.B.; Pereira, U.P.; Martins, M.L.; Mouriño, J.L.P. Sylimarin as hepatic protector and immunomodulator in Nile tilapia during *Streptococcus agalactiae* infection. *Fish Shellfish Immunol.* **2018**, *82*, 565–572. [CrossRef]
16. Pereira, M.O.; Moraes, A.V.; Rodhermel, J.C.B.; Hess, J.D.; Alves, L.; Chaaban, A.; Jatobá, A. Supplementation of *Curcuma longa* hydrolate improves immunomodulatory response in Nile tilapia reared in a recirculation aquaculture system. *Arq. Bras. Med. Vet. Zootec.* **2020**, *72*, 1805–1812. [CrossRef]
17. Pereira, M.O.; Hess, J.D.; Rodhermel, J.C.B.; Farias, D.R.; Schleder, D.D.; Alves, L.; Bertoldi, F.C.; Chaban, A.; Andrade, J.I.A.; Jatobá, A. *Curcuma longa* hydrolate improves Nile tilapia survival in a recirculation rearing system, maintaining the animal homeostasis and modulating the gut microbial community. *An. Acad. Bras. Cienc.* **2021**, *93*, e20210088. [CrossRef]
18. Marchi, J.P.; Tedesco, L.; Melo, A.D.C.; Frasson, A.C.; França, V.F.; Sato, S.W.; Lovato, E.C.W. *Curcuma longa* L., o açafrão da terra, e seus benefícios medicinais. *Arq. Ciênc. Saúde UNIPAR* **2016**, *20*, 189–194. [CrossRef]
19. Omosa, L.K.; Midiwo, J.O.; Kuete, V. Curcuma longa. In *Medicinal Spices and Vegetables from Africa*; Academic Press: Cambridge, MA, USA, 2017; pp. 425–435.
20. Jyotirmayee, B.; Mahalik, G. A review on selected pharmacological activities of *Curcuma longa* L. *Int. J. Food Prop.* **2022**, *25*, 1377–1398. [CrossRef]
21. FAO. *The State of World Fisheries and Aquaculture Blue Transformation in Action*; FAO: Rome, Italy, 2024. [CrossRef]
22. Delannoy, C.M.; Samai, H.; Labrie, L. *Streptococcus agalactiae* serotype IV in farmed tilapia. *Aquaculture* **2021**, *544*, 737033. [CrossRef]
23. Tavares-Dias, M.; Martins, M.L. An overall estimation of losses caused by diseases in the Brazilian fish farms. *J. Parasit. Dis.* **2017**, *41*, 913–918. [CrossRef] [PubMed]
24. Huang, X.; He, H.; Li, Z.; Liu, C.; Jiang, B.; Huang, Y.; Su, Y.; Li, W. Screening and effects of intestinal probiotics on growth performance, gut health, immunity, and disease resistance of Nile tilapia (*Oreochromis niloticus*) against *Streptococcus agalactiae*. *Fish Shellfish Immunol.* **2024**, *151*, 109668. [CrossRef]
25. Gatesoupe, F.J. The use of probiotics in aquaculture. *Aquaculture* **1999**, *180*, 147–165. [CrossRef]
26. Newaj-Fyzul, A.; Al-Harbi, A.H.; Austin, B. Developments in the use of probiotics for disease control in aquaculture. *Aquaculture* **2014**, *431*, 1–11. [CrossRef]
27. Moraes, A.V.; Owatari, M.S.; Silva, E.; Pereira, M.O.; Piola, M.; Ramos, C.; Farias, D.R.; Schleder, D.D.; Jesus, G.F.A.; Jatobá, A. Effects of microencapsulated probiotics-supplemented diet on growth, non-specific immunity, intestinal health and resistance of juvenile Nile tilapia challenged with *Aeromonas hydrophila*. *Anim. Feed Sci. Technol.* **2022**, *287*, 115286. [CrossRef]
28. Jatobá, A.; Vieira, F.D.N.; Buglione Neto, C.; Silva, B.C.; Mouriño, J.L.P.; Jerônimo, G.T.; Dotta, G.; Martins, M.L. Utilização de bactérias ácido-lácticas isoladas do trato intestinal de tilápia-do-nilo como probiótico. *Pesqui. Agropecu. Bras.* **2008**, *43*, 1201–1207. [CrossRef]
29. Jatobá, A.; Vieira, F.D.N.; Buglione-Neto, C.C.; Mouriño, J.L.P.; Silva, B.C.; Seiftter, W.Q.; Andreatta, E.R. Diet supplemented with probiotic for Nile tilapia in polyculture system with marine shrimp. *Fish Physiol. Biochem.* **2011**, *37*, 725–732. [CrossRef]
30. Jatobá, A.; Pereira, M.O.; Vieira, L.M.; Bitencourt, M.; Rodrigues, E.; Fachini, F.A.; Moraes, A.V. Action time and feed frequency of *Lactobacillus plantarum* for Nile tilapia. *Arq. Bras. Med. Vet. Zootec.* **2018**, *70*, 327–332. [CrossRef]
31. Yue, B.; Zong, G.; Tao, R.; Wei, Z.; Lu, Y. Crosstalk between traditional Chinese medicine-derived polysaccharides and the gut microbiota: A new perspective to understand traditional Chinese medicine. *Phytother. Res.* **2022**, *36*, 4125–4138. [CrossRef]
32. Hamed, H.S.; Amen, R.M.; Elelemi, A.H.; Mahboub, H.H.; Elabd, H.; Abdelfattah, A.M.; Abdel Moniem, H.; El-Beltagy, M.A.; Alkafafy, M.; Yassin, E.M.M.; et al. Effect of dietary *Moringa oleifera* leaves nanoparticles on growth performance, physiological, immunological responses, and liver antioxidant biomarkers in Nile tilapia (*Oreochromis niloticus*) against Zinc oxide nanoparticles toxicity. *Fishes* **2022**, *7*, 360. [CrossRef]
33. Elabd, H.; Mahboub, H.H.; Salem, S.M.; Abdelwahab, A.M.; Alwutayd, K.M.; Shaalan, M.; Ismail, S.H.; Abdelfattah, A.M.; Khalid, A.; Mansour, A.T.; et al. Nano-curcumin/chitosan modulates growth, biochemical, immune, and antioxidative profiles, and the expression of related genes in Nile tilapia, *Oreochromis niloticus*. *Fishes* **2023**, *8*, 333. [CrossRef]
34. Goda, M.N.; Shaheen, A.A.; Hamed, H.S. Potential role of dietary parsley and/or parsley nanoparticles against zinc oxide nanoparticles toxicity induced physiological, and histological alterations in Nile tilapia, *Oreochromis niloticus*. *Aquac. Rep.* **2023**, *28*, 101425. [CrossRef]
35. AOAC. *Official Methods of Analysis*, 18th ed.; Association of Official Analytical Chemists: Gaithersburg, MD, USA, 2005.
36. Goldenfarb, P.B.; Bowyer, F.P.; Hall, E.; Brosious, E. Reproducibility in the hematology laboratory: The microhematocrit determination. *Am. J. Clin. Pathol.* **1971**, *56*, 35–39. [CrossRef] [PubMed]
37. Blaxhall, P.C.; Daisley, K.W. Routine haematological methods for use with fish blood. *J. Fish Biol.* **1973**, *5*, 771–781. [CrossRef]
38. De Paiva, M.J.T.R.; Pádua, S.B.; Tavares-Dias, M.; Egami, M.I. *Métodos Para Análise Hematológica em Peixes*; Editora da Universidade Estadual de Maringá-EDUEM: Maringá, Brazil, 2013.
39. Rosenfeld, G. Corante pancrômico para hematologia e citologia clínica. Nova combinação dos componentes do May-Grünwald e do Giemsa num só corante de emprego rápido. *Mem. Inst. Butantan* **1947**, *20*, 329–334.
40. Ishikawa, N.M.; Ranzani-Paiva, M.J.T.; Lombardi, J.V. Total leukocyte counts methods in fish, *Oreochromis niloticus*. *Arch. Vet. Sci.* **2008**, *13*, 54–63. Available online: http://hdl.handle.net/11449/70526 (accessed on 2 December 2024).

41. Silva, B.C.; Martins, M.L.; Jatobá, A.; Buglione Neto, C.C.; Vieira, F.N.; Pereira, G.V.; Jerônimo, G.T.; Seiffert, W.Q.; Mouriño, J.L.P. Hematological and immunological responses of Nile tilapia after polyvalent vaccine administration by different routes. *Pesq. Vet. Bras.* **2009**, *29*, 874–880. [CrossRef]
42. Amar, E.C.; Kiron, V.; Satoh, S.; Okamoto, N.; Watanabe, T. Effects of dietary β carotene on the immune response of rainbow trout *Oncorhynchus mykiss*. *Fish. Sci.* **2000**, *66*, 1068–1075. [CrossRef]
43. Andrawes, N.G.; Nour, A.A.A.; Shaheen, A.A.; Hamed, H.S. Effect of *Nigella sativa* enriched diet on biochemical variables and antioxidant damage caused by silver nanoparticles toxicity in the African catfish, *Clarias gariepinus*. *Egypt. J. Aquat. Biol. Fish.* **2021**, *25*, 63–76. [CrossRef]
44. Mansour, A.T.; Hamed, H.S.; El-Beltagi, H.S.; Mohamed, W.F. Modulatory effect of papaya extract against chlorpyrifos-induced oxidative stress, immune suppression, endocrine disruption, and DNA damage in female *Clarias gariepinus*. *Int. J. Environ. Res. Public Health* **2022**, *19*, 4640. [CrossRef]
45. Santos, G.G.; Libanori, M.C.M.; Pereira, S.A.; Ferrarezi, J.V.S.; Ferreira, M.B.; Soligo, T.A.; Yamashita, E.; Martins, M.L.; Mouriño, J.L.P. Probiotic mix of *Bacillus* spp. and benzoic organic acid as growth promoter against *Streptococcus agalactiae* in Nile tilapia. *Aquaculture* **2023**, *566*, 739212. [CrossRef]
46. Ferreira, L.A.; Henriques, O.B.; Andreoni, A.A.; Vital, G.R.; Campos, M.M.; Habermehl, G.G.; Moraes, V.L. Antivenom and biological effects of ar-turmerone isolated from *Curcuma longa* (*Zingiberaceae*). *Toxicon* **1992**, *30*, 1211–1218. [CrossRef]
47. Lee, H.S. Antiplatelet property of *Curcuma longa* L. rhizome-derived ar-turmerone. *Bioresour. Technol.* **2006**, *97*, 1372–1376. [CrossRef] [PubMed]
48. Alonso-Amelot, M.E. Multitargeted bioactive materials of plants in the *Curcuma* genus and related compounds: Recent advances. *Stud. Nat. Prod. Chem.* **2016**, *47*, 111–200. [CrossRef]
49. Su, P.; Henriksson, A.; Nilsson, C.; Mitchell, H. Synergistic effect of green tea extract and probiotics on the pathogenic bacteria, *Staphylococcus aureus* and *Streptococcus pyogenes*. *World J. Microbiol. Biotechnol.* **2008**, *24*, 1837–1842. [CrossRef]
50. Jatobá, A.; Mouriño, J.L.P. *Lactobacillus plantarum* effect on intestinal tract of *Oreochromis niloticus* fingerlings. *Ciênc. Anim. Bras.* **2015**, *16*, 45–53. [CrossRef]
51. Bomba, A.; Nemcová, R.; Mudroňová, D.; Guba, P. The possibilities of potentiating the efficacy of probiotics. *Trends Food Sci. Technol.* **2002**, *13*, 121–126. [CrossRef]
52. Ringø, E.; Gatesoupe, F.J. Lactic acid bacteria in fish: A review. *Aquaculture* **1998**, *160*, 177–203. [CrossRef]
53. Vázquez, J.A.; González, M.; Murado, M.A. Effects of lactic acid bacteria cultures on pathogenic microbiota from fish. *Aquaculture* **2005**, *245*, 149–161. [CrossRef]
54. Ringø, E.; Hoseinifar, S.H.; Ghosh, K.; Doan, H.V.; Beck, B.R.; Song, S.K. Lactic acid bacteria in finfish—An update. *Front. Microbiol.* **2018**, *9*, 1818. [CrossRef]
55. Gill, H.S. Probiotics to enhance anti-infective defences in the gastrointestinal tract. *Best Pract. Res. Clin. Gastroenterol.* **2003**, *17*, 755–773. [CrossRef] [PubMed]
56. Owatari, M.S.; Cardoso, L.; Pereira, S.A.; Pereira, U.P.; Tachibana, L.; Martins, M.L.; Mouriño, J.L.P. Laboratory-controlled challenges of streptococcosis in Nile tilapia using the oral route (infected-feed) for infection. *Fish Shellfish Immunol.* **2022**, *120*, 295–303. [CrossRef] [PubMed]
57. Devi, K.N.; Dhayanithi, N.B.; Kumar, T.T.A.; Balasundaram, C.; Harikrishnan, R. In vitro and in vivo efficacy of partially purified herbal extracts against bacterial fish pathogens. *Aquaculture* **2016**, *458*, 121–133. [CrossRef]
58. Dawood, M.A.; Koshio, S.; Abdel-Daim, M.M.; Van Doan, H. Probiotic application for sustainable aquaculture. *Rev. Aquac.* **2019**, *11*, 907–924. [CrossRef]
59. Yousuf, S.; Tyagi, A.; Singh, R. Probiotic supplementation as an emerging alternative to chemical therapeutics in finfish aquaculture: A Review. *Probiotics Antimicrob. Proteins* **2022**, *15*, 1151–1168. [CrossRef]

fishes

MDPI

Review

Feed Additives in Aquaculture: Benefits, Risks, and the Need for Robust Regulatory Frameworks

Ekemini Okon [1,2,*], Matthew Iyobhebhe [3], Paul Olatunji [4], Mary Adeleke [2], Nelson Matekwe [5] and Reuben Okocha [2,*]

[1] Faculty of Bioscience Engineering, Ghent University, 9000 Ghent, Belgium
[2] Department of Agriculture, Landmark University, Omu-Aran 251103, Kwara State, Nigeria; themaryadeleke@gmail.com
[3] Department of Biochemistry, Landmark University, Omu-Aran 251103, Kwara State, Nigeria; iyobhebhematthew@gmail.com
[4] Freshwater and Oceanic Sciences Unit of Research, University of Liege, Sart Tilman, 4000 Liège, Belgium; olatunji.o.paul@gmail.com
[5] Department of Agriculture, Environmental Affairs, Rural Development and Land Reform, Veterinary Services, Kimberley 8300, South Africa; nmatekwe@daerl.co.za
* Correspondence: okon.ekemini@lmu.edu.ng or okon.ekeminimoses@gmail.com (E.O.); okocha.reuben@lmu.edu.ng (R.O.)

Abstract: Aquaculture currently supplies over half of the world's fish and relies heavily on feed additives to enhance growth, improve feed efficiency, and increase disease resistance. This review consolidates peer-reviewed studies identified through targeted searches of Web of Science, Scopus, and Google Scholar, focusing on aquaculture feed additives. It emphasizes the principal classes of additives employed in finfish and shrimp cultivation, such as natural immunostimulants (including beta-glucans and nucleotides), probiotics, prebiotics, synbiotics, phytogenics, enzymes, and synthetic nutrients. For each, it summarizes their mechanisms of action, commonly reported inclusion rates, production outcomes, environmental risks, and regulatory statuses. Evidence indicates that immunostimulants enhance innate defences (including phagocyte activity and cytokine responses). Probiotics and prebiotics, on the other hand, regulate gut microbiota and barrier function. Phytogenics offer antimicrobial and antioxidant effects, and synthetic additives provide targeted nutrients or functional compounds that support growth and product quality. Where data are available, typical application ranges include probiotics in the order of 10^4–10^9 CFU per gram, prebiotics at approximately 2–10 g per kilogram, and pigments or antioxidants (such as astaxanthin) at 50–100 mg per kilogram. Significant gaps exist, notably the absence of species-specific dose–response data for tropical and subtropical aquaculture species, as well as limited experimental evidence regarding additive–additive interactions under commercial rearing conditions. Additional gaps include long-term ecological fate, regional regulatory discrepancies, and species-specific dose–response relationships. It is recommended that mechanistic studies employing omics approaches, standardised dose–response trials, and harmonized risk assessments be conducted to promote the sustainable and evidence-based application of feed additives.

Keywords: aquaculture; feed additives; fish nutrition; shrimp nutrition; fish health; immune response

Key Contribution: This paper offers a comprehensive overview of traditional and emerging feed additives in aquaculture, incorporating mechanistic insights, typical application ranges, and measurable indicators of performance and health. It provides a critical assessment of factors influencing the practical adoption of these technologies, including efficacy, cost-effectiveness, availability, regulatory constraints, and environmental considerations. Notable contributions include a comparative table that details conventional additives alongside next-generation approaches (e.g., bacteriophage–probiotic cocktails, postbiotics, and nano-encapsulated bioactives). We also provide a summary of key research gaps and strategic recommendations for sustainable application across various species and production systems.

1. Introduction

Fish and shrimp aquaculture continue to grow rapidly due to increasing global demand [1,2]. According to the FAO SOFIA 2024 report, global fisheries and aquaculture production in 2022 reached a record 223.2 million tonnes, including 130.9 million tonnes from aquaculture. This marks the first time aquaculture has surpassed capture fisheries in aquatic animal production (94.4 million tonnes vs. 91.0 million tonnes) [3].

Furthermore, unlike most land-based agricultural food production systems, over 95% of global aquaculture output occurs in developing countries, with a yearly growth rate of 6.13% [3]. However, the intensification of aquaculture has raised various environmental concerns regarding its potential effects on aquatic life, amid efforts to achieve sustainability [4].

In response to the growing demand for fish and shrimp, there is an increasing interest in using diverse feed additives to enhance the health and performance of these species [5]. Additionally, fish and shrimp production has experienced significant growth due to technological advancements, improved breeding methods, and the availability of high-quality feed [6]. Specifically, various feed sources, including plant and animal products, minerals, and other additives, are utilised to meet the growing demands of this industry. These feed additives have been shown to reduce reliance on natural aquatic products, resulting in higher yields per unit of water surface area or volume at grow-out facilities [7]. This is due to their various component ingredients, which are formulated to meet the nutritional requirements of aquatic animals, support their immune system function, and promote growth [5].

Globally, aquaculture feeds account for approximately 3.6% of the total compound feed volume used in animal production [7]. Despite their relatively modest proportion, the importance of aquaculture feeds should not be overlooked [8]. Feed additives in aquaculture play a crucial role in maintaining health, reducing costs, promoting robust growth, and enhancing yield potential across various aquatic species. These feeds are formulated to provide balanced nutrition, thereby meeting the dietary needs of aquatic animals [5,9].

For this review, feed additives are defined as intentionally added, non-nutritive or functional substances. They are included in formulated aquafeeds or delivered as supplements to modulate growth, feed utilisation, health, product quality, or feed stability. Different feed additives are employed to optimise the intake, digestion, absorption, and transportation of dietary nutrients in aquatic feeds. These feed additives encompass a wide range of substances, including probiotics, prebiotics, immunostimulants, enzymes, and essential nutrients [9]. They are typically incorporated into aquafeed formulations to supplement the nutritional requirements of farmed fish and shrimp [9]. Feed additives can further enhance growth rates and feed conversion efficiency by supporting digestive health and nutrient utilisation, ultimately leading to improved production outcomes for farmers [8].

Furthermore, certain additives have demonstrated immune-stimulatory properties, enhancing the disease resistance of farmed species and reducing the need for antibiotics [10]. These additives are crucial for optimising the digestion and absorption of nutrients, leading to enhanced overall health and performance [9]. Feed additives, such as probiotics, prebiotics, enzymes, and organic acids, work together to support a balanced microbial community within the digestive systems of animals [9,11]. Probiotics introduce beneficial bacteria that assist in digestion and improve nutrient absorption. Moreover, prebiotics nourish these helpful microorganisms by boosting their growth and activity levels [12].

On the other hand, enzymes are essential for breaking down complex feed components, such as fibres and non-starch polysaccharides, into simpler forms that fish and shrimp can more easily absorb [13]. Additionally, organic acids like formic, propionic, and butyric acids help establish an optimal pH environment in the digestive systems of these aquatic animals [14]. This balanced environment not only suppresses the growth of harmful bacteria but also encourages the growth of beneficial bacteria [14].

However, the utilisation of feed additives in aquaculture, encompassing fish and shrimp farming, poses some challenges. Recent research has focused on understanding the impact of feed additives on the aquatic environment [15]. This emphasizes the importance of carefully selecting suitable additives and ensuring the correct dosage to prevent adverse environmental impacts [15,16]. Earlier studies documented that certain additives can have a negative impact on the environment when released into aquatic ecosystems. Therefore, responsible use and sustainable practices are essential, involving the careful consideration of various key factors [8,15,16].

Environmental concerns linked to the use of additives now include documented outbreaks of antibiotic-resistant infections in aquaculture systems. In Southeast Asia, the presence of multi-drug-resistant *E. coli* and other pathogens has been linked to feed containing antibiotics and poor waste management. This is particularly notable in shrimp farms in Thailand and integrated fish–chicken systems in Vietnam [17,18]. These findings highlight the need to adopt sustainable practices, implement rigorous monitoring, and enforce regulatory measures. Taking these steps is essential for tackling these challenges and ensuring the sustainability of aquaculture in the long term.

Although several additive classes have demonstrated benefits for growth and immunity, inconsistencies in comparisons across species and formulations hinder the standardization of dosages. Furthermore, the environmental fate and long-term ecological effects of many additives are not well understood, and regulatory approaches are inconsistent across regions. This limits the safe and uniform use of these additives. Therefore, this review provides a comprehensive overview of the impact of feed additives on the health and growth of fish and shrimp. It focuses on natural immunostimulants, probiotics, prebiotics, synthetic feeds, and phytogenics. The specific objectives are (1) to synthesize experimental and regulatory evidence on the main classes of aquafeed additives and their typical application ranges; (2) to summarize known mechanisms of action and measurable endpoints used to assess additive efficacy and safety; and (3) to identify key knowledge gaps and propose research and regulatory priorities for sustainable use. This study aims to assist fish and shrimp farmers, researchers, policymakers, and stakeholders in the aquaculture industry in making evidence-based decisions for sustainable aquafeed use.

To ensure a comprehensive and balanced synthesis, relevant literature was identified through searches of Scopus, Web of Science, and Google Scholar using combinations of keywords such as "aquaculture," "feed additives," "immunostimulants," "prebiotics," "probiotics," "synbiotics," "phytogenics," and "postbiotics." Studies were included if they were published in English and addressed fish or shrimp species of aquacultural relevance. Priority was given to research that reported measurable outcomes related to growth performance, feed utilisation, immune modulation, physiological responses, or environmental impact.

2. Natural Immunostimulants as Feed Additives in Fish and Shrimp Diets

Natural immunostimulants are compounds that enhance or stimulate the immune system upon exposure to foreign agents [19]. These substances are derived from diverse sources, including plants such as herbs, seaweed, fruits, and vegetables, as well as other naturally occurring compounds [20]. Among the natural immunostimulants, polysaccharides (beta-glucans) and nucleotides have undergone extensive research concerning their application in fish and shrimp to sustain or improve their health [21].

Several polysaccharides, including beta-glucans, mannan oligosaccharides (MOS), chitosan, and alginates, have demonstrated immunomodulatory properties in fish and shrimp [22–26]. Beta-glucans are complex polysaccharides found abundantly in the cell walls of bacteria, fungi, and the extracellular matrix of yeast [21]. They induce immune cells, such as neutrophils and macrophages, thereby activating and stimulating the immune response [27]. Beta-glucan binds to pattern-recognition receptors, including Dectin-1 (CLEC7A), complement receptor 3 (CR3), and Toll-like receptors (TLRs), on immune cells. This receptor engagement triggers intracellular signalling cascades, particularly via Syk kinase, PKCδ, and the CARD9–BCL10–MALT1 (CBM) complex, leading to NF-κB activation, the production of inflammatory cytokines, a respiratory burst, and enhanced phagocytosis [28,29].

Recent research on beta-glucan has revealed promising findings regarding its immunomodulatory effects in fish and shrimp [30]. In channel catfish (*Ictalurus punctatus*), exposure to beta-glucan significantly increased the phagocytic rates of neutrophils and macrophages, accompanied by the upregulation of genes related to phagocytosis and receptor signalling pathways [31]. The most important aspect of beta-glucan is its ability to substantially enhance the activity of phagocytes. This includes granulocytes and monocytes, which then differentiate into macrophages and dendritic cells. These phagocytes play a crucial role in preventing infections by ingesting potentially dangerous pathogens [32]. Stimulating phagocytes via beta-glucans triggers a series of events that improve immune defence mechanisms. This increases the immunity and resilience of aquatic organisms against various pathogenic agents [33].

The profound impact of beta-glucans on modulating immunological responses within cells derived from the intestinal mucosa cannot be overemphasised. This is important considering their significance for maintaining optimal well-being among fish and shrimp species [30]. Moreover, beta-glucan modifies diet activity levels, which can enhance fish resistance to infections transmitted through the primary route (i.e., the gut) [34,35]. Furthermore, beta-glucans have demonstrated favourable immunomodulatory effects in fish and shrimp [35,36]. This suggests that beta-glucans may be a valuable supplement for enhancing resistance to infectious diseases and improving productivity in fish and shrimp farming.

Besides beta-glucans and polysaccharides, nucleotides are essential components of DNA and RNA, playing a crucial role in the biological processes of many animals. Nucleotides substantially enhance the immune response by stimulating the production of antibodies and other essential immune cells in juveniles of the Pacific white shrimp *Litopenaeus vannamei* [37]. The immunomodulatory effects of these compounds on aquatic animals have been recently investigated, yielding promising results for their efficacy when administered at optimal dosages and for suitable species and durations that support immune enhancement without adverse effects [38,39]. Furthermore, the administration of dietary supplements containing high levels of nucleotides augmented the immune response, thereby facilitating the production of essential antibodies, known as immunoglobulins, which are active during viral episodes [39].

Research has further demonstrated that the inclusion of nucleotide supplements can enhance activity levels responsible for pathogen eradication during infection, while concurrently promoting the growth of beneficial bacteria to support balanced microbiota processes [40]. Consequently, maintaining an optimal microbiota balance promotes robust intestinal health, minimises disease susceptibility, and fosters overall well-being [12]. In aquatic species, such immunostimulants may be administered through various methods, including incorporation into feed, immersion, or injection, to attain targeted immunological effects.

Fish and Shrimp Immune Response to Natural Immunostimulants as a Feed Additive

Despite anatomical differences between vertebrate and invertebrate immune systems, dietary polysaccharides stimulate comparable innate immune mechanisms in fish and shrimp. In both groups, supplementation enhances phagocytic activity, respiratory burst, lysozyme and complement activity, and pathogen resistance [22,41–46]. In fish populations, administering beta-glucans through immersion, food inclusion, or injection is effective in enhancing immunological responses [47,48]. A strong immune system in fish protects them against diseases; therefore, boosting their immune response with beta-glucans can enhance overall health and reduce susceptibility to diseases [24,30]. Nevertheless, further research is required to ascertain the optimal dosages and methods of administration. Thus, it is necessary to investigate the long-term effects and interactions with other essential dietary components to evaluate the practical application of beta-glucans as immunostimulants specifically within the contexts of ichthyological and crustacean aquaculture. Additionally, the inclusion of specific polysaccharides in fish and shrimp feed enhances growth performance by improving nutrient uptake and stimulating immune responses [25]. These polysaccharides provide an ecologically sustainable alternative to synthetic options, owing to their low toxicity levels and easy biodegradability, making them a safe choice for use in aquaculture [23].

In shrimp, the immune response to natural immunostimulants is a relatively new research area that has gained significant global attention in recent years [49]. Beta-glucans have also shown promising results for enhancing a shrimp's immune response; however, they exhibit limitations when treating certain infections [34,50]. Despite these limitations, a significant understanding of their complex immune system is crucial for developing robust strategies that enhance disease resistance in shrimp. In addition, plant extracts and compounds derived from algae have demonstrated potential as immunostimulant sources, with studies revealing an increase in antibody production [49,51,52]. For instance, supplementation with algae extract significantly increased antibody cell synthesis in shrimp [53]. These consistent patterns suggest that mechanistic insights from one taxon can often inform application in the other, although dose optimisation remains species-specific. However, efficacy may vary across species and life stages, and long-term effects are not yet fully understood, requiring further research.

3. Probiotics and Prebiotics as Feed Additives in Fish and Shrimp Diets

There is a rapidly growing body of literature indicating the success of probiotics and prebiotics in immunomodulation (innate, cellular, and humoral immune responses). Probiotics are considered to be living microorganisms administered orally for health benefits [54]. They can alter the microflora (by implantation or colonisation) in a specific host's compartment, exerting beneficial health effects on the host [55]. On the other hand, prebiotics are indigestible fibres that enhance the growth of beneficial commensal gut bacteria, resulting in improved host health [56]. These beneficial effects of prebiotics are due to by-products derived from the fermentation of intestinal commensal bacteria [57].

Among the many health benefits attributed to probiotics and prebiotics, the modulation of the immune system is one of the most anticipated benefits (Figure 1), and it can stimulate systemic and local immunity [58] by directly enhancing the innate immune response, including phagocytosis, neutrophil activation, alternative complement system activation, and lysozyme activity [59]. In some cases, they improve growth, such as increasing size and weight gain, and could act as alternative antimicrobial compounds in fish and shrimp [60].

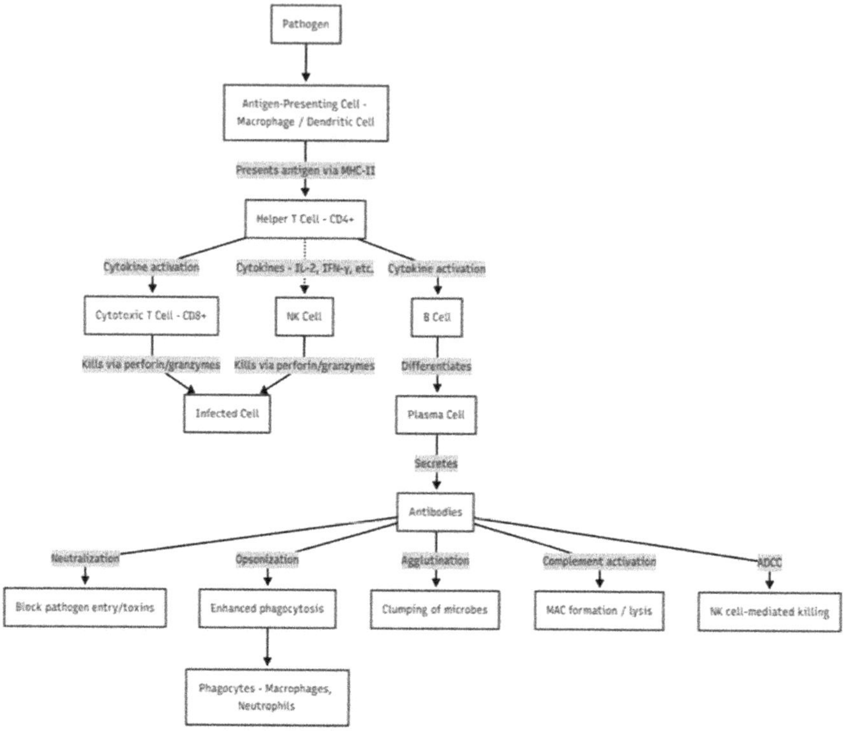

Figure 1. The cellular/humoral immunity interactions in fish and shrimp.

3.1. *Immune Response of Fish to Probiotics and Prebiotics as Feed Additives*

The integration of probiotics into the diets of fish and shrimp has been shown to have a positive impact on both cellular and humoral immunity, thereby significantly enhancing the immune system's functionality [61]. These beneficial microorganisms have the capacity to activate T and B lymphocytes, which are essential elements of the adaptive immune response [62]. The interaction between these two cell types results in an improved defence mechanism against pathogens and supports the development of long-lasting immunity [58,63]. Moreover, probiotics can augment antibody production by stimulating B cells that generate specific antibodies targeting pathogens within a humoral immune response, ultimately reducing the risk of infections in fish [64]. Furthermore, probiotics influence T-cell activity, a vital facet of the immune system. T lymphocytes are responsible for recognising and eliminating infected cells while mediating other immune responses, indicating that probiotics facilitate a more efficacious cellular immune response by modulating various T cell subsets in fish [65].

Probiotics can directly affect the immune cells and indirectly support their function by strengthening the gut barrier [66]. This is important in fish health, as the gut serves as a vital interface between the body's external environment and its immune system. Thus, having a robust intestinal barrier ensures pathogenic microorganisms cannot enter, grow, or survive in the fish's body systems [67]. Primarily, probiotics impact fish immune system through various mechanisms, including enhancing the integrity of the gut barrier by promoting mucus production, strengthening tight junctions between cells, and inhibiting the growth of harmful bacteria [58]. These actions ultimately lead to improved immune responses on both innate and adaptive levels. However, not all strains of probiotics produce identical effects when used as feed additives [68,69].

The dosage of probiotic supplements in fish may influence their efficacy in boosting immunological function [70] (Table 1). Each fish has a distinctive composition of gut microbiota, which may further influence individual responses to probiotic treatments for immune support [71]. Notably, probiotics impact humoral immunity, and certain strains of probiotics have been reported to positively influence immunoglobulin production, specifically secretory IgA [58,72]. These findings emphasize the significance of probiotics as feed additives for fish and shrimp, promoting improved health [73]. Despite promising findings regarding the potential benefits of probiotics for immune enhancement, further comprehensive studies are needed to elucidate their mechanisms and optimise their safe and effective use.

Table 1. Probiotic dosage and immunological effects in fish. Arrow facing upwards indicate an increase.

Probiotic Strain (Fish Species)	Dosage	Observed Immune Effects	Source
Bacillus subtilis E20 (*Epinephelus coioides*)	1×10^4–1×10^8 CFU g^{-1} feed	↑ Lysozyme, phagocytosis, superoxide dismutase (SOD), serum ACP	[74]
B. subtilis + fructooligosaccharides—FOS (*Trachinotus ovatus*)	1.05–5.62 $\times 10^7$ CFU g^{-1} feed + 0.2% or 0.4% FOS	↑ Specific growth rate (SGR), lysozyme, disease resistance, serum ACP	[7]
B. subtilis + *B. licheniformis* (*Oreochromis niloticus*)	0–10 g kg^{-1} feed	↑ Lysozyme, SGR protease, anti-protease, SOD, and immunoglobulin	[75]
Lactobacillus plantarum (Ep-M1) (*Litopenaeus vannamei*)	5×10^8 CFU g^{-1} feed	↑ SGR, SOD, immunometabolism, survival	[76]
Enterococcus casseliflavus (EC-001) (*Cyprinus carpio*)	1×10^7–1×10^9 CFU g^{-1} feed	↑ SGR, lysozyme, disease resistance, serum Acid Phosphatase (ACP)	[77]
Lactobacillus rhamnosus (*Oncorhynchus mykiss*)	1×10^6 CFU g^{-1} feed	↑ Weight gain, SOD, immunometabolism, lysozyme, disease resistance	[78]
Shewanella putrefaciens (Pdp11) (*Solea senegalensis*)	1×10^7 CFU g^{-1} feed	↑ Stress tolerance, disease resistance, and gut microbiota modulation	[79]

On the other hand, prebiotics serve as an energy source for the gut microbiota, thereby improving the immune system of aquatic animals and promoting general growth [24,80]. One good example of prebiotics that can stimulate an immune response is immunosaccharides [81]. Immunosaccharides can be considered an alternative to antibiotics in managing the health of aquatic animals in aquaculture [82,83].

Controlling fish diseases involves the use of vaccination rather than antibiotic treatments in aquaculture [84]. However, some diseases still lack available vaccines or are still under development. Therefore, alternative strategies have gained interest in recent years as supplements to vaccination and to reduce unnecessary antibiotic use. One strategy to supplement vaccination is the use of prebiotics, given their role in intestinal microbiota in fish and shrimp [85]. The microorganisms found within the caeca-colon ferment prebiotics, leading to the modification of the colonic microbiota and changes in the gut. This results from the utilisation of oligosaccharides by anaerobic bacteria, especially bifidobacteria, as a substrate, thereby eliminating the growth of harmful and putrefactive bacteria that can cause diseases [86].

Additional strategies to complement vaccination through the use of prebiotics encompass the stimulation of gut microbiota development or the activation of innate immune mechanisms [67]. The microbiota generate substances that activate the immune system and bolster the host's defences against infections [87]. For instance, the non-specific immune system of grass carp (*Ctenopharyngodon idella*), gilt-head seabream (*Sparus aurata*), and hybrid catfish (*Pangasianodon gigas* × *Pangasianodon hypophthalmus*) showed a positive response to the supplementation of mannan oligosaccharides in their diet [88–90]. Moreover, incorporating mannan oligosaccharides at a rate of 0.4% into the diet of gilt-head seabream improved their immune system. It increased their resistance to bacterial infections when directly introduced into the gut, a common site of infection in fish [91,92].

3.2. The Immune Response of Shrimps to Probiotics and Prebiotics as Feed Additives

The shrimp aquaculture sector is experiencing rapid growth, and the utilisation of prebiotics and probiotics is becoming increasingly popular [93]. Studies have shown the immune response of shrimps to probiotics, including the effects of short-chain fructooligosaccharides (FOS) supplementation on Pacific white prawn (*Litopenaeus vannamei*) [94]. Furthermore, Grobiotic®, a prebiotic dietary supplement for growth and health management in Pacific white shrimp (*Litopenaeus vannamei*), showed improved survival rates when cultured in a low-salinity water level of 2 ppt [95]. However, the specific mechanisms responsible for this enhanced survival under low-salinity conditions have yet to be determined [96], as the optimal salinity for their growth and survival is between 20 and 25 ppt [97]. Furthermore, prebiotics have been demonstrated to alter the microbial community in the gastrointestinal tract, improving non-specific immune responses [12].

Research has shown that probiotics can improve shrimp's immunity against pathogenic bacteria and viruses [98]. For instance, it was observed that *Lactobacillus plantarum* could enhance the non-specific immune response in *Litopenaeus vannamei* when it is exposed to *Vibrio harveyi*. Further research has demonstrated that *Bacillus subtilis* WB60 enhanced the growth performance and immune response of shrimps [99]. *Pediococcus pentosaceus* has also been recognised as a potential probiotic for shrimps, exhibiting positive impacts on growth performance and immune response [100]. This also includes minimising disease recurrence and increasing the enzymatic activities associated with feed ingestion, growth, and shrimp survival (Figure 2) [101].

A comprehensive evaluation of *Bacillus coagulans* and *Bacillus firmus*, in both live and lyophilised forms, revealed significant improvements in growth parameters, including weight gain, length increase, and specific growth rate (SGR), when these probiotics were administered either independently or in combination with spirulina and yeast. Notably, the combined probiotic diet also led to improved survival rates and feed conversion ratios (FCR) [102]. Further investigation isolated *Bacillus subtilis* and *Bacillus licheniformis* from *Penaeus monodon*, demonstrating that their inclusion in the diet significantly increased the activities of digestive enzymes, such as protease, amylase, and cellulase, which indicates enhanced digestive efficiency [103]. The incorporation of hydrolysed squid by-products alongside *Bacillus subtilis* BF12 in a plant-based diet for *P. monodon* resulted in substantial improvements in growth performance, feed efficiency, nutrient retention, and immune responses. This combination not only enhanced digestive enzyme activity but also promoted beneficial gut microbiota, suggesting a multifaceted impact on shrimp health [104].

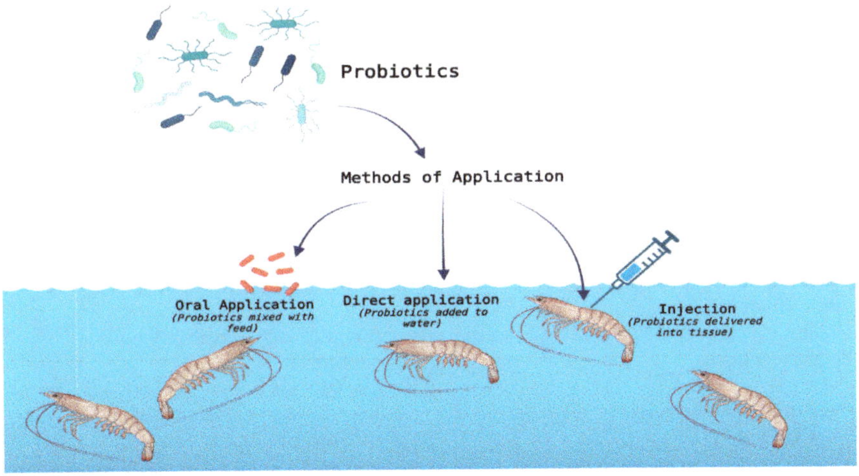

Figure 2. Methods of delivering probiotics to *Litopenaeus vannamei* shrimp. Adapted with permission from Amiin et al. [98]. Copyright 2023—under the terms of the Creative Commons Attribution 4.0 license.

A cost-effective probiotic formulation containing *Bacillus cereus* has demonstrated notable improvements in growth rates, immune parameters, and survival rates within shrimp diets [105,106]. Supplementing shrimp ponds with *Streptococcus phocae* PI80, administered through both feed and water, has been documented to significantly enhance growth and immune responses while reducing the prevalence of pathogenic bacteria [107]. Similarly, the incorporation of *Clostridium butyricum* into shrimp diets has been shown to enhance growth, increase digestive enzyme activity, and strengthen antioxidant defences, all of which contribute to increased resilience against nitrite stress [108]. Moreover, the application of *Bacillus* sp. Mk22 has shown dual benefits by promoting growth and survival while simultaneously reducing infections associated with *Vibrio* spp. and white spot syndrome virus (WSSV), highlighting its potential in disease management strategies [109]. Finally, the *Bacillus* isolate P11, identified as *Bacillus subtilis*, has demonstrated improved growth performance, feed efficiency, and an enhanced immune response against *Vibrio harveyi*, thus underscoring its effectiveness as a probiotic agent [110].

These studies have shown that when used as dietary supplements, probiotics enhance the competitive elimination of pathogens from aquaculture systems and improve the shrimp's immunological parameters. Thus, probiotics serve as a sustainable alternative to antibiotics with enhanced environmental protection and stability. However, there are some constraints to their use in shrimp aquaculture due to their expense, which leads to high production costs and complications in administering precise probiotics as dietary additives [111]. Table 2 summarises the constraints and potentials of probiotics, prebiotics, and synbiotics in shrimp aquaculture.

Table 2. Constraints and potentials of probiotics, prebiotics, and symbiotics in shrimp aquaculture.

Potentials	Constraints	References
	PROBIOTICS	
Improve gut health and nutrient uptake.	Challenges in strain selection and dosage optimisation.	[112–114]
Enhance disease resistance, immune response, and the secretion of antibacterial compounds and antitoxins.	Environmental conditions impacting efficacy.	[112–116]
Maintain a balanced microbial community in ponds.	An overdose can cause immunosuppression.	[112,113,117]
Reduce pathogen levels and enhance water quality.	There is limited understanding of mechanisms in aquaculture systems.	[112–115,117]
Enhance feed efficiency, stimulate digestive enzyme activity, and promote growth and reproduction.	Storage and maintenance of live cultures.	[118,119]
Regulate the immune system and manage allergic responses.	Potential environmental incompatibility with aquatic hosts.	[112,116,120]
	PREBIOTICS	
Improve water quality and decrease pollution.	Limited research on specific prebiotic effects in shrimp.	[121,122]
Enhance growth and survival rates, increase stress resistance and health status, and modulate enteric microbiota and immune responses.	There is a need for further studies to understand the molecular impacts.	[121,122]
	SYMBIOTICS	
Combine the benefits of probiotics and prebiotics.	Complexity in formulation and application.	[123,124]
Improve metabolic pathways and energy metabolism, boost growth performance and immune function, and decrease the severity of infections, thereby raising survival rates.	There is a need for more research on specific symbiotic combinations.	[123–125]

3.3. Growth Response of Fish to Probiotics and Prebiotics as Feed Additives

According to the Food and Agriculture Organization (FAO), probiotics are live microbial feed supplements that confer health benefits to the host by modifying the gastrointestinal microbial community [111,126]. Conversely, prebiotics are non-digestible feed additives that stimulate the activity of beneficial gut microorganisms [111,127]. Probiotics produce beneficial enzymes that facilitate digestion and safeguard the gastrointestinal tract in fish [128,129]. Additionally, they enhance the balance of intestinal microbes, resulting in increased digestive enzyme activity, improved nutrient absorption, and decreased pathogenic issues within the gastrointestinal tract when administered at appropriate dosages [130]. Probiotics function synergistically with digestive enzymes in the fish gastrointestinal tract, serving as supplements to optimise nutrition [131]. Consequently, fish feed efficiency and growth rates are augmented, while also preventing antinutritional factors in ingredients, reducing intestinal disorders, and aiding pre-digestion [129].

In Atlantic salmon (*Salmo salar*), the addition of mannanoligosaccharide, fructooligosaccharide, and galactooligosaccharide at 10 g kg^{-1} of prebiotics to a fish meal-based diet did not impact growth and digestibility. This finding suggests that adding prebiotics at a particular concentration may not enhance growth or digestive efficacy. In other studies, a diet containing 2 g kg^{-1} mannanoligosaccharide has been shown to improve fish growth, feed efficiency, and survival rates compared to those fed the basal diet [132,133]. These findings highlight the superior growth performance of mannanoligosaccharide compared to other prebiotics, such as fructooligosaccharide and galactooligosaccharide [58,91,134].

The consequences of prebiotic intake extend beyond typical biological networks, influencing the composition of fish gut microbiota; they also actively participate in this process [135]. The process goes both ways, as diverse materials present within the gut can affect microbial communities while simultaneously affecting those components directly or indirectly [119]. By feeding prebiotic oligosaccharides, such as inulin and oligofructose, populations of beneficial bacteria can be initiated, enhancing gut performance while eliminating harmful bacterial competition, as they find these substrates unsuitable [136].

Similarly, prebiotics promote the growth of beneficial bacteria without undergoing digestion [137]. By facilitating the proliferation of these microorganisms within aquatic fauna, it becomes feasible to sustain optimal gastrointestinal conditions across various fish species without dependence on costly or complex treatments typically associated with conventional methods [138]. Beneficial microorganisms flourish in environments with proper nutrition, which stimulates their growth and activity [131]. Prebiotics demonstrate a remarkable potential to uphold optimal fish health by modulating gut microbiota balance, enhancing nutrient utilisation, and promoting gastrointestinal health [139].

4. Feed Additives in Fish and Shrimp Diets

Feed additives are widely used in fish and shrimp diets to enhance growth performance, feed efficiency, and overall health [140]. These additives are specifically formulated to provide essential nutrients, vitamins, and minerals that may be deficient in conventional feed ingredients. They can be utilised to reduce the costs associated with specialized feeds (e.g., soybean meal and rice bran in fish feed pellets) and to improve the feed's taste and appeal, making it more palatable to fish and shrimp. Table 3 provides an overview of various feed additives used in fish and shrimp aquaculture, along with their respective dosages.

Table 3. Different types of synthetic feed additives in fish and shrimp diets.

Feed Additives	Function	Fish Species	Dosage Recommendation	Environmental Risk	References
Antibiotics	Growth and feed efficiency, reduced disease occurrence	Various fish species, shrimp	Varies with antibiotic type	High risk	[141]
Astaxanthin	Pigmentation, growth, and antioxidant	Atlantic salmon (*Salmo salar*), Rainbow trout (*Oncorhynchus mykiss*), Discus fish (*Symphysodon* spp.).	50–100 mg kg^{-1} of feed	Limited information, but likely biodegradable	[142,143]
Beta-carotene	Pigmentation, growth, and antioxidant	Nile tilapia (*Oreochromis niloticus*), African catfish (*Clarias gariepinus*), shrimp	50–100 mg kg^{-1} of feed	Limited information, but likely biodegradable	[142]
Betaine	Osmoregulation, nutrient utilisation, and stress resistance	Barramundi (*Lates calcarifer*), prawn	500–1000 mg kg^{-1} of feed	Limited information, but likely biodegradable	[144]
Butylated Hydroxytoluene (BHT)	Preservative	Channel catfish (*Ictalurus punctatus*), lobster (*Homarus gammarus*)	50–100 mg kg^{-1} of feed	Potential risk due to low biodegradability	[143]

Table 3. *Cont.*

Feed Additives	Function	Fish Species	Dosage Recommendation	Environmental Risk	References
Choline Chloride	Growth promoter	Various fish species, shrimp	500–1000 mg kg^{-1} of feed	Limited information, but likely biodegradable	[5]
Enzymes (e.g., Phytase)	Digestive enhancer	Various fish species, shrimp	As per the enzyme activity levels	Generally considered safe	[5,145]
Ethoxyquin	Antioxidant	Rainbow trout (*Oncorhynchus mykiss*), crab (*Brachyura* spp.)	100–200 mg kg^{-1} of feed	Potential risk due to low biodegradability	[146,147]
Mould Inhibitors (e.g., Propionic Acid)	Antifungal agent	Various species	2–4 g kg^{-1} of feed	Limited information, but likely biodegradable	[148,149]
Sodium Bicarbonate	pH regulator	Carp (*Cyprinus carpio*), shrimp	1–2 g kg^{-1} of feed	Limited information, but likely biodegradable	[150,151]
Synthetic Arginine	Amino acid supplement	Atlantic salmon (*Salmo salar*), Rainbow trout (*Oncorhynchus mykiss*), African catfish (*Clarias gariepinus*), shrimp	Varies with species	Limited information, but likely biodegradable	[46,152]
Synthetic Lysine	Amino acid supplement	Nile tilapia (*Oreochromis niloticus*), African catfish (*Clarias gariepinus*), shrimp	Varies with species	Limited information, but likely biodegradable	[153]
Synthetic Methionine	Amino acid supplement	Various fish species, shrimp	2–4 g kg^{-1} of feed	Limited information, but likely biodegradable	[154]
Synthetic Phospholipids	Emulsifiers, chemoattraction	Various fish species, shrimp	-	Limited information, but likely biodegradable	[155]
Synthetic Taurine	Cellular and physiological processes	Atlantic salmon (*Salmo salar*), Rainbow trout (*Oncorhynchus mykiss*), African catfish (*Clarias gariepinus*),	-	Limited information, but likely biodegradable	[156,157]
Vitamin C	Immune booster, antioxidant, and stress resistance	Various fish species, shrimp	100–300 mg kg^{-1} of feed	Generally considered safe	[158,159]

5. Phytogenics as Feed Additives in Fish and Shrimp Diets

Phytogenics, a novel category of feed additives, are increasingly attracting attention in the aquaculture sector [160]. Phytogenics offer several advantages as feed additives for aquatic animals, such as the potential to enhance feed digestibility, particularly with proteins and amino acids [161] (Table 4). Phytogenic feed additives (PFAs) are plant-derived substances incorporated into animal feed to enhance their performance. In aquatic feeds, aromatic plant essential oil-based feed additives are widely used as the predominant type of phytogenic product due to their antimicrobial and antioxidant properties, stability, ease of formulation, and regulatory acceptance in many regions [9,162,163].

Growing evidence supports the potential benefits of phytogenics in aquaculture, particularly for species such as fish and shrimp [9,164]. Phytogenic feed additives have been shown to positively impact the growth and health of fish and shrimp [165]. Fish and shrimp fed diets containing phytogenics experienced improved feed conversion, enhanced growth, and increased innate immunity factors [166]. *Phyllanthus niruri* leaf extracts enhanced specific and non-specific immune responses in Mozambique tilapia (*Oreochromis mossambicus*) [167]. The extract also enhanced the growth performance of the fish through enzymatic and antibody-mediated mechanisms.

Dietary supplementation with white mustard (*Sinapis alba*) seed oil at concentrations of 0.5%, 1% and 1.5% for 42–63 days significantly enhanced growth performance in rainbow trout (*Oncorhynchus mykiss*), with the 1.5% inclusion level yielding the most significant growth response [168]. The use of extracts from plants such as *Aegle marmelos, Cynodon dactylon, Withania somnifera* (ashwagandha or winter cherry), and *Zingiber officinale* (ginger), as feed additives, also improved feed efficiency and growth performance in *Oreochromis mossambicus* [169]. Table 4 summarizes the benefits of different phytogenic compounds in aquaculture.

Table 4. Phytogenics as feed additives in aquaculture.

Phytogenic Compounds	Benefit	Description	References
Ginger, oregano, thyme, garlic	Growth promotion	Enhances feed intake and digestion, leading to improved growth performance	[170,171]
Carvacrol, thymol	Antimicrobial activity, growth promotion	Inhibits pathogenic bacteria and fungi in the gut	[172]
Echinacea, garlic, turmeric	Immune system enhancement	Stimulates innate immune responses and disease resistance	[173]
Curcumin, flavonoids, polyphenols	Antioxidant properties, disease resistance, reproductive, and growth performance	Reduces oxidative stress and improves cellular health, survival, and growth	[174]
Fennel and anise essential oils	Antibacterial, antioxidant, growth performance, lipid metabolism	Enhanced growth performance and well-being	[175,176]
Liquorice	Antioxidant properties, disease-resistant, immunostimulant	Enhances growth and survival, reduces oxidative stress	[177]
Herbal blends	immune responses, antioxidants, and disease resistance	Enhanced growth and survival	[178]

Further studies have demonstrated that phytogenic additives can positively impact fish growth, although their mechanisms and effectiveness vary widely. For example, *Zingiber officinale* (ginger) has been shown to enhance the growth of African catfish (*Clarias gariepinus*), as reflected by growth metrics such as weight gain, specific growth rate, and protein efficiency ratio [179,180], with 20% supplementation producing the most notable growth increase. Including 20% as a supplement may not be economically feasible for commercial feed production compared to traditional additives, due to the high cost. However, lower supplementation levels (2–3%) have been shown to significantly improve growth performance and health metrics, offering a more cost-effective solution [181,182]. Therefore, it is recommended to use ginger at these lower levels to achieve economic and health benefits in commercial aquaculture operations. While feed additives have demonstrated benefits for growth, feed efficiency, and disease resistance, their adoption in commercial aquaculture largely depends on considerations of cost-effectiveness and environmental sustainability. For example, in fish, the inclusion levels of phytogenic additives must be economical relative to traditional feeds, and their uptake is influenced by ingredient availability and regional market dynamics [179,180]. Moreover, some additives raise environmental concerns; antibiotic-based additives, in particular, have been linked to antimicrobial resistance in intensive aquaculture, highlighting the necessity for safer alternatives [17,18]. Even natural compounds and innovative delivery methods encounter challenges such as variable efficacy under different conditions or the potential accumulation of residues in the environment. Conversely, *Aloe vera* crude polysaccharide extracts have also been found to improve growth performance and protein efficiency ratio in the same species, even at much lower levels (0.5–4.0%) [183], suggesting that certain phytogenics may be effective at lower, more cost-effective doses.

Moreover, a study on channel catfish (*Ictalurus punctatus*) demonstrated that including matrix-encapsulated phytogenics in their diets improved weight gain and reduced the feed conversion ratio, providing practical alternatives for improving feed efficiency for feed manufacturers [161]. Garlic–cinnamon blends have also been shown to enhance the immune response and overall well-being in shrimp [184]. Supplementing the diet of striped catfish (*Pangasianodon hypophthalmus*) with different cinnamon products increased the specific growth rate and protein retention [185]. Carvacrol supplementation has also resulted in significant improvements in the growth performance of Nile tilapia (*Oreochromis niloticus*) [186], emphasizing that the effectiveness and function of phytogenic additives may be species-specific.

These studies highlight the importance of PFAs in aquafeed and their potential to enhance growth performance and overall health in fish and shrimp. They also suggest that adding phytogenics as feed additives to fish and shrimp diets could reduce reliance on fish meal as a feed ingredient, potentially leading to a more sustainable and cost-effective approach [187–189]. However, the effectiveness of these extracts may vary depending on the plant type, the extraction method used, and the concentration of the extract [9].

5.1. Emerging and Next-Generation Feed Additives in Aquaculture

In addition to conventional immunostimulants and probiotics, recent years have seen the emergence of innovative additives and engineered methodologies aimed at enhancing the specificity, stability, and sustainability of health interventions in aquaculture. These advanced, next-generation feed additives are designed to enhance animal health and productivity while addressing issues related to antibiotic resistance. The developments encompass bacteriophage–probiotic combinations, postbiotics, nano- and micro-encapsulation techniques for targeted delivery, and the development of engineered microbial strains.

Bacteriophage therapy has demonstrated considerable potential in reducing pathogen carriage within animals, particularly when administered immediately prior to slaughter to target "super-shedders" [190]. However, the transient nature of bacteriophages and the possibility of phage-resistant subpopulations necessitate meticulous application. Conversely, multi-species probiotics are increasingly employed in aquaculture to enhance the health, growth, and disease resistance of aquatic organisms [191,192]. They operate through mechanisms such as competitive exclusion of pathogens and stimulation of the host immune system, thereby rendering them more effective than single-strain probiotics [193,194]. Thus, their application can be facilitated via various methods, primarily as feed additives. A summary of the comparative advantages and limitations of conventional and emerging feed additives in aquaculture is presented in Table 5.

Table 5. Comparative overview of emerging and conventional feed additives in aquaculture.

Additive	Main Action	Advantages	Limitations and Risks	Cost-Effectiveness	Environmental Impact	Reference
Conventional probiotics	Microbiome modulation, competitive exclusion	Widely available, several validated strains	Strain survival, variable effects	Moderate	Low–moderate	[195]
Bacteriophage–probiotic	Targeted pathogen lysis	Antibiotic alternative; specificity	Host range, regulatory, and environmental concerns	Moderate–high	Low–moderate	[196]
Nano-encapsulated phytogenics	Protected delivery, controlled release	Lower dose, improved bioavailability	Cost and potential nano-ecotoxicity	Moderate–high	Potential accumulation risk	[197]
Engineered/ CRISPR probiotics	Precision metabolic modulation	High specificity potential	Genetically modified organism regulation; public acceptance	Currently high	Unknown	[198]
Postbiotics	Bioactive metabolites, immune modulation	Storage stability, no live microbes	Need to identify active compounds	High	Low	[199]
Synbiotics/ microencapsulated probiotics	Enhanced survival + prebiotic support	Improved stability and colonisation	Cost, formulation complexity	High	Moderate	[197,200]

Postbiotics, comprising preparations of inanimate microorganisms and their constituents, confer health benefits while circumventing the risks associated with live probiotics. They have the capacity to improve gastrointestinal health, decrease inflammation, and exhibit antimicrobial properties [201,202]. The potential applications of postbiotics are presently being investigated within the scope of active food packaging to prolong shelf life and serve as bio-preservatives, thereby demonstrating their adaptability in both animal health and food safety sectors [203].

Innovative encapsulation methods, including hydrogels, electrospinning, and 3D bioprinting, are being developed to enhance the delivery and viability of probiotics and postbiotics [204]. These nano- and micro-encapsulation technologies serve to protect active ingredients from adverse environmental conditions and facilitate their targeted delivery within the gastrointestinal tract [204]. By ensuring the probiotics reach the intestinal environment in an active state, encapsulation can markedly enhance their stability and functional efficacy.

Engineered microbial strains, including next-generation probiotics (NGPs) and CRISPR-edited strains, represent a significant advancement in the field of feed additives [205]. Next-generation probiotics (NGPs), such as *Akkermansia muciniphila* and *Bacteroides* spp., are being developed to confer specific health benefits, including modulation of gut microbiota, reduction of inflammation, and enhancement of metabolic health [206,207]. These strains require more nutritional resources and exhibit increased sensitivity to aerobic conditions, thereby necessitating the implementation of sophisticated delivery systems [208]. Furthermore, engineered strains, such as *Lactobacillus plantarum* engineered to express antimicrobial peptides, demonstrate considerable potential in improving disease resistance and growth performance in aquaculture, through the targeted expression of beneficial compounds [205].

For industry adoption, evaluations of new additives must address key factors: (1) effectiveness and reproducibility across species and scales, as laboratory results may not translate to cages or ponds; (2) a cost–benefit analysis covering production, storage, and administration costs, as well as benefits like improved growth or lower mortality; (3) supply chain and availability for producers in low- and middle-income countries; (4) regulatory approval and consumer acceptance, especially for options like phages, nanoparticles, and genetically modified microbes; (5) ecological risks and residue concerns, such as phage persistence, nanoparticle buildup, and antibiotic-resistance gene mobilization. While current research is promising, few studies combine biological effectiveness with economic and environmental assessments. Addressing this gap should be a focus for future trials and evaluations.

5.2. Mechanisms of Action and Measurable Biomarkers

Natural immunostimulants, particularly beta-glucans and nucleotides, play a pivotal role in enhancing the immune response of fish. Beta-glucans, which are polysaccharides derived from yeast and fungi, activate the innate immune system by binding to specific receptors on immune cells, such as macrophages and neutrophils. This interaction triggers a cascade of immune responses, including increased phagocytic activity and the production of reactive oxygen species, which are crucial for combating pathogens [30,209]. Furthermore, beta-glucans have been shown to upregulate the expression of cytokines, such as interleukin-1β and interleukin-10, thereby modulating the immune response and enhancing the fish's ability to resist infections [210–212]. This immunomodulatory effect not only improves disease resistance but also contributes to overall growth performance by reducing the energy expenditure associated with immune challenges.

Nucleotides, on the other hand, serve as essential building blocks for nucleic acids and play a critical role in cellular metabolism and energy transfer. Their supplementation in fish diets has been linked to improved gut health and enhanced growth performance [213–215]. Nucleotides can stimulate the proliferation of intestinal epithelial cells, thereby improving gut integrity and nutrient absorption. They have been shown to modulate the gut microbiota, promoting a favourable microbial balance that supports digestion and immune function [216,217]. The synergistic effects of beta-glucans and nucleotides underscore the importance of a comprehensive approach to immunostimulation in aquaculture, where enhancing both innate immunity and gut health can lead to substantial improvements in fish growth and resilience (Figure 3).

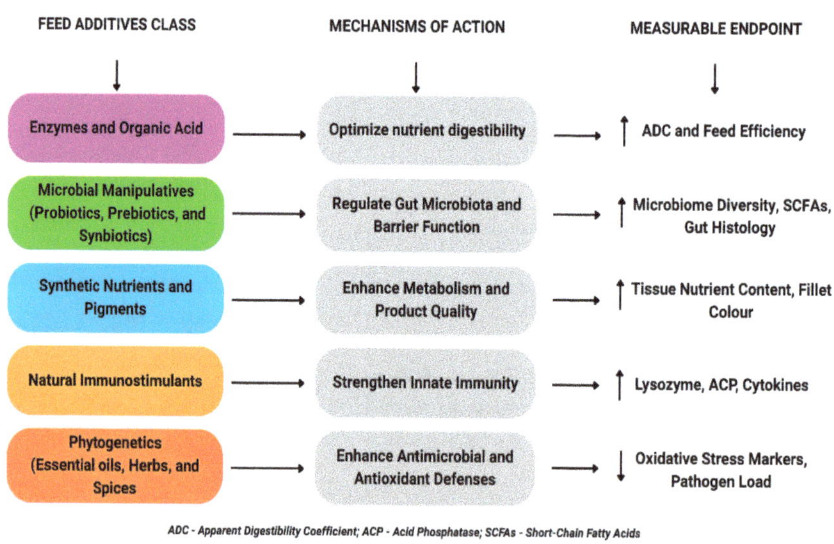

Figure 3. Mechanisms of action and measurable endpoints for feed additives in aquaculture.

Probiotics enhance nutrient utilisation and improve growth performance by increasing digestive enzyme activity and promoting a balanced gut microbiome [218,219]. The presence of probiotics in the gastrointestinal tract can inhibit the growth of pathogenic bacteria through competitive exclusion and the production of antimicrobial substances [220,221].

This not only improves the overall health of the fish but also reduces the incidence of disease, thereby minimising the need for antibiotic interventions. Prebiotics, on the other hand, selectively stimulate the growth of beneficial gut bacteria, complementing the effects of probiotics by fostering a healthy microbial environment [222,223]. The fermentation of prebiotics in the gut results in the production of short-chain fatty acids (SCFAs), which serve as an energy source for intestinal cells and exhibit anti-inflammatory properties [223,224]. Thus, the combined use of probiotics and prebiotics can significantly enhance the gut health of fish, leading to improved feed conversion ratios and growth rates. This microbiome modulation is particularly significant in aquaculture, where the stress of farming conditions can disrupt the natural balance of gut microbiota, resulting in health issues and reduced growth.

Phytogenics enhance the immune response in aquaculture species by modulating cytokine production and activating immune-related genes [225–227]. They possess antimicrobial properties that help control bacterial and parasitic infections by penetrating microbial cells and causing dysfunction [227,228]. These compounds reduce oxidative stress through their antioxidant activity, upregulating antioxidant enzymes to protect tissues from damage [229,230]. Phytogenics facilitate growth by enhancing digestive enzyme activity and nutrient absorption, thereby improving feed conversion ratios [225,229]. They additionally demonstrate anti-inflammatory properties through the downregulation of pro-inflammatory cytokines, contributing to overall health and resilience against disease [227,229,230]. Furthermore, phytogenics may influence reproductive performance through interactions with hormone receptors and modulation of hormone levels, thereby improving gonadal development and sperm quality [230].

Enzymes in aquaculture enhance protein digestibility and nutrient utilisation by breaking down proteins into smaller peptides and amino acids, facilitating better absorption [231,232]. They also improve gut health and immune response, contributing to overall fish health and reducing waste, which benefits water quality [231,232]. Organic acids primarily operate through their antimicrobial effects, suppressing the proliferation of detrimental bacteria within the digestive system and reducing stomach pH levels to augment the activity of digestive enzymes [233,234]. This facilitates improved nutrient absorption and enhances growth performance.

6. Environmental Concerns About Aquatic Feed Additives

Concerns about the environmental effects of fish and shrimp farming relate to the improper use of feed additives. Although these additives offer certain benefits, their misuse and improper disposal can lead to adverse environmental impacts [235]. For instance, the improper disposal and accumulation of florfenicol in the aquatic environment have been documented with significant environmental and health implications [236,237]. Consequently, addressing these issues requires a combination of improved practices, alternative treatments, and stringent regulatory measures to ensure the sustainability and safety of aquaculture. Antibiotics or antimicrobials are commonly used feed additives to control bacterial diseases and promote growth in fish and shrimp farming [238,239]. However, their use in aquaculture and livestock farming is increasingly restricted in regions such as the European Union and the United States due to worries about antimicrobial resistance and food safety. These resistant bacteria can spread to the surrounding environment through effluent discharge, potentially contaminating water bodies and harming other organisms that interact with them [240].

Fish feed is enriched with essential nutrients, including vitamins, minerals, and amino acids, to meet the nutritional needs of farmed fish and shrimp [241]. Improper application or excessive use of supplements in aquaculture can cause nutrient buildup and eutrophication in nearby water bodies [242]. These effects include decreased oxygen levels, the death of aquatic organisms, and disruptions to the ecosystem [243].

Several measures can effectively reduce the negative environmental impacts of feed additives [244]. Implementing strategies to reduce antibiotic use in aquatic animal feed has proven highly effective [245]. This is especially important in addressing misuse and overuse in regions like Asia, the Middle East, and Africa. Promoting alternative feed additives, such as probiotics and prebiotics, which have similar benefits for animal health and growth, can help achieve this goal [246]. Overall, developing and applying effective strategies that meet both environmental and sustainable needs for using aquatic animal feed additives is crucial. These strategies should consider environmental and public health issues, necessitating collaboration among stakeholders, including farmers, researchers, and policymakers [247], while also taking economic factors into account. A summary of the impact of antibiotics used as aqua feed additives on the aquatic environment is presented in Table 6.

Table 6. Impact of antibiotics used as aquafeed additives on the aquatic environment.

Antibiotic	Environmental Impact	References
Doxycycline	- Causes a significant decrease in glycogen and protein levels in fish tissues. - Results in scale loss, mucus hypersecretion, and abnormal swimming patterns. - Concentrations above 10 mg/L raise public health concerns due to the presence of antibioticresistant genes.	[248]
Sulfonamides	- Inhibits periphyton growth and alters microbial community structure. - Increases the abundance of resistance genes (sul1 and sul2). - Causes genotoxicity and histopathological changes in aquatic organisms.	[249,250]
Florfenicol	- Detected in high concentrations in aquaculture water, posing a significant risk to algae. - Inhibits the growth of marine microalgae *Tetraselmis suecica*.	[251,252]
Oxytetracycline	- Inhibits the growth of marine microalgae *Tetraselmis suecica*. - Accumulates in sediments and nontarget organisms, impacting biodiversity.	[251,253]
Chloramphenicol	- Inhibits the growth of marine microalgae *Tetraselmis suecica*.	[251]
Enrofloxacin; Ciprofloxacin	- High transportability from plasma to muscle and liver in fish. - Detected in high concentrations in mariculture water, posing resistance risks.	[254,255]
Erythromycin	- Detected in high concentrations in mariculture water, posing resistance risks.	[252,255]

7. Regulatory Framework for the Use of Feed Additives in Fish and Shrimp Farming

Aquatic feed additives have become crucial to ensure efficient and sustainable operations within the aquaculture industry [256]. Regulatory measures and oversight are critical in ensuring that aquatic feed additives are used safely, effectively, and sustainably [257]. Such measures include monitoring and enforcing guidelines for the use of feed additives to enhance pigmentation, growth, and antioxidant function [258]. This is essential for consumer safety and the protection of the aquatic species' environment.

7.1. International Regulatory Framework

At the international level, various organizations have played a key role in establishing regulations and standards for the proper use of feed additives in aquaculture across different countries [259]. The Food and Agriculture Organization (FAO) and the World Health Organization (WHO) are well-known global bodies involved in such efforts. They jointly established the Codex Alimentarius Commission (CAC/GL 3-1989), which carefully oversees food standards, focusing on setting guidelines related to aquafeed and other feed additives [260,261]. The regulation CAC/GL 3-1989 provides a detailed method for screening and assessing dietary exposure to food additives using accessible data, including food consumption patterns and maximum permitted levels. This helps evaluate compliance with Acceptable Daily Intake values effectively and supports risk management decisions, emphasizing the need for further evaluation or monitoring of additives [260,262].

Another key regulatory authority is the World Organisation for Animal Health (OIE), which aims to strictly enforce responsible use practices within animal production systems worldwide, especially those related to aquatic animals [263]. This organization also emphasizes the importance of maintaining animals' well-being throughout their lifecycle and ensuring that production remains sustainable and within the bounds of food safety [264]. Table 7 summarizes the different regulatory aspects of aquaculture feed additives in the European Union, the United States, and China.

Table 7. A comparison of regulatory approval timelines or safety thresholds (EU vs. US vs. China) of feed additives used in aquaculture.

Aspect	European Union (EU)	United States (US)	China	Reference
Regulatory Body	- European Food Safety Authority (EFSA)	- Food and Drug Administration (FDA) and Association of American Feed Control Officials (AAFCO)	Ministry of Agriculture and Rural Affairs (MARA)	[265–267]
Approval Process	- Comprehensive assessment of safety for animals, consumers, the environment, and handlers - Compulsory authorization for all additives - Guidance documents provided by EFSA	- FDA and AAFCO regulate safety and efficacy - Approval based on scientific evidence and safety data	- Proactive government policies - Focus on reducing feed cost and waste discharge	[265–271]
Safety Thresholds	- Specific safety assessments for each additive - Acceptable Daily Intake (ADI) levels are defined for certain additives	- Safety and efficacy must be demonstrated - Specific safety thresholds for additives like acetic acid	- Emphasis on accurate nutrient supply and reducing toxicity	[265,266,269, 271–273]
Legislation	- Regulation (EC) No 1831/2003 - Food Improvement Agent Package (FIAP) - REFIT program for evaluating legislation	- Federal Food, Drug, and Cosmetic Act (FFDCA) - Specific regulations for genetically engineered animals	- Policies for sustainable development and food safety	[265,268,270, 274–276]
Focus Areas	- Safety for target animals, consumers, the environment, and handlers - Efficacy of additives	- Safety and efficacy for animal health and productivity - Environmental impact assessments for GE animals	- Reducing feed cost and waste discharge - Ensuring food safety	[265,269–271,274]
Challenges	- Ambiguities in guidance documents - Need for better endpoints for data collection	- Concerns about the adequacy of regulatory safeguards for GE animals	- Coordination among governmental organizations - Ensuring compliance with regulations	[270,274]

7.2. Regional and National Regulatory Frameworks

The rules for incorporating feed additives in aquaculture vary significantly by region or country [263]. For instance, Regulation (EC) No. 1831/2003 oversees the management of nutritional supplements within the European Union (EU) territories, covering everything from authorization and product labelling to distribution through marketing channels [277]. Whenever a new feed additive enters the EU market, it undergoes a thorough evaluation by the European Food Safety Authority (EFSA) before approval; safety requirements also extend to post-approval issues like environmental impact and human safety, making sure no adverse effects occur once it is approved and distributed across various markets [278].

In the United States, regulations regarding feed supplements are overseen by the Food and Drug Administration (FDA), which operates under the authority of the Federal Food, Drug, and Cosmetic Act. The Association of American Feed Control Officials (AAFCO) guides each state in regulating its respective feed supplements [279]. Among other responsibilities, the FDA regulates finished dietary supplement products and dietary ingredients under rules that are different from those governing "conventional" foods and drug products [280]. The FDA also enforces a mandatory safety program for fish and fishery products, and manufacturers of dietary supplements must follow current good manufacturing practices [281].

Furthermore, in China, the Ministry of Agriculture of the State Council, now called the Ministry of Agriculture and Rural Affairs, oversees the regulation of national feed and feed additive administration. It manages the classification and approval of feed additives in accordance with the outlined legislation. At the local level, this responsibility falls to the relevant administrative departments under county governments [282].

While each country has its standards for feed supplements in the global aquaculture industry, countries in the European Union follow a unified legal framework that supports intra-EU trade. Regardless of jurisdiction, assessing the safety, potential risks, and overall effectiveness of an additive remains a key part of regulatory oversight [283]. During the approval process, scientific committees from leading regulatory agencies review 'dossiers,' which include scientific data, toxicological information, and detailed environmental risk assessments of the additives before approval.

8. Key Issues and Measurable Endpoints in the Application of Feed Additives

The practical application of feed additives in aquaculture is associated with several key issues that must be addressed to ensure their sustainable use. These include variability in additive efficacy across species and environments, inconsistent quality and availability, potential adverse effects on fish physiology, consumer safety concerns regarding residues, and ecological impacts such as antimicrobial resistance or nutrient loading.

To evaluate these challenges systematically and comparably, future studies should adopt standardized, measurable endpoints. Core growth and production metrics include feed conversion ratio (FCR), specific growth rate (SGR), protein efficiency ratio (PER), and survival rate. Health and welfare indicators encompass haematological and biochemical profiles, innate immune markers (e.g., lysozyme, complement, phagocytic activity, phenoloxidase), gut histomorphology, and stress enzyme activities (e.g., superoxide dismutase (SOD), catalase (CAT), malondialdehyde (MDA)). Environmental endpoints should encompass residue accumulation in water and sediments, shifts in microbial community composition, and quantification of antibiotic resistance genes (ARGs) in effluents. Economic outcomes, such as the incremental cost per ton of feed and the return on investment per kilogram of biomass produced, should also be consistently reported.

Integrating these measurable endpoints across trials will facilitate comparability, facilitate meta-analyses, and ultimately inform evidence-based policy on additive use in aquaculture (Table 8).

Table 8. Recommended endpoints and standard assays for feed additive trials.

Parameters	Why	Assays	When
Growth and performance	Primary production metric	WG, SGR, FCR, PER	Weekly or biweekly during trials (typical 4–12 week trials)
Survival and disease resistance	Real-world production viability	Cumulative survival, post-challenge survival curves; challenge tests with common pathogens	End of trial + challenge follow-up
Innate immunity	Mechanism and prophylactic value	Serum lysozyme activity, alternative complement (ACP), phagocytic activity, respiratory burst, SOD, and CAT	Baseline, mid-trial, end-trial, and post-challenge
Adaptive immunity	Longer-term protection	Specific antibody titres, lymphocyte proliferation, and cytokine gene expression	Later timepoints (weeks) and post-vaccination/challenge
Gut health and microbiome	Central to nutrient uptake and immunity	Intestinal histology, digestive enzyme activities, 16S rRNA gene sequencing (alpha/beta diversity), SCFA quantification	Mid and end trial
Digestibility and nutrient retention	Feed efficiency and waste output	Apparent digestibility coefficients (ADC), nutrient retention indices, and faecal nutrient analysis	Final phase ± periodic sampling
Oxidative stress and tissue health	Phytogenic and antioxidant effects	SOD, MDA, CAT in liver/muscle; histopathology	End trial
Residues and environmental fate	Public health and environmental risk (esp. antibiotics, antioxidants)	Antibiotic residues in tissue/water/sediment; ARGs (qPCR); chemical residuals (LC-MS)	During and after the feeding period, sediment sampling for benthic accumulation
Product quality and composition	Market value	Fillet colour, lipid profile, proximate composition	Harvest
Environmental metrics	eutrophication/ecosystem impact	Dissolved inorganic nitrogen and phosphorus, chlorophyll a, BOD, and benthic oxygen demand	Regular pond/pen/effluent monitoring
Economics/LCA	Adoption decisions	Cost per kg gain, feed cost ratio, basic LCA endpoints if available	End of trial and scenario modelling

Legend: Weight gain (WG), Specific Growth Rate (SGR), Feed Conversion Ratio (FCR), Protein Efficiency Ratio (PER), Life cycle assessment (LCA), Short-Chain Fatty Acids (SCFA), Biochemical Oxygen Demand (BOD).

9. Knowledge Gaps and Research Priorities

Our review identifies several gaps: (1) a deficiency in long-term and field-scale research concerning ecological fate and accumulation, as most studies are limited to short laboratory experiments or small pond trials; (2) an absence of systematic studies on additive–additive and additive–ingredient interactions that could reveal synergy or antagonism; (3) inadequate dose–response data for many commercially relevant species on a species-specific basis; (4) a paucity of mechanistic studies linking additive application to molecular pathways, such as those employing omics approaches; and (5) regional disparities in regulation and enforcement that impede harmonised efforts and recommendations. It is therefore recommended to prioritise (i) standardised, multi-site dose–response trials with agreed endpoints; (ii) integrative mechanistic studies (transcriptomics, metabolomics, microbiomics) to establish modes of action and biomarkers of efficacy; (iii) long-term ecotoxicology and residue studies (water, sediment, non-target organisms); (iv) economic and life-cycle assessments for promising additives; (v) development of harmonised, evidence-based guidance documents and registries of authorised additives with species-specific dose ranges.

10. Conclusions and Future Outlook

Fish feed additives provide numerous benefits in aquaculture, including promoting growth, controlling diseases, and enhancing nutritional value. Natural immunostimulants have shown great potential as feed additives, boosting disease resistance and improving the innate immune response in fish and shrimp. Probiotics and prebiotics also have positive effects on the health of fish and shrimp in aquaculture. Synthetic feed additives provide a well-balanced diet tailored to meet the specific nutritional needs of various fish and shrimp species. These meals supply targeted nutrition to support the health of aquatic animals. However, the unregulated or improper use of these additives can lead to harmful environmental effects. Therefore, responsible use of aquatic feed additives should be a priority in sustainable aquaculture practices. This includes promoting environmentally friendly alternatives and protecting the long-term health of aquatic ecosystems.

Given the increasing interest in sustainable and eco-friendly practices, future research may explore innovative feed additives derived from natural sources, such as plant extracts, prebiotics, probiotics, and synbiotics. Studying their effectiveness, safety, and mechanisms of action can offer alternatives to traditional feed additives, thereby decreasing reliance on synthetic compounds (Figure 4). Researchers might focus on enhancing nutrient utilisation and minimising waste in aquaculture systems. This could involve examining how feed additives affect nutrient digestion, absorption, and retention, as well as their impact on waste composition and nutrient excretion. The goal is to optimise feed formulas and feeding strategies for improved efficiency and reduced environmental impact.

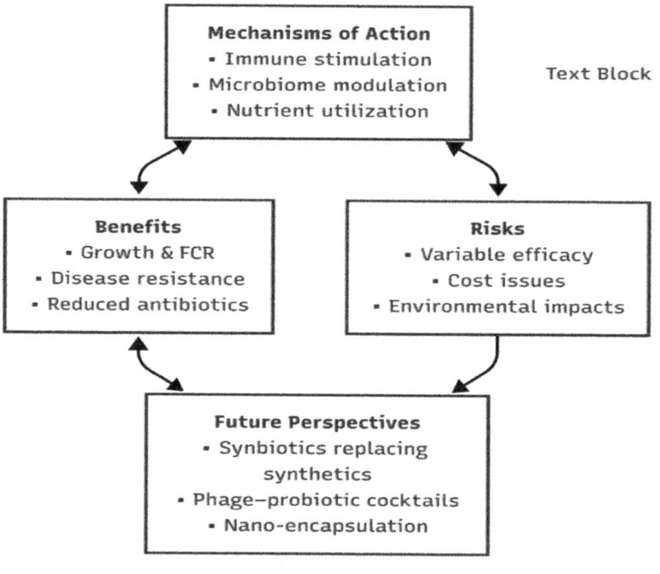

Figure 4. Schematic summary of feed additive mechanisms, benefits, risks, and future perspectives in aquaculture.

This review highlights new opportunities, but advancing requires a shift from broad speculation to targeted, empirical research. Future studies should test hypotheses supporting the use of sustainable feed additives in aquaculture. By 2030, well-developed symbiotic formulations could reduce synthetic additives by over 50% without compromising growth or survival, as confirmed through controlled farm trials that evaluated growth, immune markers, microbiome, and costs. Bacteriophage–probiotic combinations are expected to reduce *Vibrio* and *Aeromonas* infections by at least 40% and lower antibiotic resistance genes in effluent by 1 log10, as validated through farm studies on outbreaks, survival, antibiotic use, and resistance profiles. Nano-encapsulated phytogenic blends may offer growth and health benefits at half the standard dose, while reducing environmental residues by at least 30%, as demonstrated in pond trials and chemical residue analyses.

Author Contributions: E.O.: Conceptualization, Writing—Original draft preparation, Writing—Reviewing and Editing, Visualization. M.I.: Writing—Original draft preparation. P.O.: Visualization, Writing—Original draft preparation, Writing—Reviewing and Editing. M.A.: Writing—Original draft preparation. N.M.: Writing—Reviewing and Editing. R.O.: Conceptualization; Writing—Reviewing and Editing. All authors have read and agreed to the published version of the manuscript.

Funding: This research received no external funding.

Institutional Review Board Statement: Not applicable.

Informed Consent Statement: Not applicable.

Data Availability Statement: Not applicable.

Conflicts of Interest: The authors declare no conflict of interest.

References

1. Edwards, P.; Zhang, W.; Belton, B.; Little, D.C. Misunderstandings, myths and mantras in aquaculture: Its contribution to world food supplies has been systematically over reported. *Mar. Policy* **2019**, *106*, 103547. [CrossRef]
2. Kidane, D.G.; Brækkan, E.H. Global seafood demand growth differences across regions, income levels, and time. *Mar. Resour. Econ.* **2021**, *36*, 289–305. [CrossRef]
3. Food and Agriculture Organization of the United Nations. The state of world fisheries and aquaculture 2024: Blue transformation in action. In *The State of World Fisheries and Aquaculture (SOFIA)*; FAO: Rome, Italy, 2024. [CrossRef]
4. Stentiford, G.D.; Bateman, I.J.; Hinchliffe, S.J.; Bass, D.; Hartnell, R.; Santos, E.M.; Devlin, M.J.; Feist, S.W.; Taylor, N.G.H.; Verner-Jeffreys, D.W.; et al. Sustainable aquaculture through the One Health lens. *Nat. Food* **2020**, *1*, 468–474. [CrossRef]
5. Bai, S.C.; Hamidoghli, A.; Bae, J. Feed additives: An overview. In *Feed and Feeding Practices in Aquaculture*; Woodhead Publishing: Cambridge, UK, 2022; pp. 195–229. [CrossRef]
6. Tacon, A.G. Trends in global aquaculture and aquafeed production: 2000–2017. *Rev. Fish. Sci. Aquac.* **2020**, *28*, 43–56. [CrossRef]
7. Boyd, C.E.; McNevin, A.A. Overview of aquaculture feeds: Global impacts of ingredient production, manufacturing, and use. In *Feed and Feeding Practices in Aquaculture*; Woodhead Publishing: Cambridge, UK, 2022; pp. 3–28.
8. Marimuthu, V.; Shanmugam, S.; Sarawagi, A.D.; Kumar, A.; Kim, I.H.; Balasubramanian, B. A glimpse on influences of feed additives in aquaculture. *eFood* **2022**, *3*, e6. [CrossRef]
9. Encarnação, P. Functional feed additives in aquaculture feeds. In *Aquafeed Formulation*; Academic Press: Cambridge, MA, USA, 2016; pp. 217–237.
10. Goh, J.X.H.; Tan, L.T.H.; Law, J.W.F.; Ser, H.L.; Khaw, K.Y.; Letchumanan, V.; Lee, L.H.; Goh, B.H. Harnessing the potentialities of probiotics, prebiotics, synbiotics, paraprobiotics, and postbiotics for shrimp farming. *Rev. Aquac.* **2022**, *14*, 1478–1557. [CrossRef]
11. Ochoa-Romo, J.P.; Cornejo-Granados, F.; Lopez-Zavala, A.A.; Viana, M.T.; Sánchez, F.; Gallardo-Becerra, L.; Luque-Villegas, M.; Valdez-López, Y.; Sotelo-Mundo, R.R.; Cota-Huízar, A.; et al. Agavin induces beneficial microbes in the shrimp microbiota under farming conditions. *Sci. Rep.* **2022**, *12*, 1–16. [CrossRef] [PubMed]
12. Butt, U.D.; Lin, N.; Akhter, N.; Siddiqui, T.; Li, S.; Wu, B. Overview of the latest developments in the role of probiotics, prebiotics and synbiotics in shrimp aquaculture. *Fish Shellfish Immunol.* **2021**, *114*, 263–281. [CrossRef]
13. Liang, Q.; Yuan, M.; Xu, L.; Lio, E.; Zhang, F.; Mou, H.; Secundo, F. Application of enzymes as a feed additive in aquaculture. *Mar. Life Sci. Technol.* **2022**, *4*, 208–221. [CrossRef]
14. Busti, S.; Rossi, B.; Volpe, E.; Ciulli, S.; Piva, A.; D'Amico, F.; Soverini, M.; Candela, M.; Gatta, P.P.; Bonaldo, A.; et al. Effects of dietary organic acids and nature identical compounds on growth, immune parameters and gut microbiota of European sea bass. *Sci. Rep.* **2020**, *10*, 21321. [CrossRef] [PubMed]
15. EFSA Panel on Additives and Products or Substances used in Animal Feed (EFSA FEEDAP Panel); Bampidis, V.; Azimonti, G.; Bastos, M.L.; Christensen, H.; Dusemund, B.; Durjava, M.; Kouba, M.; López-Alonso, M.; Puente, S.L.; et al. Guidance on the assessment of the safety of feed additives for the environment. *EFSA J.* **2019**, *17*, e05648.
16. Pandey, A.K.; Kumar, P.; Saxena, M.J. Feed additives in animal health. *Nutraceuticals Vet. Med.* **2019**, 345–362. [CrossRef]
17. Hedberg, N.; Stenson, I.; Pettersson, M.N.; Warshan, D.; Nguyen-Kim, H.; Tedengren, M.; Kautsky, N. Antibiotic use in Vietnamese fish and lobster sea cage farms; implications for coral reefs and human health. *Aquaculture* **2018**, *495*, 366–375. [CrossRef]

18. Pham, T.T.H.; Rossi, P.; Dinh, H.D.K.; Pham, N.T.A.; Tran, P.A.; Ho, T.T.K.M.; Dinh, Q.T.; De Alencastro, L.F. Analysis of antibiotic multi-resistant bacteria and resistance genes in the effluent of an intensive shrimp farm (Long An, Vietnam). *J. Environ. Manag.* **2018**, *214*, 149–156. [CrossRef]
19. Nikitin, A.V.; Navashin, S.M. Natural immunostimulants. *Antibiotiki* **1983**, *28*, 702–716.
20. Zebeaman, M.; Tadesse, M.G.; Bachheti, R.K.; Bachheti, A.; Gebeyhu, R.; Chaubey, K.K. Plants and Plant-Derived Molecules as Natural Immunomodulators. *BioMed Res. Int.* **2023**, *2023*, 7711297. [CrossRef]
21. Pooljun, C.; Jariyapong, P.; Wongtawan, T.; Hirono, I.; Wuthisuthimethavee, S. Effect of feeding different types of β-glucans derived from two marine diatoms (*Chaetoceros muelleri* and *Thalassiosira weissflogii*) on growth performance and immunity of banana shrimp (*Penaeus merguiensis*). *Fish Shellfish Immunol.* **2022**, *130*, 512–519. [CrossRef] [PubMed]
22. Yin, M.; Zhang, Y.; Li, H. Advances in research on immunoregulation of macrophages by plant polysaccharides. *Front. Immunol.* **2019**, *10*, 145. [CrossRef] [PubMed]
23. Servin Arce, K.; de Souza Valente, C.; do Vale Pereira, G.; Shapira, B.; Davies, S.J. Modulation of the gut microbiota of Pacific white shrimp (*Penaeus vannamei* Boone, 1931) by dietary inclusion of a functional yeast cell wall-based additive. *Aquac. Nutr.* **2021**, *27*, 1114–1127. [CrossRef]
24. Mohammadi, G.; Hafezieh, M.; Karimi, A.A.; Azra, M.N.; Van Doan, H.; Tapingkae, W.; Abdelrahman, H.A.; Dawood, M.A. The synergistic effects of plant polysaccharide and Pediococcus acidilactici as a synbiotic additive on growth, antioxidant status, immune response, and resistance of Nile tilapia (*Oreochromis niloticus*) against Aeromonas hydrophila. *Fish Shellfish Immunol.* **2022**, *120*, 304–313. [CrossRef] [PubMed]
25. Mohan, K.; Ravichandran, S.; Muralisankar, T.; Uthayakumar, V.; Chandirasekar, R.; Seedevi, P.; Rajan, D.K. Potential uses of fungal polysaccharides as immunostimulants in fish and shrimp aquaculture: A review. *Aquaculture* **2019**, *500*, 250–263. [CrossRef]
26. Li, H.; Liu, Y.; Teng, Y.; Zheng, Y.; Zhang, M.; Wang, X.; Cheng, H.; Xu, J.; Chen, X.; Zhao, X.; et al. Enhancement of seaweed polysaccharides (fucoidan and laminarin) on the phagocytosis of macrophages via activation of intelectin in blunt snout bream (*Megalobrama amblycephala*). *Front. Mar. Sci.* **2023**, *10*, 1124880. [CrossRef]
27. Kim, H.S.; Hong, J.T.; Kim, Y.; Han, S.B. Stimulatory effect of β-glucans on immune cells. *Immune Netw.* **2011**, *11*, 191–195. [CrossRef] [PubMed]
28. Sahasrabudhe, N.M.; Dokter-Fokkens, J.; de Vos, P. Particulate β-glucans synergistically activate TLR4 and Dectin-1 in human dendritic cells. *Mol. Nutr. Food Res.* **2016**, *60*, 2514–2522. [CrossRef]
29. Elder, M.J.; Webster, S.J.; Chee, R.; Williams, D.L.; Hill Gaston, J.S.; Goodall, J.C. β-Glucan size controls dectin-1-mediated immune responses in human dendritic cells by regulating IL-1β production. *Front. Immunol.* **2017**, *8*, 791. [CrossRef] [PubMed]
30. Hadiuzzaman, M.; Moniruzzaman, M.; Shahjahan, M.; Bai, S.C.; Min, T.; Hossain, Z. β-Glucan: Mode of Action and Its Uses in Fish Immunomodulation. *Front. Mar. Sci.* **2022**, *9*, 905986. [CrossRef]
31. Petrie-Hanson, L.; Peterman, A.E. Trained immunity provides long-term protection against bacterial infections in channel catfish. *Pathogens* **2022**, *11*, 1140. [CrossRef]
32. Lim, J.J.; Grinstein, S.; Roth, Z. Diversity and versatility of phagocytosis: Roles in innate immunity, tissue remodeling, and homeostasis. *Front. Cell. Infect. Microbiol.* **2017**, *7*, 191. [CrossRef]
33. Stier, H.; Ebbeskotte, V.; Gruenwald, J. Immune-modulatory effects of dietary Yeast Beta-1, 3/1, 6-D-glucan. *Nutr. J.* **2014**, *13*, 38. [CrossRef]
34. Lin, S.; Pan, Y.; Luo, L.; Luo, L. Effects of dietary β-1, 3-glucan, chitosan or raffinose on the growth, innate immunity and resistance of koi (*Cyprinus carpio koi*). *Fish Shellfish Immunol.* **2011**, *31*, 788–794. [CrossRef]
35. Meena, D.K.; Das, P.; Kumar, S.; Mandal, S.C.; Prusty, A.K.; Singh, S.K.; Akhtar, M.S.; Behera, B.K.; Kumar, K.; Pal, A.K.; et al. Beta-glucan: An ideal immunostimulant in aquaculture (a review). *Fish Physiol. Biochem.* **2013**, *39*, 431–457. [CrossRef]
36. Ochoa-Álvarez, N.A.; Casillas-Hernández, R.; Magallón-Barajas, F.J.; Ramirez-Orozco, J.M.; Carbajal-Millan, E. Protector effect of beta-glucans from shrimp pond-related yeasts in *Penaeus vannamei* rearing under white spot syndrome virus presence. *Lat. Am. J. Aquat. Res.* **2021**, *49*, 18–28. [CrossRef]
37. Andrino, K.G.S.; Augusto, E.; Serrano, J.; Valeriano, L.; Corre, J. Effects of dietary nucleotides on the immune response and growth of juvenile Pacific white shrimp *Litopenaeus vannamei* (Boone, 1931). *Asian Fish. Sci.* **2012**, *25*, 180–192. [CrossRef]
38. Gil, A. Modulation of the immune response mediated by dietary nucleotides. *Eur. J. Clin. Nutr.* **2002**, *56*, S1–S4. [CrossRef] [PubMed]
39. Wei, Z.; Yi, L.; Xu, W.; Zhou, H.; Zhang, Y.; Zhang, W.; Mai, K. Effects of dietary nucleotides on growth, non-specific immune response and disease resistance of sea cucumber *Apostichopus japonicas*. *Fish Shellfish Immunol.* **2015**, *47*, 1–6. [CrossRef]
40. Xie, S.; Yin, P.; Tian, L.; Yu, Y.; Liu, Y.; Niu, J. Dietary supplementation of astaxanthin improved the growth performance, antioxidant ability and immune response of juvenile largemouth bass (*Micropterus salmoides*) fed high-fat diet. *Mar. Drugs* **2020**, *18*, 642. [CrossRef] [PubMed]

41. Hamidoghli, A.; Lee, Y.; Hwang, S.; Choi, W.; Choi, Y.H.; Bai, S.C. Evaluation of Yeast Hydrolysate in a Low-Fishmeal Diet for Whiteleg Shrimp (*Litopenaeus vannamei*). *Animals* **2023**, *13*, 1877. [CrossRef]
42. Elumalai, P.; Kurian, A.; Lakshmi, S.; Faggio, C.; Esteban, M.A.; Ringø, E. Herbal immunomodulators in aquaculture. *Rev. Fish. Sci. Aquac.* **2020**, *29*, 33–57. [CrossRef]
43. Terzi, E.; Kucukkosker, B.; Bilen, S.; Kenanoglu, O.N.; Corum, O.; Özbek, M.; Parug, S.S. A novel herbal immunostimulant for rainbow trout (*Oncorhynchus mykiss*) against *Yersinia ruckeri*. *Fish Shellfish Immunol.* **2021**, *110*, 55–66. [CrossRef]
44. Akbar Ali, I.; Radhakrishnan, D.K.; Kumar, S. Immunostimulants and Their Uses in Aquaculture. In *Aquaculture Science and Engineering*; Springer Nature: Singapore, 2022; pp. 291–322.
45. Muahiddah, N.; Diamahesa, W.A. Potential use of brown algae as an immunostimulant material in the aquaculture field to increase non-specific immunity and fight disease. *J. Fish Health* **2022**, *2*, 109–115. [CrossRef]
46. Li, M.; Zhang, M.; Jiang, H.; Qin, C. Comparison of dietary arginine or/and inulin supplementation on growth, digestive ability and ammonia tolerance of juvenile yellow catfish *Pelteobagrus fulvidraco*. *Aquac. Rep.* **2023**, *30*, 101543. [CrossRef]
47. Vetvicka, V.; Vannucci, L.; Sima, P. The effects of β–glucan on fish immunity. *N. Am. J. Med. Sci.* **2013**, *5*, 580. [CrossRef]
48. Vallejos-Vidal, E.; Reyes-López, F.; Teles, M.; MacKenzie, S. The response of fish to immunostimulant diets. *Fish Shellfish Immunol.* **2016**, *56*, 34–69. [CrossRef] [PubMed]
49. Kumar, S.; Verma, A.K.; Singh, S.P.; Awasthi, A. Immunostimulants for shrimp aquaculture: Paving pathway towards shrimp sustainability. *Environ. Sci. Pollut. Res.* **2023**, *30*, 25325–25343. [CrossRef] [PubMed]
50. Smith, V.J.; Brown, J.H.; Hauton, C. Immunostimulation in crustaceans: Does it really protect against infection? *Fish Shellfish Immunol.* **2003**, *15*, 71–90. [CrossRef] [PubMed]
51. dos, S.; Filho, L.G.A.; Diniz, F.M.; Pereira, A.M. Immunostimulants derived from plants and algae to increase resistance of pacific white shrimp (*Litopenaeus vannamei*) against vibriosis. *Stud. Nat. Prod. Chem.* **2023**, *77*, 297–337.
52. Hossain, M.S.; Small, B.C.; Kumar, V.; Hardy, R. Utilisation of functional feed additives to produce cost-effective, ecofriendly aquafeeds high in plant-based ingredients. *Rev. Aquac.* **2023**, *16*, 121–153. [CrossRef]
53. Kiataramgul, A.; Maneenin, S.; Purton, S.; Areechon, N.; Hirono, I.; Brocklehurst, T.W.; Unajak, S. An oral delivery system for controlling white spot syndrome virus infection in shrimp using transgenic microalgae. *Aquaculture* **2020**, *521*, 735022. [CrossRef]
54. Mikucka, A.; Deptuła, A.; Bogiel, T.; Chmielarczyk, A.; Nurczyńska, E.; Gospodarek-Komkowska, E. Bacteraemia Caused by Probiotic Strains of *Lacticaseibacillus rhamnosus*—Case Studies Highlighting the Need for Careful Thought before Using Microbes for Health Benefits. *Pathogens* **2022**, *11*, 977. [CrossRef]
55. Kaur, G.; Kaur, P. Green tea (*Camellia sinensis*): It's promising health benefits for the welfare of humans. *J. Pharmacogn. Phytochem.* **2019**, *8*, 299–302.
56. Zaib, S.; Hayat, A.; Khan, I. Probiotics and their beneficial health effects. *Mini Rev. Med. Chem.* **2023**, *24*, 110–125. [CrossRef]
57. Miquel, S.; Martín, R.; Bridonneau, C.; Robert, V.; Sokol, H.; Bermúdez-Humarán, L.G.; Thomas, M.; Langella, P. Ecology and metabolism of the beneficial intestinal commensal bacterium *Faecalibacterium prausnitzii*. *Gut Microbes* **2014**, *5*, 146–151. [CrossRef]
58. Ren, C.; Faas, M.M.; de Vos, P. Disease managing capacities and mechanisms of host effects of lactic acid bacteria. *Crit. Rev. Food Sci. Nutr.* **2020**, *61*, 1365–1393. [CrossRef] [PubMed]
59. Yvon, S.; Schwebel, L.; Belahcen, L.; Tormo, H.; Peter, M.; Haimoud-Lekhal, D.A.; Eutamène, H.; Jard, G. Effects of thermized donkey milk with lysozyme activity on altered gut barrier in mice exposed to water-avoidance stress. *J. Dairy Sci.* **2019**, *102*, 7697–7706. [CrossRef]
60. Aji, M.B. Antimicrobial Compounds Activities of d2. 2 Biocontrol Bacteria Against Bacterial Pathogens on Shrimp and Fish In Vitro Aktifitas Senyawa Antimikroba Dari Bakteri Biokontrol d2. 2 Terhadap Bakteri PatogenPada Udang Dan Ikan Secara In Vitro. 2014. Available online: https://digilib.unila.ac.id/3667/1/ABSTRACT.pdf (accessed on 7 May 2025).
61. Kurashima, Y.; Kiyono, H. Mucosal ecological network of epithelium and immune cells for gut homeostasis and tissue healing. *Annu. Rev. Immunol.* **2017**, *35*, 119–147. [CrossRef] [PubMed]
62. Didierlaurent, A.M.; Collignon, C.; Bourguignon, P.; Wouters, S.; Fierens, K.; Fochesato, M.; Dendouga, N.; Langlet, C.; Malissen, B.; Lambrecht, B.N.; et al. Enhancement of adaptive immunity by the human vaccine adjuvant AS01 depends on activated dendritic cells. *J. Immunol.* **2014**, *193*, 1920–1930. [CrossRef]
63. Tang, H.G.; Wu, T.X.; Zhao, Z.Y.; Pan, X.D. Effects of fish protein hydrolysate on growth performance and humoral immune response in large yellow croaker (*Pseudosciaena crocea* R.). *J. Zhejiang Univ. Sci. B* **2008**, *9*, 684–690. [CrossRef] [PubMed]
64. Rahe, M.C.; Murtaugh, M.P. Mechanisms of adaptive immunity to porcine reproductive and respiratory syndrome virus. *Viruses* **2017**, *9*, 148. [CrossRef]
65. Fletcher, T.C.; Secombes, C.J. Immunology of fish. *eLS* **2015**, 1–9. [CrossRef]
66. Wang, J.; Ji, H.; Wang, S.; Liu, H.; Zhang, W.; Zhang, D.; Wang, Y. Probiotic *Lactobacillus plantarum* promotes intestinal barrier function by strengthening the epithelium and modulating gut microbiota. *Front. Microbiol.* **2018**, *9*, 1953. [CrossRef]
67. Corbel, M.J. The immune response in fish: A review. *J. Fish Biol.* **1975**, *7*, 539–563. [CrossRef]

68. Van Zijderveld, S.M.; Fonken, B.; Dijkstra, J.; Gerrits, W.J.J.; Perdok, H.B.; Fokkink, W.; Newbold, J.R. Effects of a combination of feed additives on methane production, diet digestibility, and animal performance in lactating dairy cows. *J. Dairy Sci.* **2011**, *94*, 1445–1454. [CrossRef] [PubMed]
69. Küçükyılmaz, K.; Bozkurt, M.; Çatlı, A.U.; Çınar, M.; Bİntaş, E. The effect of dietary supplementation of yeast and humate on broiler performance and some slaughter characteristics. *Ziraat Fakültesi Derg. Süleyman Demirel Üniversitesi* **2012**, *7*, 83–92.
70. Zheng, X.; Liu, B.; Wang, N.; Yang, J.; Zhou, Q.; Sun, C.; Zhao, Y. Low fish meal diet supplemented with probiotics ameliorates intestinal barrier and immunological function of *Macrobrachium rosenbergii* via the targeted modulation of gut microbes and derived secondary metabolites. *Front. Immunol.* **2022**, *13*, 1074399. [CrossRef]
71. Zakostelska, Z.; Kverka, M.; Klimesova, K.; Rossmann, P.; Mrazek, J.; Kopecny, J.; Hornova, M.; Srutkova, D.; Hudcovic, T.; Ridl, J.; et al. Lysate of probiotic *Lactobacillus casei* DN-114 001 ameliorates colitis by strengthening the gut barrier function and changing the gut microenvironment. *PLoS ONE* **2011**, *6*, e27961. [CrossRef] [PubMed]
72. Purwandari, A.R. Sari DNRSystem of Leukocytes Respiratory Burst Activity (RBA) in Grouper (*Epinephelus coioides*). *J. Biota* **2022**, *8*, 25–32. [CrossRef]
73. Lim, C.; Lee, C.S.; Webster, C.D. (Eds.) *Alternative Protein Sources in Aquaculture Diets*; CRC Press: Boca Raton, FL, USA, 2023.
74. Liu, C.H.; Chiu, C.H.; Wang, S.W.; Cheng, W. Dietary administration of the probiotic, *Bacillus subtilis* E20, enhances the growth, innate immune responses, and disease resistance of the grouper, *Epinephelus coioides*. *Fish Shellfish Immunol.* **2012**, *33*, 699–706. [CrossRef]
75. Abarike, E.D.; Cai, J.; Lu, Y.; Yu, H.; Chen, L.; Jian, J.; Tang, J.; Jun, L.; Kuebutornye, F.K. Effects of a commercial probiotic BS containing *Bacillus subtilis* and *Bacillus licheniformis* on growth, immune response and disease resistance in Nile tilapia, *Oreochromis niloticus*. *Fish Shellfish Immunol.* **2018**, *82*, 229–238. [CrossRef]
76. Du, Y.; Xu, W.; Wu, T.; Li, H.; Hu, X.; Chen, J. Enhancement of growth, survival, immunity and disease resistance in *Litopenaeus vannamei*, by the probiotic, *Lactobacillus plantarum* Ep-M17. *Fish Shellfish Immunol.* **2022**, *129*, 36–51. [CrossRef]
77. Akbari, H.; Shekrabi, S.P.H.; Soltani, M.; Mehrgan, M.S. Effects of potential probiotic *Enterococcus casseliflavus* (EC-001) on growth performance, immunity, and resistance to *Aeromonas hydrophila* infection in common carp (*Cyprinus carpio*). *Probiotics Antimicrob. Proteins* **2021**, *13*, 1316–1325. [CrossRef]
78. Hooshyar, Y.; Abedian Kenari, A.; Paknejad, H.; Gandomi, H. Effects of *Lactobacillus rhamnosus* ATCC 7469 on different parameters related to health status of rainbow trout (*Oncorhynchus mykiss*) and the protection against *Yersinia ruckeri*. *Probiotics Antimicrob. Proteins* **2020**, *12*, 1370–1384. [CrossRef]
79. Tapia-Paniagua, S.T.; Vidal, S.; Lobo, C.; Prieto-Álamo, M.J.; Jurado, J.; Cordero, H.; Cerezuela, R.; Banda, G.; Esteban, M.A.; Balebona, M.C.; et al. The treatment with the probiotic *Shewanella putrefaciens* Pdp11 of specimens of Solea senegalensis exposed to high stocking densities to enhance their resistance to disease. *Fish Shellfish Immunol.* **2014**, *41*, 209–221. [CrossRef]
80. Carbone, D.; Faggio, C. Importance of prebiotics in aquaculture as immunostimulants. Effects on immune system of *Sparus aurata* and *Dicentrarchus labrax*. *Fish Shellfish. Immunol.* **2016**, *54*, 172–178. [CrossRef] [PubMed]
81. Nawaz, A.; Irshad, S.; Hoseinifar, S.H.; Xiong, H. The functionality of prebiotics as immunostimulant: Evidences from trials on terrestrial and aquatic animals. *Fish Shellfish Immunol.* **2018**, *76*, 272–278. [CrossRef]
82. Hoseinifar, S.H.; Sun, Y.Z.; Zhou, Z.; Van Doan, H.; Davies, S.J.; Harikrishnan, R. Boosting immune function and disease bio-control through environment-friendly and sustainable approaches in finfish aquaculture: Herbal therapy scenarios. *Rev. Fish. Sci. Aquac.* **2020**, *28*, 303–321. [CrossRef]
83. Vazirzadeh, A.; Marhamati, A.; Rabiee, R.; Faggio, C. Immunomodulation, antioxidant enhancement and immune genes up-regulation in rainbow trout (*Oncorhynchus mykiss*) fed on seaweeds included diets. *Fish Shellfish Immunol.* **2020**, *106*, 852–858. [CrossRef] [PubMed]
84. Lillehaug, A.; Børnes, C.; Grave, K. A pharmaco-epidemiological study of antibacterial treatments and bacterial diseases in Norwegian aquaculture from 2011 to 2016. *Dis. Aquat. Org.* **2018**, *128*, 117–125. [CrossRef] [PubMed]
85. Savaş, S.; Kubilay, A.; Basmaz, N. Effect of bacterial load in feeds on intestinal microflora of seabream (*Sparus surata*) larvae and juveniles. *Isr. J. Aquac. Bamidgeh* **2005**, *57*, 3–9. [CrossRef]
86. Mussatto, S.I.; Mancilha, I.M. Non-digestible oligosaccharides: A review. *Carbohydr. Polym.* **2007**, *68*, 587–597. [CrossRef]
87. Anjum, J.; Quach, A.; Wongkrasant, P.; Nazir, S.; Tariq, M.; Barrett, K.E.; Zaidi, A. Potentially probiotic *Limosilactobacillus reuteri* from human milk strengthens the gut barrier in T84 cells and a murine enteroid model. *J. Appl. Microbiol.* **2023**, *134*, lxac029. [CrossRef]
88. Hahor, W.; Thongprajukaew, K.; Suanyuk, N. Effects of dietary supplementation of oligosaccharides on growth performance, gut health and immune response of hybrid catfish (*Pangasianodon gigas* × *Pangasianodon hypophthalmus*). *Aquaculture* **2019**, *507*, 97–107. [CrossRef]
89. Lu, Z.Y.; Jiang, W.D.; Wu, P.; Liu, Y.; Kuang, S.Y.; Tang, L.; Yang, J.; Zhou, X.; Feng, L. Mannan oligosaccharides supplementation enhanced head-kidney and spleen immune function in on-growing grass carp (*Ctenopharyngodon idella*). *Fish Shellfish Immunol.* **2020**, *106*, 596–608. [CrossRef]

90. Khosravi-Katuli, K.; Mohammadi, Y.; Ranjbaran, M.; Ghanaatian, H.; Khazaali, A.; Paknejad, H.; Santander, J. Effects of mannan oligosaccharide and synbiotic supplementation on growth performance and immune response of Gilthead Sea Bream (*Sparus aurata*) before and after thermal stress. *Aquac. Res.* **2021**, *52*, 3745–3756. [CrossRef]
91. Lu, J.; Qi, C.; Limbu, S.M.; Han, F.; Yang, L.; Wang, X.; Xin, J.G.; Chen, L. Dietary mannan oligosaccharide (MOS) improves growth performance, antioxidant capacity, non-specific immunity and intestinal histology of juvenile Chinese mitten crabs (*Eriocheir sinensis*). *Aquaculture* **2019**, *510*, 337–346. [CrossRef]
92. Torrecillas, S.; Makol, A.; Caballero, M.J.; Montero, D.; Robaina, L.; Real, F.; Sweetman, J.; Tort, L.; Izquierdo, M.S. Immune stimulation and improved infection resistance in European sea bass (*Dicentrarchus labrax*) fed mannan oligosaccharides. *Fish Shellfish Immunol.* **2007**, *23*, 969–981. [CrossRef]
93. Alak, G.; Atamanalp, M. Usage of probiotics and prebiotics in aquaculture. *Yüzüncü Yil Üniversitesi J. Agric. Sci.* **2012**, *22*, 62–68.
94. Mustafa, A.; Buentello, A.; Gatlin, D., III; Lightner, D.; Hume, M.; Lawrence, A. Effects of fructooligosaccharides (FOS) on growth, survival, gut microflora, stress, and immune response in Pacific white shrimp, *Litopenaeus vannamei*, cultured in a recirculating system. *J. Immunoass. Immunochem.* **2020**, *41*, 45–59. [CrossRef] [PubMed]
95. Li, P.; Wang, X.; Murthy, S.; Gatlin, D.M., III; Castille, F.L.; Lawrence, A.L. Effect of dietary supplementation of brewer's yeast and Grobiotic®-A on growth, immune responses, and low-salinity tolerance of Pacific white shrimp *Litopenaeus vannamei* cultured in recirculating systems. *J. Appl. Aquac.* **2009**, *21*, 110–119. [CrossRef]
96. Gatlin, D.M., III; Li, P.; Wang, X.; Burr, G.S.; Castille, F.; Lawrence, A.L. Potential application of prebiotics in aquaculture. In *Avances en Nutrición Acuicola*; Universidad Autónoma de Nuevo León: Monterrey, Mexico, 2006; ISBN 970-694-333-5.
97. Li, T.; Chu, C.; Zhang, Y.; Ju, M.; Wang, Y. Contrasting eutrophication risks and countermeasures in different water bodies: Assessments to support targeted watershed management. *Int. J. Environ. Res. Public Health* **2017**, *14*, 695. [CrossRef]
98. Amiin, M.K.; Lahay, A.F.; Putriani, R.B.; Reza, M.; Putri, S.M.E.; Sumon, M.A.A.; Jamal, M.T.; Santanumurti, M.B. The role of probiotics in vannamei shrimp aquaculture performance–A review. *Vet. World* **2023**, *16*, 638. [CrossRef]
99. Javahery, S.; Noori, A.; Hoseinifar, S.H. Growth performance, immune response, and digestive enzyme activity in Pacific white shrimp, *Penaeus vannamei* Boone, 1931, fed dietary microbial lysozyme. *Fish Shellfish Immunol.* **2019**, *92*, 528–535. [CrossRef]
100. Preethi, R.; Thanigaivel, S. Analysis and Evaluation of Probiotic (*Lactococcus*) Based Feed Supplements to Improve the Growth and Immune Related Responses of Rohu Fishes. *ECS Trans.* **2022**, *107*, 14051.
101. Eissa, E.S.H.; Ahmed, R.A.; Abd Elghany, N.A.; Elfeky, A.; Saadony, S.; Ahmed, N.H.; Sakr, S.E.; Tolenada, C.P.; Atienza, A.A.; Mabrok, M.; et al. Potential Symbiotic Effects of β-1, 3 Glucan, and Fructooligosaccharides on the Growth Performance, Immune Response, Redox Status, and Resistance of Pacific White Shrimp, *Litopenaeus vannamei* to *Fusarium solani* Infection. *Fishes* **2023**, *8*, 105. [CrossRef]
102. Kolanchinathan, P.; Kumari, P.R.; Raja, K.; John, G.; Balasundaram, A. Analysis of feed composition and growth parameters of *Penaeus monodon* supplemented with two probiotic species and formulated diet. *Aquaculture* **2022**, *549*, 737740. [CrossRef]
103. Wang, Y.; Al Farraj, D.A.; Vijayaraghavan, P.; Hatamleh, A.A.; Biji, G.D.; Rady, A.M. Host associated mixed probiotic bacteria induced digestive enzymes in the gut of tiger shrimp *Penaeus monodon*. *Saudi J. Biol. Sci.* **2020**, *27*, 2479–2484. [CrossRef] [PubMed]
104. Pan, M.V.; Ferriols, V.M.E.N.; Traifalgar, R.F.M. Synergistic influence of hydrolyzed squid processing by-products and Bacillus probiotics as dietary supplements on growth performance, immunological responses, and gut health of juvenile black tiger shrimp fed fishmeal-free diets. *Aquac. Int.* **2024**, *32*, 4551–4580. [CrossRef]
105. NavinChandran, M.; Iyapparaj, P.; Moovendhan, S.; Ramasubburayan, R.; Prakash, S.; Immanuel, G.; Palavesam, A. Influence of probiotic bacterium *Bacillus cereus* isolated from the gut of wild shrimp *Penaeus monodon* in turn as a potent growth promoter and immune enhancer in *P. monodon*. *Fish Shellfish Immunol.* **2014**, *36*, 38–45. [CrossRef] [PubMed]
106. Chandran, M.N.; Suganya, A.M.; Immanuel, G.; Palavesam, A. Immunomodulatory and growth-promoting potential of low-cost probiotic product in *Penaeus monodon* culture system. *Croat. J. Fish.* **2017**, *75*.
107. Pattukumar, V.; Kanmani, P.; Satish Kumar, R.; Yuvaraj, N.; Paari, A.; Arul, V. Enhancement of innate immune system, survival and yield in *Penaeus monodon* reared in ponds using *Streptococcus phocae* PI 80. *Aquac. Nutr.* **2014**, *20*, 505–513. [CrossRef]
108. Duan, Y.; Zhang, J.; Huang, J.; Jiang, S. Effects of dietary Clostridium butyricum on the growth, digestive enzyme activity, antioxidant capacity, and resistance to nitrite stress of *Penaeus monodon*. *Probiotics Antimicrob. Proteins* **2019**, *11*, 938–945. [CrossRef]
109. Sekar, A.; Packyam, M.; Kim, K. Growth enhancement of shrimp and reduction of shrimp infection by *Vibrio parahaemolyticus* and white spot syndrome virus with dietary administration of *Bacillus* sp. Mk22. *Microbiol. Biotechnol. Lett.* **2016**, *44*, 261–267. [CrossRef]
110. Utiswannakul, P.; Sangchai, S.; Rengpipat, S. Enhanced growth of black tiger shrimp *Penaeus monodon* by dietary supplementation with Bacillus (BP11) as a probiotic. *J. Aquac. Res. Dev.* **2011**, *2*, 6. [CrossRef]
111. Amenyogbe, E.; Chen, G.; Wang, Z.; Huang, J.; Huang, B.; Li, H. The exploitation of probiotics, prebiotics and synbiotics in aquaculture: Present study, limitations and future directions.: A review. *Aquac. Int.* **2020**, *28*, 1017–1041. [CrossRef]

112. El-Saadony, M.T.; Swelum, A.A.; Ghanima, M.M.A.; Shukry, M.; Omar, A.A.; Taha, A.E.; Salem, H.M.; El-Tahan, A.M.; El-Terabily, K.A.; Abd El-Hack, M.E. Shrimp production, the most important diseases that threaten it, and the role of probiotics in confronting these diseases: A review. *Res. Vet. Sci.* **2022**, *144*, 126–140. [CrossRef]
113. Muthu, C.M.; Vickram, A.S.; Sowndharya, B.B.; Saravanan, A.; Kamalesh, R.; Dinakarkumar, Y. A comprehensive review on the utilisation of probiotics in aquaculture towards sustainable shrimp farming. *Fish Shellfish Immunol.* **2024**, *147*, 109459. [CrossRef] [PubMed]
114. Tamilselvan, M.; Raja, S. Exploring the role and mechanism of potential probiotics in mitigating the shrimp pathogens. *Saudi J. Biol. Sci.* **2024**, *31*, 103938. [CrossRef] [PubMed]
115. Soltani, M.; Ahmadivand, S.; Ringø, E. Bacillus as Probiotics in Shellfish Culture. In *Bacillus Probiotics for Sustainable Aquaculture*; CRC Press: Boca Raton, FL, USA, 2024; pp. 192–207.
116. Abo-Al-Ela, H.G.; Mahdi, S.; Angthong, P.; Rungrassamee, W. Probiotic modulation of key immune macromolecules in shrimp. *Microb. Pathog.* **2025**, *203*, 107463. [CrossRef] [PubMed]
117. Van Hai, N.; Fotedar, R. A review of probiotics in shrimp aquaculture. *J. Appl. Aquac.* **2010**, *22*, 251–266. [CrossRef]
118. Jamal, M.T.; Abdulrahman, I.A.; Al Harbi, M.; Chithambaran, S. Probiotics as alternative control measures in shrimp aquaculture: A review. *J. Appl. Biol. Biotechnol.* **2019**, *7*, 69–77.
119. Zheng, X.; Duan, Y.; Dong, H.; Zhang, J. The effect of *Lactobacillus plantarum* administration on the intestinal microbiota of whiteleg shrimp *Penaeus vannamei*. *Aquaculture* **2020**, *526*, 735331. [CrossRef]
120. Aruta, K.J.C.; Caipang, C.M.A. Local aquatic microflora as a potential source of probionts in biofloc technology for whiteleg shrimp, *Penaeus vannamei*. *Environ. Exp. Biol.* **2025**, *23*, 1–8. [CrossRef]
121. Trujillo, L.E.; Rivera, L.; Hardy, E.; Llumiquinga, E.M.; Garrido, F.; Chávez, J.A.; Abril, V.H.; País-Chanfrau, J.M. Estrategias Naturales para Mejorar el Crecimiento y la Salud en los Cultivos Masivas de Camarón en Ecuador. *Rev. Bionatura* **2017**, *2*, 318–325. [CrossRef]
122. Noman, M.; Kazmi, S.S.U.H.; Saqib, H.S.A.; Fiaz, U.; Pastorino, P.; Barcelò, D.; Tayyab, M.; Liu, W.; Wang, Z.; Yaseen, Z.M. Harnessing probiotics and prebiotics as eco-friendly solution for cleaner shrimp aquaculture production: A state of the art scientific consensus. *Sci. Total Environ.* **2024**, *915*, 169921. [CrossRef]
123. Kari, Z.A.; Wee, W.; Hamid, N.K.A.; Dawood, M.A.; Zakaria, N.N.A.B.; Wei, L.S. The roles of polysaccharides in tilapia farming: A review. *Aquac. Fish.* **2024**, *9*, 20–27. [CrossRef]
124. Kumar, R.; Dharumadurai, D. Growth and immunomodulatory postbiotic effects in shrimp. In *Postbiotics*; Academic Press: Cambridge, MA, USA, 2025; pp. 565–573.
125. Cheng, A.C.; Chang, H.T.; Lee, T.Y.; Lin, J.S.; Liu, C.H. SYNLAC prime probiotics enhances growth performance, and resistance of white shrimp, *Penaeus vannamei* to *Enterocytozoon hepatopenaei* and *Vibrio alginollyticus*: Insights into immune and metabolic pathway modulations. *Fish Shellfish Immunol.* **2024**, *155*, 110016. [CrossRef] [PubMed]
126. Food and Agricultural Organization of the United Nations and World Health Organization. Health and nutritional properties of probiotics in food including powder milk with live lactic acid bacteria. In *Report of a Joint FAO/WHO Expert Consultation on Evaluation of Health and Nutritional Properties of Probiotics in Food Including Powder Milk with Live Lactic Acid Bacteria*; FAO: Rome, Italy, 2021.
127. Bozkurt, M.; Aysul, N.; Küçükyilmaz, K.; Aypak, S.; Ege, G.; Catli, A.U.; Akşit, H.; Çöven, F.; Seyrek, K.; Çınar, M. Efficacy of in-feed preparations of an anticoccidial, multienzyme, prebiotic, probiotic, and herbal essential oil mixture in healthy and *Eimeria* spp.-infected broilers. *Poult. Sci.* **2014**, *93*, 389–399. [CrossRef]
128. Muhammad, Z.; Ramzan, R.; Zhang, R.; Zhang, M. Resistant starch-based edible coating composites for spray-dried microencapsulation of *Lactobacillus acidophilus*, comparative assessment of thermal protection, in vitro digestion and physicochemical characteristics. *Coatings* **2021**, *11*, 587. [CrossRef]
129. Nikiforov-Nikishin, A.; Nikiforov-Nikishin, D.; Kochetkov, N.; Smorodinskaya, S.; Klimov, V. The influence of probiotics of different microbiological composition on histology of the gastrointestinal tract of juvenile *Oncorhynchus mykiss*. *Microsc. Res. Tech.* **2022**, *85*, 538–547. [CrossRef] [PubMed]
130. Weinrauch, A.M.; Hoogenboom, J.L.; Anderson, W.G. A review of reductionist methods in fish gastrointestinal tract physiology. *Comp. Biochem. Physiol. Part B Biochem. Mol. Biol.* **2021**, *254*, 110571. [CrossRef] [PubMed]
131. Sparagon, W.J.; Gentry, E.C.; Minich, J.J.; Vollbrecht, L.; Laurens, L.M.; Allen, E.E.; Sims, N.A.; Dorrestein, P.C.; Kelly, L.W. Nelson, C.E. Fine scale transitions of the microbiota and metabolome along the gastrointestinal tract of herbivorous fishes. *Anim. Microbiome* **2022**, *4*, 33. [CrossRef]
132. Kanani, H.; Javadian, S.R.; Bahram, S. Evaluation of feed efficiency, growth and biochemical parameters of rainbow trout (*Oncorhynchus mikyss*) juveniles fed with different levels of Alphamune prebiotic. *Sustain. Aquac. Health Manag. J.* **2021**, *7*, 47–55. [CrossRef]

133. Maytorena-Verdugo, C.I.; Peña-Marín, E.S.; Alvarez-Villagómez, C.S.; Pérez-Jiménez, G.M.; Sepúlveda-Quiroz, C.A.; Alvarez-González, C.A. Inclusion of mannan-oligosaccharides in diets for tropical gar atractosteus tropicus larvae: Effects on growth, digestive enzymes, and expression of intestinal barrier genes. *Fishes* **2022**, *7*, 127. [CrossRef]
134. Yousefi, M.; Farsani, M.N.; Ghafarifarsani, H.; Raeeszadeh, M. Dietary Lactobacillus helveticus and Gum Arabic improves growth indices, digestive enzyme activities, intestinal microbiota, innate immunological parameters, antioxidant capacity, and disease resistance in common carp. *Fish Shellfish Immunol.* **2023**, *135*, 108652. [CrossRef]
135. Romero, J.; Ringø, E.; Merrifield, D.L. The gut microbiota of fish. In *Aquaculture Nutrition: Gut Health, Probiotics and Prebiotics*; John Willie and the Suns: Hoboken, NJ, USA, 2014; pp. 75–100.
136. Legrand, T.P.; Wynne, J.W.; Weyrich, L.S.; Oxley, A.P. Investigating both mucosal immunity and microbiota in response to gut enteritis in yellowtail kingfish. *Microorganisms* **2020**, *8*, 1267. [CrossRef] [PubMed]
137. Li, X.; Zhou, S.; Zhang, J.; Zhou, Z.; Xiong, Q. Directional changes in the intestinal bacterial community in black soldier fly (*Hermetia illucens*) larvae. *Animals* **2021**, *11*, 3475. [CrossRef]
138. Tsao, J.C.; Zeltzer, L.K. Complementary and alternative medicine approaches for pediatric pain: A review of the state-of-the-science. *Evid. Based Complement. Altern. Med.* **2005**, *2*, 149–159. [CrossRef]
139. Gupta, V.; Garg, R. Probiotics. *Indian J. Med. Microbiol.* **2009**, *27*, 202–209. [CrossRef]
140. Ogunkalu, O. Effects of feed additives in fish feed for improvement of aquaculture. *Eurasian J. Food Sci. Technol.* **2019**, *3*, 49–57.
141. Cuvin-Aralar, M.L.A.; Ricafort, C.H.; Salvacion, A. *An Overview of Agricultural Follution in the Philippines: The Fisheries Sector*; World Bank: Washington, DC, USA, 2016.
142. Rahman, M.M.; Khosravi, S.; Chang, K.H.; Lee, S.M. Effects of dietary inclusion of astaxanthin on growth, muscle pigmentation and antioxidant capacity of juvenile rainbow trout (*Oncorhynchus mykiss*). *Prev. Nutr. Food Sci.* **2016**, *21*, 281. [CrossRef] [PubMed]
143. Zhao, W.; Guo, Y.C.; Huai, M.Y.; Li, L.; Man, C.; Pelletier, W.; Wei, H.L.; Yao, R.; Niu, J. Comparison of the Retention Rates of Synthetic and Natural Astaxanthin in Feeds and Their Effects on Pigmentation, Growth, and Health in Rainbow Trout (*Oncorhynchus mykiss*). *Antioxidants* **2022**, *11*, 2473. [CrossRef]
144. Song, X.; Wang, L.; Li, X.; Chen, Z.; Liang, G.; Leng, X. Dietary astaxanthin improved the body pigmentation and antioxidant function, but not the growth of discus fish (*Symphysodon* spp.). *Aquac. Res.* **2017**, *48*, 1359–1367. [CrossRef]
145. Lall, S.P.; Kaushik, S.J. Nutrition and metabolism of minerals in fish. *Animals* **2021**, *11*, 2711. [CrossRef]
146. Liu, Z.L.; Zhao, W.; Hu, W.S.; Zhu, B.; Xie, J.J.; Liu, Y.J.; Tian, L.X.; Niu, J. Lipid metabolism, growth performance, antioxidant ability and intestinal morphology of rainbow trout (*Oncorhynchus mykiss*) under cage culture with flowing water were affected by dietary lipid levels. *Aquac. Rep.* **2021**, *19*, 100593. [CrossRef]
147. Wen, B.; Jiang, Y.; Yuan, R.; Heqiu, Y.; Liu, Q.; Li, X.; Zuo, R. Effects of dietary lipid sources on the survival, growth, body composition, antioxidant capacity and expression of antioxidant and pro-inflammatory genes in juvenile Chinese mitten crab (*Eriocheir sinensis*) reared under three salinities. *Aquac. Res.* **2021**, *52*, 5307–5320. [CrossRef]
148. Ibrahim, S.A.; Ayivi, R.D.; Zimmerman, T.; Siddiqui, S.A.; Altemimi, A.B.; Fidan, H.; Esatbeyoglu, T.; Bakhshayesh, R.V. Lactic acid bacteria as antimicrobial agents: Food safety and microbial food spoilage prevention. *Foods* **2021**, *10*, 3131. [CrossRef]
149. Nasrollahzadeh, A.; Mokhtari, S.; Khomeiri, M.; Saris, P. EAntifungal preservation of food by lactic acid bacteria. *Foods* **2022**, *11*, 395. [CrossRef] [PubMed]
150. Luo, Z.Z.; Sun, H.M.; Guo, J.W.; Luo, P.; Hu, C.Q.; Huang, W.; Shu, H. Molecular characterisation of a RNA polymerase (RNAP) II (DNA directed) polypeptide H (POLR2H) in Pacific white shrimp (*Litopenaeus vannamei*) and its role in response to high-pH stress. *Fish Shellfish Immunol.* **2020**, *96*, 245–253. [CrossRef]
151. Korkmaz, K. The Effect of Sodium Bicarbonate Injection on the Physico-Chemical Quality of Post-Harvest Trout. *Foods* **2023**, *12*, 2437. [CrossRef]
152. Li, X.; Zheng, S.; Wu, G. Nutrition and metabolism of glutamate and glutamine in fish. *Amino Acids* **2020**, *52*, 671–691. [CrossRef] [PubMed]
153. Rachmawati, D.; Nurhayati, D. Effect of dietary lysine on the growth performance of Pangasius hypophthalmus. *Depik* **2022**, *11*, 111–116. [CrossRef]
154. Guo, J.; Duan, M.; Qiu, X.; Masagounder, K.; Davis, D.A. Characterisation of methionine uptake and clearance in the hemolymph of Pacific white shrimp *Litopenaeus vannamei*. *Aquaculture* **2020**, *526*, 735351. [CrossRef]
155. Ahmmed, M.K.; Ahmmed, F.; Tian, H.; Carne, A.; Bekhit, A.E.D. Marine omega-3 (n-3) phospholipids: A comprehensive review of their properties, sources, bioavailability, and relation to brain health. *Compr. Rev. Food Sci. Food Saf.* **2020**, *19*, 64–123. [CrossRef]
156. Aragão, C.; Teodósio, R.; Colen, R.; Richard, N.; Rønnestad, I.; Dias, J.; Conceição, L.E.C.; Ribeiro, L. Taurine Supplementation to Plant-Based Diets Improves Lipid Metabolism in Senegalese Sole. *Animals* **2023**, *13*, 1501. [CrossRef]
157. Koven, W.; Bracha, C.; Nixon, O.; Israeli, D.; Tandler, A.; Meiri-Ashkenazi, I.; Rosenfeld, H. The effect of dietary taurine and its potential biosynthesis on juvenile grey mullet (*Mugil cephalus*) performance. *Isr. J. Aquac. Bamidgeh* **2023**, *75*, 1–15. [CrossRef]

158. Chen, S.; Yu, Y.; Gao, Y.; Yin, P.; Tian, L.; Niu, J.; Liu, Y. Exposure to acute ammonia stress influences survival, immune response and antioxidant status of pacific white shrimp (*Litopenaeus vannamei*) pretreated with diverse levels of inositol. *Fish Shellfish Immunol.* **2019**, *89*, 248–256. [CrossRef]
159. Kong, F.; Zhu, Y.; Yu, H.; Wang, X.; Azm, F.R.A.; Yuan, J.; Tan, Q. Effect of dietary vitamin C on the growth performance, non-specific immunity and antioxidant ability of red swamp crayfish (*Procambarus clarkii*). *Aquaculture* **2021**, *541*, 736785. [CrossRef]
160. Sreekanth, G.B.; Varghese, T.; Mishal, P.; Sandeep, K.P.; Praveen, K.V. Food security in India: Is aquaculture a solution in the offing. *Int. J. Sci. Res.* **2015**, *4*, 553–560.
161. Gaponov, N.V.; Gamko, L.N. Nutrient digestibility of fishmeal rations in primates. *Vet. Sci. Today* **2021**, *10*, 239–242. [CrossRef]
162. Souza, C.D.F.; Baldissera, M.D.; Baldisserotto, B.; Heinzmann, B.M.; Martos-Sitcha, J.A.; Mancera, J.M. Essential oils as stress-reducing agents for fish aquaculture: A review. *Front. Physiol.* **2019**, *10*, 785. [CrossRef] [PubMed]
163. Zheng, X.; Bossier, P. Toxicity assessment and anti-Vibrio activity of essential oils: Potential for application in shrimp aquaculture. *Rev. Aquac.* **2023**, *15*, 1554–1573. [CrossRef]
164. Mbokane, E.M.; Moyo, N.A.G. Use of medicinal plants as feed additives in the diets of Mozambique tilapia (*Oreochromis mossambicus*) and the African Sharptooth catfish (*Clarias gariepinus*) in Southern Africa. *Front. Vet. Sci.* **2022**, *9*, 1072369. [CrossRef]
165. Yousefian, M.; Amiri, M.S. A review of the use of prebiotic in aquaculture for fish and shrimp. *Afr. J. Biotechnol.* **2009**, *8*, 7313–7318.
166. Sitjà-Bobadilla, A. Nutritional Interventions to Mitigate Parasitic Enteritis. *Aquac. Eur.* 2017, 17. Available online: http://hdl.handle.net/10261/191297 (accessed on 7 May 2025).
167. Muthulakshmi, M.; Subramani, P.A.; Michael, R.D. Immunostimulatory effect of the aqueous leaf extract of Phyllanthus niruri on the specific and non-specific immune responses of *Oreochromis mossambicus* Peters. *Iran. J. Vet. Res.* **2016**, *17*, 200. [PubMed]
168. Salem, M.O.A.; Taştan, Y.; Bilen, S.; Terzi, E.; Sönmez, A.Y. Effects of white mustard (*Sinapis alba*) oil on growth performance, immune response, blood parameters, digestive and antioxidant enzyme activities in rainbow trout (*Oncorhynchus mykiss*). *Fish Shellfish Immunol.* **2022**, *131*, 283–299. [CrossRef]
169. Immanuel, G.; Uma, R.P.; Iyapparaj, P.; Citarasu, T.; Punitha Peter, S.M.; Michael Babu, M.; Palavesam, A. Dietary medicinal plant extracts improve growth, immune activity and survival of tilapia *Oreochromis mossambicus*. *J. Fish Biol.* **2009**, *74*, 1462–1475. [CrossRef]
170. Jahanjoo, V.; Yahyavi, M.; Akrami, R.; Bahri, A.H. Influence of adding garlic (*Allium sativum*), Ginger (*Zingiber officinale*), thyme (*Thymus vulgaris*) and their combination on the growth performance, haematoimmunological parameters and disease resistance to Photobacterium damselae in sobaity sea bream (*Sparidentex hasta*) Fry. *Turk. J. Fish. Aquat. Sci.* **2024**, *18*, 633–645.
171. Alam, S.; Afzal, G.; Siddique, A.B.; Afzal, M.; Shahid, M.; Ramzan, A.; Iqbal, Z.; Ali, H.M.; Hira Ahsan, H.; Khan, M.S.; et al. Essential Oil-Based Functional Feeds for Promoting Growth in Aquaculture Species. *Complement. Altern. Med. Essent. Oils* **2024**, 199–206. [CrossRef]
172. Dawood, M.A.; El Basuini, M.F.; Yilmaz, S.; Abdel-Latif, H.M.; Alagawany, M.; Kari, Z.A.; Razab, M.K.; Hamid, N.K.; Moonmanee, T.; Van Doan, H. Exploring the roles of dietary herbal essential oils in aquaculture: A review. *Animals* **2022**, *12*, 823. [CrossRef]
173. Awad, E.; Awaad, A. Role of medicinal plants on growth performance and immune status in fish. *Fish Shellfish Immunol.* **2017**, *67*, 40–54. [CrossRef]
174. Ahmadifar, E.; Yousefi, M.; Karimi, M.; Fadaei Raieni, R.; Dadar, M.; Yilmaz, S.; Dawood, M.A.O.; Abdel-Latif, H.M. Benefits of dietary polyphenols and polyphenol-rich additives to aquatic animal health: An overview. *Rev. Fish. Sci. Aquac.* **2021**, *29*, 478–511. [CrossRef]
175. Ashry, A.M.; Habiba, M.M.; El-Zayat, A.M.; Hassan, A.M.; Moonmanee, T.; Van Doan, H.; Shadrack, R.S.; Dawood, M.A. Dietary anise (*Pimpinella anisum* L.) enhances growth performance and serum immunity of European sea bass (*Dicentrarchus labrax*). *Aquac. Rep.* **2022**, *23*, 101083. [CrossRef]
176. Ashry, A.M.; Habiba, M.M.; Abdel-Wahab, A.; Younis, E.M.; Davies, S.J.; Elnakeeb, M.A.; Abdelghany, M.F.; El-Zayat, A.M.; El-Sebaey, A.M. Dietary effect of powdered herbal seeds on zootechnical performance, hemato-biochemical indices, immunological status, and intestinal microbiota of European sea bass (*Dicentrarchus labrax*). *Aquac. Rep.* **2024**, *36*, 102074. [CrossRef]
177. Kalaiselvan, P.; Malarvizhi, K.; Ranjan, A. Exploring phytobiotics in aquaculture: Sources, mode of action, effects, administration, and its bioavailability in fish. *Aquac. Int.* **2024**, *32*, 5737–5799. [CrossRef]
178. Chakraborty, S.B.; Hancz, C. Application of phytochemicals as immunostimulant, antipathogenic and antistress agents in finfish culture. *Rev. Aquac.* **2011**, *3*, 103–119. [CrossRef]
179. Olaniyi, C.O.; Atoyebi, M.O.; Obafunmiso, H.T.; Salaam, K.A. Effect of ginger (*Zingiber officinale*) in the nutrition of African catfish-A cholesterol reducer and fertility enhancer. *Int. J. Aquac. Fish. Sci.* **2020**, *6*, 21–28. [CrossRef]
180. Oparaku, N.F.; Ijem, N.; Nduka, J.C. Growth performance of Clarias gariepinus (*African catfish*) juveniles fed Zingiber officinale (*Ginger*) and Moringa oleifera bark supplemented diets. *Acad. J. Med. Plants* **2021**, *9*, 147–153.

181. Wei, L.S.; Kari, Z.A.; Kabir, M.A.; Khoo, M.I.; Azra, M.N.; Wee, W. Promoting growth and health of African catfish, Clarias gariepinus, through dietary novel supplement, ginger, zingiber officinale rosc, leaf powder. *Aquac. Stud.* **2023**, *24*, AQUAST1719. [CrossRef]
182. Wijayanto, D. The effect of ginger (*Zingiber officinale*) enrichment in artificial feed on the growth, survival, and profitability of giant gourami (*Osphronemus goramy*) cultivation. *Aquac. Aquar. Conserv. Legis.* **2025**, *18*, 269–276.
183. Gabriel, N.N.; Wilhelm, M.R.; Habte-Tsion, H.M.; Chimwamurombe, P.; Omoregie, E.; Iipinge, L.N.; Shimooshili, K. Effect of dietary Aloe vera polysaccharides supplementation on growth performance, feed utilisation, hemato-biochemical parameters, and survival at low pH in African catfish (*Clarias gariepinus*) fingerlings. *Int. Aquat. Res.* **2019**, *11*, 57–72. [CrossRef]
184. Jaafar, N.; Musa, S.M.; Azfaralariff, A.; Mohamed, M.; Yusoff, A.H.; Lazim, A.M. Improving the efficiency of post-digestion method in extracting microplastics from gastrointestinal tract and gills of fish. *Chemosphere* **2020**, *260*, 127649. [CrossRef]
185. Tartila, S.S.Q.; Jusadi, D.; Setiawati, M.; Fauzi, I.A. Evaluation of dietary supplementation with cinnamon products on growth, blood composition, liver structure, and meat quality of striped catfish (*Pangasianodon hypophthalmus*). *Aquac. Int.* **2021**, *29*, 2243–2257. [CrossRef]
186. Mahboub, H.H.; Tartor, Y.H. Carvacrol essential oil stimulates growth performance, immune response, and tolerance of Nile tilapia to Cryptococcus uniguttulatus infection. *Dis. Aquat. Org.* **2020**, *141*, 1–14. [CrossRef]
187. Kazempoor, R.; Alavinezhad, S.S.; Pargari, M.M.; Shakeri, Y.S.; Haghighi, M.M. A review on the application of phytogenics as feed additives for aquatic animals. *Int. J. Aquat. Res. Environ. Stud.* **2022**, *2*, 46–78. [CrossRef]
188. Onomu, A.J.; Okuthe, G.E. The role of functional feed additives in enhancing aquaculture sustainability. *Fishes* **2024**, *9*, 167. [CrossRef]
189. Tignani, M.V.; Santolini, E.; Secci, G.; Bovo, M.; Parisi, G.; Barbaresi, A. Assessing environmental sustainability of substitute feeding formulas for gilthead seabream (*Sparus aurata*) using Life Cycle Assessment. *Sci. Total Environ.* **2024**, *954*, 176689. [CrossRef]
190. Doyle, M.P.; Erickson, M.C. Opportunities for mitigating pathogen contamination during on-farm food production. *Int. J. Food Microbiol.* **2012**, *152*, 54–74. [CrossRef] [PubMed]
191. Puvanasundram, P.; Chong, C.M.; Sabri, S.; Yusoff, M.S.; Karim, M. Multi-strain probiotics: Functions, effectiveness and formulations for aquaculture applications. *Aquac. Rep.* **2021**, *21*, 100905. [CrossRef]
192. Dima, M.F.; Sîrbu, E.; Patriche, N.; Cristea, V.; Coadă, M.T.; Plăcintă, S. Effects of multi-strain probiotics on the growth and hematological profile in juvenile carp (*Cyprinus carpio*, Linnaeus 1758). *Carpathian J. Food Sci. Technol.* **2022**, *14*, 5–20. [CrossRef]
193. Kuebutornye, F.K.; Abarike, E.D.; Lu, Y.; Hlordzi, V.; Sakyi, M.E.; Afriyie, G.; Wang, Z.; Li, Y.; Xie, C.X. Mechanisms and the role of probiotic Bacillus in mitigating fish pathogens in aquaculture. *Fish Physiol. Biochem.* **2020**, *46*, 819–841. [CrossRef]
194. Midhun, S.J.; Arun, D.; Jyothis, M. Probiotic application of beneficial bacteria for improved health and disease control. In *Recent Advances in Aquaculture Microbial Technology*; Academic Press: Cambridge, MA, USA, 2023; pp. 275–289.
195. Calcagnile, M.; Tredici, S.M.; Alifano, P. A comprehensive review on probiotics and their use in aquaculture: Biological control, efficacy, and safety through the genomics and wet methods. *Heliyon* **2024**, *10*, e40892. [CrossRef]
196. Moon, K.; Ryu, S.; Song, S.H.; Chun, S.W.; Lee, N.; Lee, A.H. Characterization of novel bacteriophages for effective phage therapy against Vibrio infections in aquaculture. *J. Microbiol.* **2025**, *63*, e2502009. [CrossRef]
197. Hashem, N.M.; Hosny, N.S.; El-Desoky, N.; Soltan, Y.A.; Elolimy, A.A.; Sallam, S.M.; Abu-Tor, E.S.M. Alginate nanoencapsulated synbiotic composite of pomegranate peel phytogenics and multi-probiotic species as a potential feed additive: Physicochemical, antioxidant, and antimicrobial activities. *Animals* **2023**, *13*, 2432. [CrossRef] [PubMed]
198. MA, A.; Sarasan, M.; Kachiprath, B.; Sukumaran, V.; Singh, I.B.; Puthumana, J. Engineering the fish gut microbiome: Could it serve as future-proof strategy for sustainable aquaculture? *Blue Biotechnol.* **2025**, *2*, 6. [CrossRef]
199. Liu, C.; Ma, N.; Feng, Y.; Zhou, M.; Li, H.; Zhang, X.; Ma, X. From probiotics to postbiotics: Concepts and applications. *Anim. Res. One Health* **2023**, *1*, 92–114. [CrossRef]
200. Dawood, M.A.; Abo-Al-Ela, H.G.; Hasan, M.T. Modulation of transcriptomic profile in aquatic animals: Probiotics, prebiotics and synbiotics scenarios. *Fish Shellfish Immunol.* **2020**, *97*, 268–282. [CrossRef] [PubMed]
201. Quintanilla-Pineda, M.; Achou, C.G.; Díaz, J.; Gutiérrez-Falcon, A.; Bravo, M.; Herrera-Muñoz, J.I.; Peña-Navarro, N.; Alvarado, C.; Ibañez, F.C.; Marzo, F. In vitro evaluation of postbiotics produced from bacterial isolates obtained from rainbow trout and Nile tilapia against the pathogens *Yersinia ruckeri* and *Aeromonas salmonicida* subsp. *salmonicida*. *Foods* **2023**, *12*, 861. [CrossRef]
202. Tao, L.T.; Lu, H.; Xiong, J.; Zhang, L.; Sun, W.W.; Shan, X.F. The application and potential of postbiotics as sustainable feed additives in aquaculture. *Aquaculture* **2024**, *592*, 741237. [CrossRef]
203. Gervasoni, L.F.; Gervasoni, K.; de Oliveira Silva, K.; Mendes, M.E.F.; Maddela, N.R.; Prasad, R.; Winkelstroter, L.K. Postbiotics in active food packaging: The contribution of cellulose nanocomposites. *Sustain. Chem. Pharm.* **2023**, *36*, 101280. [CrossRef]
204. Rovelli, R.; Cecchini, B.; Zavagna, L.; Azimi, B.; Ricci, C.; Esin, S.; Milazzo, M.; Batoni, G.; Danti, S. Emerging Multiscale Biofabrication Approaches for Bacteriotherapy. *Molecules* **2024**, *29*, 533. [CrossRef]

205. Zhang, C.L.; Chen, G.; Ji, X.S.; Teng, J.; Yao, Z.L.; Hu, C.L.; Zhao, Y. Construction and evaluation of recombinant *Lactobacillus plantarum* expressing *Micropterus salmoides* hepcidin. *Microb. Cell Factories* **2025**, *24*, 148. [CrossRef]
206. Jiao, S.; Fu, Y.Z.; Zhang, N.F. Progress on mechanism and functional characteristics of next generation probiotics. *Chin. J. Anim. Nutr.* **2022**, *34*, 4836–4846.
207. Loo, K.Y.; Thong, J.Y.H.; Tan, L.T.H.; Letchumanan, V.; Chan, K.G.; Lee, L.H.; Law, J.W.F. A Current Overview of Next-Generation Probiotics and Their Prospects in Health and Disease Management. *Prog. Microbes Mol. Biol.* **2024**, *7*, 7. [CrossRef]
208. Barbosa, J.C.; Machado, D.; Almeida, D.; Andrade, J.C.; Brandelli, A.; Gomes, A.M.; Freitas, A.C. Next-generation probiotics. In *Probiotics*; Academic Press: Cambridge, MA, USA, 2022; pp. 483–502.
209. Amphan, S.; Unajak, S.; Printrakoon, C.; Areechon, N. Feeding-regimen of β-glucan to enhance innate immunity and disease resistance of Nile tilapia, Oreochromis niloticus Linn., against Aeromonas hydrophila and Flavobacterium columnare. *Fish Shellfish Immunol.* **2019**, *87*, 120–128. [CrossRef]
210. Schmitt, P.; Wacyk, J.; Morales-Lange, B.; Rojas, V.; Guzmán, F.; Dixon, B.; Mercado, L. Immunomodulatory effect of cathelicidins in response to a β-glucan in intestinal epithelial cells from rainbow trout. *Dev. Comp. Immunol.* **2015**, *51*, 160–169. [CrossRef] [PubMed]
211. Wen, C.; Gan, N.; Zeng, T.; Zhang, N.; Zhou, H.; Zhang, A.; Wang, X. Regulation of Il-10 gene expression by Il-6 via Stat3 in grass carp head kidney leucocytes. *Gene* **2020**, *741*, 144579. [CrossRef]
212. Zhou, N.; Chen, L.L.; Chen, J.; Guo, Z.P. Molecular characterization and expression analysis of IL-1β and two types of IL-1 receptor in barbel steed (*Hemibarbus labeo*). *Comp. Biochem. Physiol. Part B Biochem. Mol. Biol.* **2020**, *241*, 110393. [CrossRef] [PubMed]
213. Meng, Y.; Ma, R.; Ma, J.; Han, D.; Xu, W.; Zhang, W.; Mai, K. Dietary nucleotides improve the growth performance, antioxidative capacity and intestinal morphology of turbot (*Scophthalmus maximus*). *Aquac. Nutr.* **2017**, *23*, 585–593. [CrossRef]
214. Magouz, F.I.; Abdel-Rahim, M.M.; Lotfy, A.M.; Mosbah, A.; Alkafafy, M.; Sewilam, H.; Dawood, M.A. Dietary nucleotides enhanced growth performance, carcass composition, blood biochemical, and histology features of European sea bass, *Dicentrarchus labrax* L. *Aquac. Rep.* **2021**, *20*, 100738.
215. Taklu, M.; Islami, H.R.; Mousavi, S.A.; Jourdehi, A.Y. Nucleotide supplementation in the diet of Sterlet sturgeon (*Acipenser ruthenus*): Improved zootechnical performance, biochemical indices, and immune responses. *Anim. Feed Sci. Technol.* **2022**, *288*, 115322. [CrossRef]
216. Doo, E.H.; Chassard, C.; Schwab, C.; Lacroix, C. Effect of dietary nucleosides and yeast extracts on composition and metabolic activity of infant gut microbiota in PolyFermS colonic fermentation models. *FEMS Microbiol. Ecol.* **2017**, *93*, fix088. [CrossRef]
217. Zhou, W.; Ramachandran, D.; Mansouri, A.; Dailey, M.J. Glucose stimulates intestinal epithelial crypt proliferation by modulating cellular energy metabolism. *J. Cell. Physiol.* **2018**, *233*, 3465–3475. [CrossRef]
218. Ringø, E.; Van Doan, H.; Lee, S.H.; Soltani, M.; Hoseinifar, S.H.; Harikrishnan, R.; Song, S.K. Probiotics, lactic acid bacteria and bacilli: Interesting supplementation for aquaculture. *J. Appl. Microbiol.* **2020**, *129*, 116–136. [CrossRef]
219. Madhulika Ngasotter, S.; Meitei, M.M.; Kara, T.; Meinam, M.; Sharma, S.; Rathod, S.K.; Singh, S.B.; Singh, S.K.; Bhat, R.A.H. Multifaceted Role of Probiotics in Enhancing Health and Growth of Aquatic Animals: Mechanisms, Benefits, and Applications in Sustainable Aquaculture—A Review and Bibliometric Analysis. *Aquac. Nutr.* **2025**, *1*, 5746972. [CrossRef] [PubMed]
220. Corr, S.C.; Hill, C.; Gahan, C.G. Understanding the mechanisms by which probiotics inhibit gastrointestinal pathogens. *Adv. Food Nutr. Res.* **2009**, *56*, 1–15.
221. Aleman, R.S.; Yadav, A. Systematic review of probiotics and their potential for developing functional nondairy foods. *Appl. Microbiol.* **2023**, *4*, 47–69. [CrossRef]
222. Alsufyani, M.O.; Asiri, A.A.; Asiri, Y.M.; Alsughayyir, I.A.; Alzahrani, R.A.; Al Essa, T.A.; Lajhar, A.M.; Alshhrani, M.A.; Rabeh, M.A. Correlation between Probiotic and Prebiotic: A Systematic review. *Egypt. J. Chem.* **2024**, *67*, 2227–2244. [CrossRef]
223. Yoo, S.; Jung, S.C.; Kwak, K.; Kim, J.S. The role of prebiotics in modulating gut microbiota: Implications for human health. *Int. J. Mol. Sci.* **2024**, *25*, 4834. [CrossRef] [PubMed]
224. Kumari, T.; Bag, K.K.; Das, A.B.; Deka, S.C. Synergistic role of prebiotics and probiotics in gut microbiome health: Mechanisms and clinical applications. *Food Bioeng.* **2024**, *3*, 407–424. [CrossRef]
225. Caipang, C.M.A.; Suharman, I.; Avillanosa, A.L.; Gonzales-Plasus, M.M. Influence of phytogenic feed additives on the health status in the gut and disease resistance of cultured fish. In *IOP Conference Series: Earth and Environmental Science*; IOP Publishing: Bristol, UK, 2021; Volume 695, p. 012024.
226. Firmino, J.P.; Galindo-Villegas, J.; Reyes-López, F.E.; Gisbert, E. Phytogenic bioactive compounds shape fish mucosal immunity. *Front. Immunol.* **2021**, *12*, 695973. [CrossRef]
227. Abdel-Latif, M.A.; Alsenosy, A.A.; Manaa, E.A.; Abaza, S.; Elshenawi, M.A.; Aboelnour, A.; Alagawany, M. Phytobiotics and their application in poultry and aquaculture industry. In *Organic Feed Additives for Livestock*; Academic Press: Cambridge, MA, USA, 2025; pp. 1–16.

228. Dawood, M.A.; El Basuini, M.F.; Zaineldin, A.I.; Yilmaz, S.; Hasan, M.T.; Ahmadifar, E.; El Asely, A.M.; Abdel-Latif, H.M.R.; Alagawany, M.; Abu-Elala, N.M.; et al. Antiparasitic and antibacterial functionality of essential oils: An alternative approach for sustainable aquaculture. *Pathogens* **2021**, *10*, 185. [CrossRef]
229. Ibrahim, D.; Shahin, S.E.; Alqahtani, L.S.; Hassan, Z.; Althobaiti, F.; Albogami, S.; Soliman, M.M.; El-Malt, R.M.S.; Al-Harthi, H.F.; Alqadri, N.; et al. Exploring the interactive effects of thymol and thymoquinone: Moving towards an enhanced performance, gross margin, immunity and Aeromonas sobria resistance of Nile Tilapia (*Oreochromis niloticus*). *Animals* **2022**, *12*, 3034. [CrossRef] [PubMed]
230. Sepehrfar, D.; Sudagar, M.; Paknejad, H.; Yousefi Siahkalroodi, S.; Norouzitallab, P. Role of phytochemicals in farmed fish reproductive performance: A review. *Iran. J. Fish. Sci.* **2023**, *22*, 1039–1068.
231. Zheng, C.C.; Wu, J.W.; Jin, Z.H.; Ye, Z.F.; Yang, S.; Sun, Y.Q.; Fei, H. Exogenous enzymes as functional additives in finfish aquaculture. *Aquac. Nutr.* **2020**, *26*, 213–224. [CrossRef]
232. Gopalraaj, J.; Velayudhannair, K.; Arockiasamy, J.P.; Radhakrishnan, D.K. The effect of dietary supplementation of proteases on growth, digestive enzymes, oxidative stress, and intestinal morphology in fishes–A review. *Aquac. Int.* **2024**, *32*, 745–765. [CrossRef]
233. Ng, W.K.; Koh, C.B. The utilisation and mode of action of organic acids in the feeds of cultured aquatic animals. *Rev. Aquac.* **2017**, *9*, 342–368. [CrossRef]
234. Sardar, P.; Shamna, N.; Sahu, N.P. Acidifiers in aquafeed as an alternate growth promoter: A short review. *Anim. Nutr. Feed Technol.* **2020**, *20*, 353–366. [CrossRef]
235. Flecker, A.S.; Shi, Q.; Almeida, R.M.; Angarita, H.; Gomes-Selman, J.M.; García-Villacorta, R.; Sethi, S.A.; Thomas, S.A.; Poff, N.L.; Forsberg, B.R.; et al. Reducing adverse impacts of Amazon hydropower expansion. *Science* **2022**, *375*, 753–760. [CrossRef] [PubMed]
236. Zhang, T.; Ding, Y.; Peng, J.; Dai, Y.; Luo, S.; Liu, W.; Ma, Y. Effects of broad-spectrum antibiotic (florfenicol) on resistance genes and bacterial community structure of water and sediments in an aquatic microcosm model. *Antibiotics* **2022**, *11*, 1299. [CrossRef]
237. Zhang, T.; Peng, J.; Dai, Y.; Xie, X.; Luo, S.; Ding, Y.; Ma, Y. Effect of florfenicol on nirS-type denitrifying communities structure of water in an aquatic microcosm model. *Front. Vet. Sci.* **2023**, *10*, 1205394. [CrossRef]
238. Khairy, W.M.; El-Ashmawy, N.; Nofal, E.R. Analysing the evolution of environmental impacts due to fish farms expansion. *Int. J. Eng. Tech. Res.* **2020**, *9*, 689–701.
239. Gonzalez Parrao, C.; Shisler, S.; Moratti, M.; Yavuz, C.; Acharya, A.; Eyers, J.; Snilstveit, B. Aquaculture for improving productivity, income, nutrition and women's empowerment in low-and middle-income countries: A systematic review and meta-analysis. *Campbell Syst. Rev.* **2021**, *17*, e1195. [CrossRef] [PubMed]
240. Won, J. Investigating Gasoline Contamination Effects on Daphnia's Heartbeat in a Simulated Lake. *J. Glob. Ecol. Environ.* **2023**, *17*, 20–31. [CrossRef]
241. Reis, J.V.T. The Production Cycle of *Litopenaeus vannamei* in Outdoor Ponds and Tank Culture of *Trachinotus carolinus*. Internship at Auburn University's Fish and Shrimp Nutrition Lab, PQDT-Global. 2017. Source: Repositório Aberto da Universidade do Porto. Available online: https://share.google/NuKzLHbDz0ofNx6CH (accessed on 10 April 2025).
242. Gazi-Khan, L.; Haque, S.E. A review of the current status of water quality and eutrophication in Dhaka's water bodies. *Management* **2022**, *10*, 1–9. [CrossRef]
243. Lenzi, M.; Persiano, M.; Gennaro, P.; Rubegni, F. Wind mitigating action on effects of eutrophication in coastal eutrophic water bodies. *Int. J. Mar. Sci. Ocean Technol.* **2016**, *3*, 14–20. [CrossRef]
244. Hu, Y.; Cheng, H.; Tao, S.; Schnoor, J.L. China's ban on phenylarsonic feed additives, a major step toward reducing the human and ecosystem health risk from arsenic. *Environ. Sci. Technol.* **2019**, *53*, 12177–12187. [CrossRef]
245. El-Seidy, E.; Mohamed, K. General Tit-For-Tat Strategy in The Three Players Prisoner's Dilemma Game. *World Sci. Res.* **2015**, *2*, 1–9.
246. Alayande, K.A.; Aiyegoro, O.A.; Ateba, C.N. Probiotics in animal husbandry: Applicability and associated risk factors. *Sustainability* **2020**, *12*, 1087. [CrossRef]
247. Johnson, M.D.; Dubeux, J.C.; Franzluebbers, A.J. 2 Conducting and Communicating Environmental Impacts of Research: Forage Production, Soil Health, Sustainability. *J. Anim. Sci.* **2022**, *100* (Suppl. S1), 39. [CrossRef]
248. Vutukuru, S.; Bodapati, A.K.; Bhimavarapu, V. Doxycycline induced alterations in the glycogen and protein content of the zebra fish, Danio rerio. *Asian J. Microbiol. Biotechnol. Environ. Sci* **2022**, *1*, 271220031.
249. Zhou, J.; Yun, X.; Wang, J.; Li, Q.; Wang, Y. A review on the ecotoxicological effect of sulphonamides on aquatic organisms. *Toxicol. Rep.* **2022**, *9*, 534–540. [CrossRef]
250. Peng, X.; Zhang, X.; Zhang, S.; Li, Z.; Zhang, H.; Zhang, L.; Wu, Z.; Liu, B. Revealing the response characteristics of periphyton biomass and community structure to sulfamethoxazole exposure in aquaculture water: The perspective of microbial network relationships. *Environ. Pollut.* **2024**, *344*, 123301. [CrossRef]

251. Seoane, M.; Rioboo, C.; Herrero, C.; Cid, Á. Toxicity induced by three antibiotics commonly used in aquaculture on the marine microalga *Tetraselmis suecica* (Kylin) Butch. *Mar. Environ. Res.* **2014**, *101*, 1–7. [CrossRef]
252. Song, C.; Zhang, C.; Fan, L.; Qiu, L.; Wu, W.; Meng, S.; Hu, G.; Kamira, B.; Chen, J. Occurrence of antibiotics and their impacts to primary productivity in fishponds around Tai Lake, China. *Chemosphere* **2016**, *161*, 127–135. [CrossRef]
253. González-Gaya, B.; Cherta, L.; Nozal, L.; Rico, A. An optimised sample treatment method for the determination of antibiotics in seawater, marine sediments and biological samples using LC-TOF/MS. *Sci. Total Environ.* **2018**, *643*, 994–1004. [CrossRef]
254. Chen, H.; Liu, S.; Xu, X.R.; Diao, Z.H.; Sun, K.F.; Hao, Q.W.; Liu, S.S.; Ying, G.G. Tissue distribution, bioaccumulation characteristics and health risk of antibiotics in cultured fish from a typical aquaculture area. *J. Hazard. Mater.* **2018**, *343*, 140–148. [CrossRef]
255. He, L.X.; He, L.Y.; Gao, F.Z.; Wu, D.L.; Ye, P.; Cheng, Y.X.; Chen, Z.Y.; Hu, L.X.; Liu, Y.S.; Chen, J.; et al. Antibiotics, antibiotic resistance genes and microbial community in grouper mariculture. *Sci. Total Environ.* **2022**, *808*, 152042. [CrossRef]
256. Bjelland, H.V.; Føre, M.; Lader, P.; Kristiansen, D.; Holmen, I.M.; Fredheim, A.; Grøtli, E.I.; Fathi, D.E.; Oppedal, F.; Utne, I.B.; et al. Exposed aquaculture in Norway. In Proceedings of the OCEANS 2015-MTS/IEEE Washington, Washington, DC, USA, 19–22 October 2015; pp. 1–10.
257. Saleh, H.M.; Bondouk, I.I.; Salama, E.; Mahmoud, H.H.; Omar, K.; Esawii, H.A. Asphaltene or polyvinylchloride waste blended with cement to produce a sustainable material used in nuclear safety. *Sustainability* **2022**, *14*, 3525. [CrossRef]
258. Gray, W.B.; Shimshack, J.P. The effectiveness of environmental monitoring and enforcement: A review of the empirical evidence. *Rev. Environ. Econ. Policy* **2011**, *5*, 3–24. [CrossRef]
259. Hornsletten, H. Optimisation Model Aimed for the Aquaculture Industry for Fleet Composition and Routing of Wellboats. Master's Thesis, NTNU, Trondheim, Norway, 2017.
260. Codex Alimentarius Commission. Guidelines for the Simple Evaluation of Dietary Exposure to Food Additives (CAC/GL 3-1989) (Originally adopted in 1989, revised in 2014). Food and Agriculture Organization of the United Nations (FAO) and World Health Organization (WHO). Available online: https://www.fao.org/fao-who-codexalimentarius/sh-proxy/en/?lnk=1&url=https%3A%2F%2Fworkspace.fao.org%2Fsites%2Fcodex%2FStandards%2FCXG%2B3-1989%2FCXG_003e.pdf (accessed on 17 October 2024).
261. Forcella, S.; Tantawy, N.; Yilma, J. The development of a four-way linking framework in Egypt: An example of the FAO, OIE and WHO joint activities to facilitate national risk assessment. *Vet. Ital.* **2015**, *51*, 45–50. [PubMed]
262. Mantovani, A.; Aquilina, G.; Cubadda, F.; Marcon, F. Risk-benefit assessment of feed additives in the one health perspective. *Front. Nutr.* **2022**, *9*, 843124. [CrossRef]
263. Munguti, J.; Obiero, K.; Odame, H.; Kirimi, J.; Kyule, D.; Ani, J.; Liti, D. Key limitations of fish feeds, feed management practices, and opportunities in Kenya's aquaculture enterprise. *Afr. J. Food Agric. Nutr. Dev.* **2021**, *21*, 17415–17434. [CrossRef]
264. Cobb, M.L.; Otto, C.M.; Fine, A.H. The animal welfare science of working dogs: Current perspectives on recent advances and future directions. *Front. Vet. Sci.* **2021**, *8*, 666898. [CrossRef]
265. Chesson, A.; Gropp, J.; Mantovani, A.; Roncancio, C. Ten years of EFSA's FEEDAP Panel and its main achievements. *EFSA J.* **2012**, *10*, s1005. [CrossRef]
266. Renshaw, D.W. *Animal Feed Additives. Issues in Toxicology: Regulatory Toxicology in the European Union*; Royal Society of Chemistry: London, UK, 2018.
267. Pauly, T.; Wyss, U. Efficacy testing of silage additives—Methodology and existing schemes. *Grass Forage Sci.* **2019**, *74*, 201–210. [CrossRef]
268. Zhang, X.G. Aquaculture in China. *Species Syst. Sel. Sustain. Aquac.* **2007**, 131–144. [CrossRef]
269. Kumar, P.; Srivastava, A. The Use of Feed and Food Additives in United States. In *Sustainable Use of Feed Additives in Livestock: Novel Ways for Animal Production*; Springer International Publishing: Cham, Switzerland, 2023; pp. 245–281.
270. Upton, H.F.; Cowan, T. *Genetically Engineered Salmon*; US Congressional Research Service: Washington, DC, USA, 2015.
271. Xie, S.; Han, D.; Yang, Y.; Zhang, S. Feed Developments in Freshwater Aquaculture. In *Aquaculture in China: Success Stories and Modern Trends*; John Willie and the Suns: Hoboken, NJ, USA, 2018; pp. 431–450.
272. Winger, R.J. Phosphorus Food Additive Use in the European Union. In *Dietary Phosphorus*; CRC Press: Boca Raton, FL, USA, 2017; pp. 279–312.
273. EFSA Panel on Additives and Products or Substances used in Animal Feed (FEEDAP); Bampidis, V.; Azimonti, G.; Bastos, M.D.L.; Christensen, H.; Durjava, M.; Dusemund, B.; Kouba, M.; López-Alonso, M.; Puente, S.L.; et al. Assessment of the safety of the feed additives acetic acid, calcium acetate and sodium diacetate for fish (FEFANA asbl). *EFSA J.* **2023**, *21*, e08176.
274. Broughton, E.I.; Walker, D.G. Policies and practices for aquaculture food safety in China. *Food Policy* **2010**, *35*, 471–478. [CrossRef]
275. Debeuckelaere, W.; Berbejal, R.P.; Rosell, M.A.G. Food additives, enzymes, and flavourings legislation in the European Union. In *Food Additives and Packaging*; American Chemical Society: Washington, DC, USA, 2014; pp. 41–56.
276. De Angelis, F.; Carreno, I. Achieving sustainability of the EU food chain with feed additives as a key tool: The European Commission intends to modernise the legal framework to foster innovation. *Eur. J. Risk Regul.* **2021**, *12*, 866–870. [CrossRef]

277. Varchenko, O.; Artimonova, I. Formation of organisational and economic mechanism for regulation of agricultural market. *Sci. J. Cahul State Univ. Bogdan Petriceicu Hasdeu Econ. Eng. Stud.* **2018**, *4*, 4–15.
278. Petković, G.; Užar, D. Marketing channels in value creation and delivery of cheese in the Republic of Serbia. *Anal. Ekon. Fak. U Subotici* **2020**, *56*, 101–115. [CrossRef]
279. Rossi, M.A.; Basiri, M.L.; McHenry, J.A.; Kosyk, O.; Otis, J.M.; Van Den Munkhof, H.E.; Bryois, J.; Hübel, C.; Breen, G.; Guo, W.; et al. Obesity remodels activity and transcriptional state of a lateral hypothalamic brake on feeding. *Science* **2019**, *364*, 1271–1274. [CrossRef] [PubMed]
280. Kempf, M.; Reinhard, A.; Beuerle, T. Pyrrolizidine alkaloids (PAs) in honey and pollen-legal regulation of PA levels in food and animal feed required. *Mol. Nutr. Food Res.* **2010**, *54*, 158–168. [CrossRef] [PubMed]
281. Grundmann, O.; Kumar, P.; Rogge, M. ACCPPublic Policy Committee Regulation of dietary supplements nutraceutical products in the United States: An argument for greater oversight uniform standards. *J. Clin. Pharmacol.* **2022**, *62*, 14–16. [CrossRef]
282. Nunes, C.S.; Kunamneni, A.; Kumar, V.; Habte-Tsion, H.M. Registration of food and feed additives (enzymes) in the United States, Canada, and China. In *Enzymes in Human and Animal Nutrition*; Academic Press: Cambridge, MA, USA, 2018; pp. 457–480.
283. Lammersfeld, C.A.; Levin, M.D.; Reilly, P.; Coyne, J.W.; Birdsall, T.C.; Markman, M. Assuring quality of dietary supplements for cancer patients: An integrative formulary systems approach. *Integr. Med. A Clin. J.* **2017**, *16*, 38.

MDPI AG
Grosspeteranlage 5
4052 Basel
Switzerland
Tel.: +41 61 683 77 34

Fishes Editorial Office
E-mail: fishes@mdpi.com
www.mdpi.com/journal/fishes

www.ingramcontent.com/pod-product-compliance
Lightning Source LLC
Chambersburg PA
CBHW061605210926
51850CB00009B/2223
* 9 7 8 3 7 2 5 8 6 8 3 6 0 *